T0145231

NEUROMETHODS

Series Editor
Wolfgang Walz
University of Saskatchewan,
Saskatoon, SK, Canada

For further volumes:
http://www.springer.com/series/7657

Neuromethods publishes cutting-edge methods and protocols in all areas of neuroscience as well as translational neurological and mental research. Each volume in the series offers tested laboratory protocols, step-by-step methods for reproducible lab experiments and addresses methodological controversies and pitfalls in order to aid neuroscientists in experimentation. *Neuromethods* focuses on traditional and emerging topics with wide-ranging implications to brain function, such as electrophysiology, neuroimaging, behavioral analysis, genomics, neurodegeneration, translational research and clinical trials. *Neuromethods* provides investigators and trainees with highly useful compendiums of key strategies and approaches for successful research in animal and human brain function including translational "bench to bedside" approaches to mental and neurological diseases.

Cerebrovascular Disorders

Edited by

Fawaz Al-Mufti

Associate Chair of Neurology for Research, New York Medical College, Associate Professor of Neurology, Neurosurgery and Radiology, New York Medical College, Director of the Neuroendovascular Surgery Fellowship, Neuroendovascular Surgery (Interventional Neurologist) Attending, Medical Director of Neurocritical Care, Westchester Medical Center at New York Medical College, Valhalla, NY, USA

Krishna Amuluru

Division of Neurointerventional Radiology, Goodman Campbell Brain and Spine, Indianapolis, IN, USA; Division of Neurointerventional Radiology, Ascension St. Vincent Medical Center, Indianapolis, IN, USA; Department of Neurosurgery, Marian College of Osteopathic Medicine, Indianapolis, IN, USA

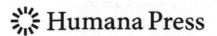 Humana Press

Editors
Fawaz Al-Mufti
Associate Chair of Neurology for
Research, New York Medical College
Associate Professor of Neurology
Neurosurgery and Radiology, New York
Medical College, Director of the
Neuroendovascular Surgery Fellowship
Neuroendovascular Surgery
(Interventional Neurologist) Attending
Medical Director of Neurocritical Care
Westchester Medical Center at New York
Medical College
Valhalla, NY, USA

Krishna Amuluru
Division of Neurointerventional Radiology
Goodman Campbell Brain and Spine
Indianapolis, IN, USA

Division of Neurointerventional Radiology
Ascension St. Vincent Medical Center
Indianapolis, IN, USA

Department of Neurosurgery
Marian College of Osteopathic Medicine
Indianapolis, IN, USA

ISSN 0893-2336   ISSN 1940-6045   (electronic)
Neuromethods
ISBN 978-1-0716-1532-4   ISBN 978-1-0716-1530-0   (eBook)
https://doi.org/10.1007/978-1-0716-1530-0

This Humana imprint is published by the registered company Springer Science+Business Media, LLC, part of Springer Nature.
The registered company address is: 1 New York Plaza, New York, NY 10004, U.S.A.

Preface to the Series

Experimental life sciences have two basic foundations: concepts and tools. The *Neuromethods* series focuses on the tools and techniques unique to the investigation of the nervous system and excitable cells. It will not, however, shortchange the concept side of things as care has been taken to integrate these tools within the context of the concepts and questions under investigation. In this way, the series is unique in that it not only collects protocols but also includes theoretical background information and critiques which led to the methods and their development. Thus it gives the reader a better understanding of the origin of the techniques and their potential future development. The *Neuromethods* publishing program strikes a balance between recent and exciting developments like those concerning new animal models of disease, imaging, in vivo methods, and more established techniques, including, for example, immunocytochemistry and electrophysiological technologies. New trainees in neurosciences still need a sound footing in these older methods in order to apply a critical approach to their results.

Under the guidance of its founders, Alan Boulton and Glen Baker, the *Neuromethods* series has been a success since its first volume published through Humana Press in 1985. The series continues to flourish through many changes over the years. It is now published under the umbrella of Springer Protocols. While methods involving brain research have changed a lot since the series started, the publishing environment and technology have changed even more radically. *Neuromethods* has the distinct layout and style of the Springer Protocols program, designed specifically for readability and ease of reference in a laboratory setting.

The careful application of methods is potentially the most important step in the process of scientific inquiry. In the past, new methodologies led the way in developing new disciplines in the biological and medical sciences. For example, Physiology emerged out of Anatomy in the nineteenth century by harnessing new methods based on the newly discovered phenomenon of electricity. Nowadays, the relationships between disciplines and methods are more complex. Methods are now widely shared between disciplines and research areas. New developments in electronic publishing make it possible for scientists that encounter new methods to quickly find sources of information electronically. The design of individual volumes and chapters in this series takes this new access technology into account. Springer Protocols makes it possible to download single protocols separately. In addition, Springer makes its print-on-demand technology available globally. A print copy can therefore be acquired quickly and for a competitive price anywhere in the world.

Saskatoon, SK, Canada   *Wolfgang Walz*

Preface

Humanity has long possessed both an anxious curiosity and a mystification regarding the consequences of scientific advancement. In January 1900, Jack London wrote in *The Shrinkage of the Planet*, "What [has] this planet of ours become!...The telegraph annihilates space and time. Each morning every part knows what every other part is thinking, contemplating, or doing. A discovery in a German laboratory is being demonstrated in San Francisco within 24 hours." In today's digital world with social media sharing, interest-based networks, interactive internet-based applications, and multiple robust online scientific forums, the 24-hour turnaround time that once amazed Jack London now seems like an eternity.

Like many other scientific fields, the discipline of neurointerventional surgery is not immune to such rapid information relay, ultimately resulting in clinical advancement. In fact, our field has experienced major developments in the past several years owing to a "shrunken world" of communication, including landmark progress in endovascular management of acute ischemic stroke, advancements in artificial intelligence, a complete re-examination of arterial access routes, and novel endovascular treatments of diseases heretofore thought to be outside the realm of neurointervention, such as pseudotumor cerebrii and chronic subdural hemorrhage, to name just a few.

Within such an environment, the utility of a textbook inevitably comes into question; however, when considering the dynamic field of neurointerventional surgery, we recognized the need for a concise body of evidence that was dedicated towards a fundamental tenet of our field—the importance of *methodology*.

This book is designed to focus on three segments of neurointervention: aneurysms, non-aneurysmal cerebrovascular malformations, and stroke. In each chapter, the authors combine evidence-based research with personal clinical experience in order to guide the reader through the treatment of complex patients. Each chapter provides pertinent—though not exhaustive—summaries of anatomy, physiology, and clinical presentation. The chapters are further concentrated to provide specific bullet-point methodology of diagnosis, clinical workup, imaging, and management of each pathological entity. This provides the reader with the ability to quickly obtain up-to-date information through a very user-friendly format. Each author is a current or emerging leader of his or her field, yet emphasis was stressed on practical take-home knowledge. Although some overlap exists among related chapters, the editors have made every effort to minimize duplicate discussion, except when controversial or state-of-the-art issues are presented.

A field such as neurointerventional surgery incorporates multiple facets, including complex medical, procedural, technical, and ethical components. Consequently, this book is certainly not meant to be a comprehensive lexicon on our subject matter. Rather, it is meant to provide a snapshot of methodology to a readership beyond the neurointerventional surgeon. We intend this book to appeal to medical students, residents, fellows, nurse practitioners, and any member of the medical community who desires an easily digestible primer on the methods of disease management that we employ on a day-to-day basis.

The growth of neurointerventional surgery is both inevitable and welcomed. Our field possesses a unique and symbiotic relationship with technology, which inevitably accelerates towards innovation, seemingly without bounds. Not only do we now have the knowledge and devices to treat diseases that were previously determined to be "inoperable," we are also now witnessing the birth of robotic assistance in neuroendovascular procedures. This technology, even in its preliminary form, holds enormous potential for our field and all of medicine. The use of robotic technology to deliver advanced care in rural areas may be a solution to the largest challenge facing patients suffering from acute ischemic stroke, a disease that is particularly burdensome to underserved populations across the planet.

Much like the flourishing technology of the Industrial Revolution in which Jack London found himself, we are in the midst of a scientific and medical revolution within our field. Throughout these exciting and innovative times, rigorous methodology to the workup, diagnostic, and therapeutic management of our patients will always remain a core principle of our field. Whether the reader is seeing a patient in a clinical setting tomorrow or even if they are performing a robotic procedure on a remote patient in the not-too-distant future, we hope they can carry the application of the doctrines expressed in this book along with them. We challenge the reader to learn and grow with us throughout this exciting journey.

Valhalla, NY, USA   *Fawaz Al-Mufti*
Indianapolis, IN, USA   *Krishna Amuluru*

Acknowledgments

The editors are grateful to have been entrusted with the responsibility of developing this book for the community of neurointerventional surgery, vascular neurology, neurocritical care, neurosurgery, neurology, and neuroradiology. The organization and publication of this text required the knowledge and assistance of many individuals. We would like to thank all of our physician contributing authors, fellows, residents, and medical students, and all the universities in which they trained and practice; the editors owe a tremendous intellectual debt. We appreciate their time, reviews, constant encouragement, and well-deserved criticism towards the formulation of *Cerebrovascular Disorders* in the *Neuromethods* series.

We would like to acknowledge the significant support offered by our executive editors, Patrick J. Marton and Dr. Wolfgang Walz, in their incipient support for allowing us to launch this project. The editors would also like to thank individual editorial assistants, who provided the continuous contacts, oversight, and management of the manuscripts we received. We specifically acknowledge Anna Rakovsky in this monumental task. Additional thanks goes to Vishnu Prakash, Amelie Vonzumbusch and Johncy Debeorah Jesu Mary for bringing this project to completion. The editors would also like to give personal acknowledgements:

Firstly, to my patients and their families for their trust and for giving me the opportunity to master the art of medicine and be a better physician.

To my parents for always having faith in me, teaching me humility and grit, to always persevere, and seek excellence in whatever task I am given.

To my brother and sister for always being my anchors, my grounding force and believing in me.

To my mentors, colleagues, and students for always challenging me and forging me into an eternal student of medicine.

To my closest friends, whose compassion, generosity, and dedication have inspired and motivated me to become a better person.

To my co-editor, mentor, friend, and selfless brother, Krishna Amuluru, you put more into the world than you take out. Thank you for being my partner on this journey.

Lastly but by no means least, to my dearest Rolla, Jude, Jana, Adam, and Loujane for the unconditional love, support, and encouragement. You have been my rock and lifeline. All that I am, or hope to be, I owe to you.

Fawaz Al-Mufti

An immense amount of gratitude goes to my parents Dr. Prabhakar Rao Amuluru and Dr. Jaladurga Amuluru, who sacrificed so much for their children, and impressed upon me the importance of dedication and hard work. An equal amount of indebtedness goes to my sisters for their infinite support and strength as role models. Thank you to my mentors Dr. Charles Prestigiacomo, Dr. Chirag Gandhi, Dr. Paul Singh, and Dr. Charles Romero for the invaluable lessons you have given me both in medicine and in life. To my co-editor Dr. Al-Mufti who has challenged me from the very first day we met and who truly embodies the essence of a gentleman and scholar. And finally, to my love Julia, who has shown me the true meaning of courage in difficult times.

Krishna Amuluru

Finally, the editors would like to thank all of our patients and their respective families, past, present, and future, for their courage and trust in allowing us to participate in their care. While this book may, in small part, represent all that you have taught us, there is simply no amount of text that could convey our gratitude to you.

Contents

Contributors

CATHERINE ALBIN • *Department of Neurology, Brigham and Women's Hospital, Massachusetts General Hospital, Boston, MA, USA*

MAIS N. AL-KAWAZ • *Department of Neurology, Weill Cornell Medical College, New York, NY, USA*

FAWAZ AL-MUFTI • *Associate Chair of Neurology for Research, New York Medical College, Associate Professor of Neurology, Neurosurgery and Radiology, New York Medical College, Director of the Neuroendovascular Surgery Fellowship, Neuroendovascular Surgery (Interventional Neurologist) Attending, Medical Director of Neurocritical Care, Westchester Medical Center at New York Medical College, Valhalla, NY, USA*

ABDULRAHMAN Y. ALTURKI • *Endovascular and Operative Neurovascular Surgery, BIDMC Neurosurgery & Brain Aneurysm Institute, Harvard Medical School, Boston, MA, USA; Department of Adult Neurosurgery, National Neurosciences Institute, King Fahad Medical City, Riyadh, Saudi Arabia*

KRISHNA AMULURU • *Division of Neurointerventional Radiology, Goodman Campbell Brain and Spine, Indianapolis, IN, USA; Division of Neurointerventional Radiology, Ascension St. Vincent Medical Center, Indianapolis, IN, USA; Department of Neurosurgery, Marian College of Osteopathic Medicine, Indianapolis, IN, USA*

SAMEER A. ANSARI • *Department of Radiology, Feinberg School of Medicine, Northwestern University, Chicago, IL, USA; Department of Neurology, Feinberg School of Medicine, Northwestern University, Chicago, IL, USA; Department of Neurological Surgery, Feinberg School of Medicine, Northwestern University, Chicago, IL, USA*

SALMAN ASSAD • *Shifa College of Medicine Pakistan, Shifa Tameer-e-Millat University, Islamabad, Pakistan*

GARY L. BERNARDINI • *Department of Neurology, Weill Cornell Medical College, New York, NY, USA; NewYork-Presbyterian Queens, Flushing, NY, USA*

SOHYUN BOO • *Department of Neuroradiology, Robert C. Byrd Health Sciences Center, West Virginia University School of Medicine, Morgantown, WV, USA*

DONALD R. CANTRELL • *Department of Radiology, Feinberg School of Medicine, Northwestern University, Chicago, IL, USA*

KEVIN COCKROFT • *Neurosurgery Department, Penn State Hershey Medical Center, Hershey, PA, USA*

ERIC R. COHEN • *Department of Neurosurgery, Rutgers Robert Wood Johnson Medical School, New Brunswick, NJ, USA; Division of Interventional Neuroradiology, Ohio Imaging Associates, Kent, OH, USA*

MICHAEL COHEN • *Eastern Maine Medical Center, Northern Light Health, Bangor, ME, USA*

MICHAEL CRIMMINS • *Interventional Neurology and Neurointensive Care, Walter Reed National Military Medical Center, Bethesda, MD, USA*

ELIE DANCOUR • *Department of Neurology, Vassar Brothers Medical Center, Poughkeepsie, NY, USA*

NEHA S. DANGAYACH • *Department of Neurology and Neurosurgery, Mount Sinai Hospital, New York, NY, USA*

MOHAMMAD EL-GHANEM • *Department of Neurology, Neurosurgery and Medical Imaging, University of Arizona, Tucson, AZ, USA*

YUVAL ELKUN • *Department of Neurosurgery, Westchester Medical Center at New York Medical College, Valhalla, NY, USA*

NASIR FAKHRI • *Department of Neurology, Warren Alpert School of Medicine at Brown University, Providence, RI, USA*

CHIRAG D. GANDHI • *Department of Neurosurgery, Westchester Medical Center at NY Medical College, Valhalla, NY, USA*

FRANCISCO EDUARDO GOMEZ III • *Division of Neurocritical Care, Department of Neurology, University of Pennsylvania, Philadelphia, PA, USA*

RAM GOWDA • *Section of Neurocritical Care and Emergency Neurology, Program in Trauma, Department of Neurology, University of Maryland School of Medicine, Baltimore, MD, USA*

BRIAN MAC GRORY • *Division of Vascular Neurology, Department of Neurology, Rhode Island Hospital, Providence, RI, USA*

GAURAV GUPTA • *Department of Neurosurgery, Rutgers Robert Wood Johnson Medical School, New Brunswick, NJ, USA*

AARON M. GUSDON • *Department of Neurology, Weill Cornell Medical College, New York, NY, USA*

DANIEL IKEDA • *Interventional Neurology and Neurointensive Care, Walter Reed National Military Medical Center, Bethesda, MD, USA*

SAEF IZZY • *Divisions of Stroke, Cerebrovascular and Critical Care Neurology, Department of Neurology, Brigham and Women's Hospital, Harvard Medical School, Boston, MA, USA*

SAMEEN JAFARI • *Department of Neurocritical Care, Brigham and Women's Hospital, Massachusetts General Hospital, Boston, MA, USA*

JEREMY KARLIN • *Interventional Neurology and Neurointensive Care, Walter Reed National Military Medical Center, Bethesda, MD, USA*

CHRISTOPHER KELLNER • *Department of Neurosurgery, Mount Sinai Hospital, New York, NY, USA*

ALEXANDER W. KORUTZ • *Department of Radiology, Feinberg School of Medicine, Northwestern University, Chicago, IL, USA*

SHOURI LAHIRI • *Department of Neurology, Cedars Sinai Medical Center, Los Angeles, CA, USA*

EHUD LAVI • *Department of Pathology, Weill Cornell Medical College, New York, NY, USA*

JAMES LEE • *Division of Neurocritical Care, Hospital of the University of Pennsylvania, Philadelphia, PA, USA*

MACKENZIE P. LERARIO • *Department of Neurology, Weill Cornell Medical College, New York, NY, USA; NewYork-Presbyterian Queens, Flushing, NY, USA; Fordham University Graduate School of Social Service, New York, NY, USA*

NEIL MAJMUNDAR • *Department of Neurological Surgery, Rutgers New Jersey Medical School, Newark, NJ, USA*

STEPHAN A. MAYER • *Department of Neurology, Henry Ford Health System, Detroit, MI, USA*

ALEXANDER E. MERKLER • *Clinical and Translational Neuroscience Unit, Feil Family Brain and Mind Research Institute, Weill Cornell Medical College, New York, NY, USA*

PHILLIP M. MEYER • *Department of Radiology, The Neurological Institute of New York, New York, NY, USA; Department of Neurological Surgery, The Neurological Institute of New York, New York, New York, NY, USA*

TIMOTHY R. MILLER • *Section of Interventional Neuroradiology, Department of Radiology and Nuclear Medicine, University of Maryland School of Medicine, Baltimore, MD, USA*

NICHOLAS A. MORRIS • *Section of Neurocritical Care and Emergency Neurology, Program in Trauma, Department of Neurology, University of Maryland School of Medicine, Baltimore, MD, USA*

MUHAMMAD NIAZI • *Endovascular and Operative Neurovascular Surgery, BIDMC Neurosurgery & Brain Aneurysm Institute, Harvard Medical School, Boston, MA, USA*

CHRISTOPHER S. OGILVY • *Neurosurgery, Harvard Medical School, BIDMC Brain Aneurysm Institute, Endovascular and Operative Neurovascular Surgery, Boston, MA, USA*

MATTHEW M. PADRICK • *Department of Neurology, Cedars Sinai Medical Center, Los Angeles, CA, USA*

SAMMY PISHANIDAR • *Department of Neurology, Weill Cornell Medical College, New York, NY, USA*

MICHAEL E. REZNIK • *Division of Neurocritical Care, Warren Alpert Medical School at Brown University, Providence, RI, USA*

DAVID J. ROH • *Division of Neurocritical Care, New York-Presbyterian/Columbia University Medical Center, New York, NY, USA*

SUDIPTA ROYCHOWDHURY • *Department of Neurointerventional Radiology, Rutgers Robert Wood Johnson Medical School, New Brunswick, NJ, USA*

IGOR RYBINNIK • *Department of Neurology, Rutgers University-Robert Wood Johnson University Hospital, New Brunswick, NJ, USA*

KONRAD SCHLICK • *Department of Neurology, Cedars Sinai Medical Center, Los Angeles, CA, USA*

AJITH THOMAS • *Harvard Medical School, BIDMC Brain Aneurysm Institute, Boston, MA, USA*

RYAN C. TURNER • *PGY-6, Department of Neurosurgery, West Virginia University School of Medicine, Morgantown, WV, USA*

SHADI YAGHI • *Division of Vascular Neurology, Department of Neurology, Rhode Island Hospital, Providence, RI, USA; Department of Neurology, Warren Alpert School of Medicine at Brown University, Providence, RI, USA*

Part I

Cerebral Aneurysms: Workup and Treatment of Cerebral Aneurysms

Chapter 1

Cerebral Aneurysms: Formation, Growth, and Rupture

Krishna Amuluru and Fawaz Al-Mufti

Abstract

The prevalence of unruptured intracranial aneurysms in the adult population is 2-5%, and is increasing due to modern imaging technologies. Cerebral aneurysms are not static entities; rather they may exhibit prominent dynamic features of inflammation and tissue degeneration. Additional hemodynamic, genetic, hormonal, and environmental factors may also play an important role in aneurysmal evolution and rupture. Knowledge of the pathophysiology of cerebral aneurysms may lead the way for development of novel imaging and therapeutic therapies. The purpose of this chapter is to summarize current data on the molecular mechanisms, genetics, and risk factors for aneurysm formation, growth, and rupture, and to provide a framework to implement that data into clinical practice.

Key words Cerebral aneurysm, Pathophysiology, Risk factors, Subarachnoid hemorrhage, Ischemia, Headache, Inflammation, genetics, Coil embolization, Aneurysm clipping, ISUIA , PHASES , Screening

1 Introduction

Cerebral aneurysms (CAs) occur in 2% to 5% of the general population and are characterized by localized structural deterioration of the arterial wall, with loss of the internal elastic lamina and disruption of the media [1]. Aneurysmal rupture leading to subarachnoid hemorrhage (SAH) is the single most feared complication of a CA, the likelihood of which is related to several modifiable and nonmodifiable risk factors. Despite advances in surgical and endovascular techniques and neurocritical care management, the overall fatality of aneurysmal SAH remains remarkably high at approximately 40% [2].

In order to prevent the devastating consequences of aneurysmal SAH, invasive treatment options for unruptured CAs must be carefully weighed against the risk of rupture, as these procedures

The original version of this chapter was revised. The correction to this chapter is available at https://doi.org/10.1007/978-1-0716-1530-0_23

Fawaz Al-Mufti and Krishna Amuluru (eds.), *Cerebrovascular Disorders*, Neuromethods, vol. 170, https://doi.org/10.1007/978-1-0716-1530-0_1, © Springer Science+Business Media, LLC, part of Springer Nature 2021, Corrected Publication 2021

carry considerable risks [3, 4]. Consensus does not exist on the most efficacious screening or treatment paradigm for unruptured aneurysms [2, 5, 6]. These issues will continue to persist as the discovery of incidental aneurysms increases due to the rising availability of magnetic resonance (MR) angiography and three-dimensional computed tomograghy (CT) angiography of the brain [4]. In fact, the number of hospitalizations and the total costs associated with the diagnosis and treatment of cerebral aneurysms has increased by 75% and 200%, respectively, in the United States during the past decade [2].

In recent years, several advancements have been made in the basic scientific knowledge of aneurysm pathology. Prominent inflammatory, hemodynamic, genetic, and environmental factors contribute to their clinical presentation. Knowledge of the pathophysiology of CAs may further advance predictive risk models of aneurysmal formation, growth and rupture. The purpose of this chapter is to summarize current data on the molecular mechanisms, genetics, and risk factors for aneurysm formation, growth, and rupture, and to provide a framework to implement that data into clinical practice.

2   Pathophysiology

2.1   Aneurysm Formation: Inflammation

Inflammatory processes have been implicated as leading factors in the pathogenesis of CAs. Macrophages and smooth muscle cells (SMCs) are the two main constituents of the inflammatory process [7]. Tissue-infiltrating macrophages are responsible for the release of proinflammatory cytokines, propagating additional inflammatory processes, as well as releasing matrix metalloproteinases (MMPs) that digest the arterial wall extracellular matrix [7, 8]. Aoki et al. demonstrated that macrophage expression and macrophage-derived MMPs were closely associated with aneurysm growth, and that selective inhibition of these MMPs blocked aneurysm progression [9].

SMCs, mostly concentrated in the media, are the predominant matrix-synthesizing cells of the vascular wall. During the early stages of aneurysm formation, SMCs migrate into the intima in response to endothelial injury and proliferate to produce myointimal hyperplasia. Subsequently, SMCs undergo phenotypic modulation and their ability to synthesize collagen is severely impaired [7]. This loss of SMC function and subsequent thinning of the media is a common observation in ruptured aneurysms [8]. Sakaki et al. showed that SMCs in the wall of ruptured aneurysms were much more degenerated/apoptotic than those in the wall of unruptured aneurysms [10]. Several additional inflammatory mediators including mast cells, cytokines such as IL-1β and TNFα, chemokines, and reactive oxygen species are also thought to be involved in the pathophysiology of CA formation [8, 11].

2.2 Aneurysm Formation: Hemodynamics

In addition to the microscopic cellular pathophysiology of CA genesis, macroscopic hemodynamics are also major factors in CA formation [8]. Clinical presentations and animal models have shown that CAs occur predominantly at sites of complex arterial geometries with excessive hemodynamic stress and elevated wall shear stress (WSS), such as junctions, bifurcations, or acute vascular angles. Using a combination of animal modeling, histopathology, and computational fluid dynamics, Meng et al. demonstrated that elevated WSS, as well as gradients in WSS, strongly predispose to CA formation [12]. Furthermore, significant correlation exists between the magnitude of WSS and internal elastic lamina loss, medial thinning, and luminal bulging, thus predisposing to aneurysm genesis [13, 14].

However, controversy exists regarding the extent, or lack thereof, of WSS responsible for subsequent growth and rupture of cerebral aneurysms. Some studies suggest that elevated WSS directly causes endothelial injury and thus propagates degeneration of arterial media [12]. Other studies, however, show that low WSS within aneurysms causes localized flow stagnation, flow-induced nitric oxide dysfunction, and endothelial apoptosis leading to further inflammation and mural degradation [15].

2.3 Aneurysm Formation: Genetics

Genetic factors play an important part in the pathogenesis of SAH and CAs; first-degree relatives of patients with aneurysmal SAH have up to seven times greater risk than the general population [16]. Certain heritable connective tissue disorders demonstrate well-known associations with CAs. Autosomal dominant polycystic kidney disease is the most common heritable disease associated with CAs, with 10–13% of patients developing CAs. Ehlers-Danlos type IV, fibromuscular dysplasia, COL4A1-related disorders and autosomal dominant hyper-IgE syndrome are some additional heritable disorders associated with CAs [11, 17].

In addition to uncertainty surrounding genetic contribution to CA formation, the role of genetics in CA rupture and SAH also remains inconclusive. The largest twin cohort to date, the population-based Nordic Twin Cohort, followed 79,644 complete twin pairs of Danish, Finnish, and Swedish origin for 6.01 million person-years, and did not show a significant degree of genetic contribution to SAH. The authors suggest that only a minority of the population-attributable risk for SAH is due to clustering of susceptible genes, and that familial clustering of confounding risk factors (smoking, hypertension, alcohol consumption) is an overlooked contributor to incidence rates of familial SAHs [18].

However, genome-wide linkage studies have identified genetic loci for CAs, such as 7q11, 14q22, and 5q22-31 in a Japanese study and 19q13.3 in a Finnish study [19, 20]. These loci include genes coding for structural proteins of the extracellular matrix such as elastin and collagen, which are promising candidates for allelic association with CAs [16]. Extensive research continues to search for the exact role of both genetics and environmental factors in CA formation and rupture.

2.4 Aneurysm Formation: Risk Factors

Several risk factors exist for the presence of CAs (many of which are similar risk factors for both subsequent growth and rupture, which will be discussed later). A recent, large meta-analysis that included 68 studies involving 1450 unruptured CAs in 94,912 patients from 21 countries identified autosomal dominant polycystic kidney disease, a positive family history of CA or SAH, female sex, and older age as significant risk factors for harboring CAs [21]. While the aforementioned meta-analysis could not separately assess the prevalence of CAs in smokers and patients with hypertension, active smoking and hypertension are well-known strong risk factors for CA formation, with odds ratios of 3.0 and 2.9 respectively. The combination of the two has an important additive effect, with an odds ratio of 8.3 [22]. Although female sex is a well-known risk factor for CA formation, this becomes evident only in the perimenopausal and postmenopausal periods [11].

3 Clinical Presentations

3.1 Clinical Signs and Symptoms of Unruptured Cerebral Aneurysms

Although the most feared complication of CAs are those related to SAH, CAs may also cause clinical presentations other than rupture. The majority of aneurysms that do cause symptoms are large; typically greater than 10 mm [23]. Such signs and symptoms may include the following.

- Ischemia.
- Headache.
- Seizure.
- Cranial neuropathy.
- Visual loss.
- Pyramidal tract dysfunction.
- Facial pain.

Since the majority of unruptured CAs (65–85%) are small (<7 mm in diameter), such clinical symptomatology from mass effect or ischemia is rare [2, 6, 21, 24]. The International Study of Unruptured Intracranial Aneurysms (ISUIA), the largest natural history study of CAs performed to date, demonstrated aneurysmal symptoms and signs other than rupture in only 11% of untreated patients [6].

As research on the natural history of incidental unruptured CAs continues, identifying indicators of impending rupture has been an area of considerable interest. In lieu of clinical signs or symptomatology such as ischemia or mass effect, clinicians are left with only patient-specific and aneurysm-specific variables when making decisions on whether to treat or manage conservatively.

3.2 Radiographic Presentation; Aneurysm Size and Location

Aneurysm size has been a frequently studied variable in relation to risk of rupture. In a recent well-designed Japanese cohort examining 6697 aneurysms, 91% of which were discovered incidentally, increasing CA size was associated with a significantly increased risk of rupture according to multivariate analysis [25]. The annual rupture rate by size was as follows:

- 3–4 mm—0.36%.
- 5–6 mm—0.50%.
- 7–9 mm—1.67%.
- 10–24 mm—4.37%.
- ≥25 mm—33.4%.

In the same cohort, the annual rupture rate by location was as follows:

- Paraclinoid ICA—0.26%.
- Middle cerebral artery—0.67%.
- Anterior communicating artery—1.31%.
- Posterior communicating artery—1.72%.
- Basilar artery—1.90%.

Although this cohort was limited to a Japanese population, a similar association between size and risk of rupture was observed in the ISUIA data, the largest natural history study performed to date [6, 25]. Phase 1 of ISUIA, which was a retrospective study and reported an annual rupture rate of 0.05% for CAs <10 mm in diameter, received heavy criticism for recruitment and selection bias, thus prompting the second phase [26]. Phase 2 of ISUIA involved a large-scale prospective cohort study following 1692 patients (1077 without a prior history of SAH) with unruptured CAs that were 2 mm or larger. Phase 2 of ISUIA documented an overall annual rupture rate of 0.7%, and showed that aneurysm size and location were powerful predictors of future rupture [6]. The 5-year cumulative rupture rates for patients who did not have a history of subarachnoid hemorrhage with aneurysms located in the ICA, anterior communicating or anterior cerebral artery, or middle cerebral artery were 0%, 2.6%, 14.5%, and 40% for aneurysms <7 mm, 7–12 mm, 13–24 mm, and ≥25 mm, respectively, compared with rates of 2.5%, 14.5%, 18.4%, and 50%, respectively, for the same size categories involving posterior circulation and posterior communicating artery aneurysms. Furthermore, a meta-analysis performed by Wermer et al. of more than 4700 patients and

6500 CAs taken from 19 studies worldwide revealed that CAs sized 5–10 mm and >10 mm had a 2.8 times and 5.2 times greater risk of rupture, respectively, than aneurysms <5 mm [27].

While size is clearly a significant factor in rupture risk, the inconsistent statistical analysis among several studies precludes the use of a definitive size threshold for the treatment of CAs. Aneurysm location, however, is a clear determinant of natural history. In fact, the only clear, independent predictor of future rupture among patients with a history of SAH in the ISUIA trial was basilar tip location. Likewise, the meta-analysis performed by Wermer et al. in 2007 found the posterior circulation location to be a statistically significant risk factor for aneurysmal rupture [27]. On the other hand, CAs in the cavernous ICA are much more benign and rupture of these extradural aneurysms rarely causes SAH. Further, their natural history suggests that a large number of such aneurysms will remain clinically asymptomatic [27].

3.3 Radiographic Presentation: Aneurysm Geometry

In addition to CA size, several geometric indices of CAs have been studied as determinants of rupture risk, and are taken into consideration when evaluating the imaging presentation of an aneurysm. In a practical setting, simple calculable indices, which can be derived from 2- and 3-dimensional imaging, remain the most useful.

The aspect ratio (AR), defined as CA height divided by neck diameter, is the most studied and perhaps the most useful geometric parameter. Ujiie et al. reported that 80% of ruptured aneurysms had an AR of >1.6, whereas 90% of unruptured aneurysms had an aspect ratio of <1.6 [28]. While AR should not be relied on as a sole predictor of CA rupture risk, a treating physician may consider intervention rather than observation in questionable cases with AR >1.6.

The size ratio (CA height divided by the parent vessel diameter) is another simple calculable index that has been associated with risk of CA rupture. Kashiwazaki et al. studied 854 ruptured and 180 unruptured CAs, and showed that size ratio strongly correlated with aneurysm rupture and was found to be the only predictive factor for rupture of small CAs [29]. Additional geometric indices such as nonsphericity index (the level to which an aneurysm surface deviates from a spherical shape), the undulation index (a quantitative measure of surface irregularity), and the ellipticity index have also been shown to be predictors of rupture risk, although these indices are not as easily calculated [30].

3.4 Aneurysm Growth: Risk Factors

A CA grows by expansion of the wall due to either cellular proliferation and production of extracellular matrix, or by hemodynamic pressure causing mural distension, or a combination thereof [11]. In the search for risk predictors of CA rupture, interval growth of a CA may be a powerful candidate because the same factors that drive aneurysm growth, such as inflammation and matrix degeneration, may be the same factors that lead to subsequent rupture. This is supported by studies that show highly increased risks of CA rupture in growing aneurysms [31, 32]. In a recent prospective study that followed 258 unruptured aneurysms, the annual risk of rupture was found to be 12-times greater in aneurysms with growth compared to those without growth [31].

A large meta-analysis of 18 studies, described 3990 patients with 4972 unruptured CAs, and identified several aneurysm-specific and patient-specific risk factors for CA growth.

- Female sex.
- Hypertension.
- Smoking at baseline.
- Multiple intracranial aneurysms.
- Posterior circulation location.
- Larger initial size.
- Irregular shape.

Any aneurysm that grows during the follow-up period should be considered for treatment, since interval growth suggests an active inflammatory or a hemodynamically stressed process.

3.5 Aneurysm Rupture: Risk Factors

While most CAs that grow will eventually rupture, many will not. This suggests that the processes which drive aneurysm progression and rupture are complex, but not entirely identical. Certain patient- and aneurysm- specific characteristics have been consistently reported as risk factors for CA rupture [2, 4, 6, 25].

- Larger initial size.
- CA location (i.e., posterior circulation, posterior communicating artery).
- Female sex.
- Young age.
- Previous SAH.
- Family history of intracranial aneurysms.
- Heavy alcohol consumption.
- Finnish, Japanese populations.

Interestingly, some differences exist between risk factors for aneurysm growth and risk factors for aneurysm rupture. For example, multiple intracranial aneurysms and smoking at baseline are strong risk factors for aneurysm growth, but have been shown to be of limited predictive value for aneurysm rupture in recently published prediction tools [4, 33]. For some investigators, hypertension is a risk factor for CA formation and growth, but does not affect the risk of rupture once the aneurysm has been formed [11]. Other risk factors such as a previous SAH and a positive family history of CAs are of limited predictive value for aneurysm growth, but are strongly related to aneurysm rupture [4, 33].

One must note that determining true risk factors for CA rupture is extremely difficult, as demonstrated by the fact that some of the aforementioned elements are reported as risk factors in some studies, but not in others. This is most likely due to differences in natural history of CAs between populations, study design, treatment guidelines, or imbalanced distribution of other risk factors between the cohorts. While the above list merely represents the most commonly reported risk factors, pooling of individual patient data will be mandatory to assess independent predictors of CA rupture.

3.6   Risk Stratification of Aneurysm Growth and Rupture

Estimation of absolute risk of aneurysm growth and/or rupture is extremely complex, and thus clinicians and patients are left to make difficult decisions on the basis of several factors. Greving et al. performed a systematic review and pooled analysis of data from 6 prospective cohort studies including 8382 participants with 10,272 aneurysms and 29,166 patient years of follow-up and developed a practical risk score (PHASES) that predicts a patient's 5-year risk of aneurysm rupture on the basis of six patient-specific and aneurysm-specific characteristics (Table 1, Fig. 1) [4]. In addition to stratifying the 5-year risk of CA rupture, increasing PHASES scores have also been shown to be associated with increasing hazard ratios for CA growth [3].

The PHASES score has several limitations, which must be mentioned. Many patient subgroups, such as patients with familial aneurysms or young smokers, were under-represented and thus the score may not apply to these patients. The score is validated for only the initial 5 years after CA detection because of limited long-term follow-up data. Finally, when contemplating invasive treatment of an unruptured CA, the risk of intervention must be considered, which the PHASES score does not account for.

An alternative risk-stratification model is the Unruptured Intracranial Aneurysm Treatment Score (UIATS), which involves a scoring strategy on the management of unruptured CAs based on consensus data derived from a multidisciplinary group of 69 specialists [34]. The UIATS is neither a prognostic study nor a predictive model for CA rupture, as it is derived from consensus on CA management among cerebrovascular specialists using the Delphi method and only indirectly from published data.

Table 1
Predictors composing the PHASES aneurysm rupture risk score

PHASES Aneurysm Risk Score	Points
(P) Population	
North American, European (other than Finnish)	0
Japanese	3
Finnish	5
(H) Hypertension	
No	0
Yes	1
(A) Age	
<70 years	0
≥70 years	1
(S) Size of aneurysm	
<7.0 mm	0
7.0–9.9 mm	3
10.0–19.9 mm	6
≥20 mm	10
(E) Earlier SAH from another aneurysm	
No	0
Yes	1
(S) Site of aneurysm	
ICA	0
MCA	2
ACA/Pcomm/posterior	4

To calculate the PHASES risk score for an individual, the number of points associated with each indicator can be added up to obtain the total risk score

SAH subarachnoid hemorrhage, *ICA* internal carotid artery, *MCA* middle cerebral artery, *ACA* anterior cerebral arteries (including the anterior cerebral artery, anterior communicating artery, and pericallosal artery), *Pcom* posterior communicating artery, *posterior* posterior circulation (including the vertebral artery, basilar artery, cerebellar arteries, and posterior cerebral artery)

4   Cerebral Aneurysm Screening

Currently, screening for CAs is recommended only in patients with a positive family history. Experts recommend screening in all those with a family history of autosomal dominant polycystic kidney disease, particularly with family history of CA or SAH. Screening

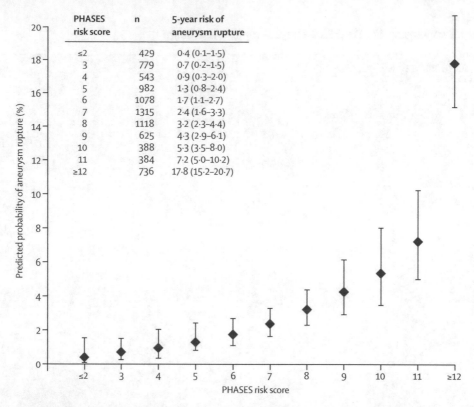

PHASES risk score	n	5-year risk of aneurysm rupture
≤2	429	0·4 (0·1–1·5)
3	779	0·7 (0·2–1·5)
4	543	0·9 (0·3–2·0)
5	982	1·3 (0·8–2·4)
6	1078	1·7 (1·1–2·7)
7	1315	2·4 (1·6–3·3)
8	1118	3·2 (2·3–4·4)
9	625	4·3 (2·9–6·1)
10	388	5·3 (3·5–8·0)
11	384	7·2 (5·0–10·2)
≥12	736	17·8 (15·2–20·7)

Fig. 1 Predicted 5-year risk of aneurysm rupture according to PHASES score (From Greving et al. Development of the PHASES score for prediction of risk of rupture of intracranial aneurysms: a pooled analysis of six prospective cohort studies. Lancet Neurol. 2014 Jan;13(1):59–66; with permission)

is also recommended for first-degree family members in families with 2 or more members with history of CA or SAH [11]. Screening in those with only one affected family member is optional if they possess other risk factors for CA formation such as the following:

- Female sex.
- Older age.
- Active smoking.
- Hypertension.
- Sibling of the affected relative.
- Affected relative harbors multiple aneurysms at a young age.

5   Management

Invasive management options for unruptured cerebral aneurysms include surgical clipping, or endovascular treatment, each of which will be discussed in subsequent chapters. Conservative management is very reasonable, and for many patients, may be the safest option. Currently, prospective randomized controlled trial data for guidance is lacking, particularly in comparing intervention with conservative management. Many publications are retrospective in nature, and lack objective short- and long-term assessment of outcomes [35].

5.1 Conservative Management

Risk factor reduction should be recommended for all patients with unruptured CAs, including optimal blood pressure control, avoidance of cigarette smoking, and moderation of alcohol use.

Few studies exist regarding the safety of antiplatelet agents and anticoagulants in patients with unruptured aneurysms. A case–control study utilizing the ISUIA data suggested that aspirin use may have a protective effect against rupture, and use of aspirin prior to rupture does not appear to worsen outcomes from SAH [36]. Anticoagulation has been associated with a poorer outcome from SAH but is not clearly associated with an increased risk for aneurysm rupture [6]. In patients with unruptured CAs, antiplatelet agents be used as necessary based on the specific indication. Anticoagulants may be necessary due to specific indications, but risks and benefits must be considered. Definitive invasive management of a CA may be considered in some situations where anticoagulation is imperative.

5.2 Treatment Recommendations

Currently the best information regarding management of unruptured CAs is based on observed rates of treatment complications compared to the natural history of unruptured CAs [3, 4]. Based on various treatment guidelines such as the American Heart Association's Stroke Council, the Mayo Clinic College of Medicine and Columbia University, generally accepted guidelines for the management of unruptured CAs include:

Strongly consider invasive treatment in the following cases:

- CA that is ≥12 mm.
- Symptomatic CA.
- Enlarging CA.
- CA in a patient with history of prior aneurysmal SAH.

Possibly consider invasive treatment in the following cases.

- CA that is ≥7 mm or <12 mm and any of the following:
  - Younger patient.
  - High-risk CA location such as posterior circulation or anterior/posterior communicating artery.
  - CA with a daughter sac.
  - Family history of SAH.
- CA that is <7 mm in a younger patient, and any of the following:

- High-risk CA location such as posterior circulation or anterior/posterior communicating artery.
- CA with a daughter sac.
- Family history of SAH.

Treat conservatively in the following cases.

- Anterior circulation CA that is <7 mm, without any high risk features such as family history of SAH or daughter sac.
- Asymptomatic cavernous ICA aneurysm.
- If a CA is treated conservatively, repeat noninvasive imaging is suggested at 6 months after diagnosis, and then annually for 3–5 years, with subsequent decrease in intervals if stability is demonstrated.

References

1. Brisman JL, Song JK, Newell DW (2006) Cerebral aneurysms. N Engl J Med 355:928–939

2. Juvela S, Poussa K, Lehto H, Porras M (2013) Natural history of unruptured intracranial aneurysms: a long-term follow-up study. Stroke 44:2414–2421

3. Backes D, Vergouwen MD, Tiel Groenestege AT, Bor AS, Velthuis BK, Greving JP et al (2015) Phases score for prediction of intracranial aneurysm growth. Stroke 46:1221–1226

4. Greving JP, Wermer MJ, Brown RD Jr, Morita A, Juvela S, Yonekura M et al (2014) Development of the phases score for prediction of risk of rupture of intracranial aneurysms: a pooled analysis of six prospective cohort studies. Lancet Neurol 13:59–66

5. Steiner T, Juvela S, Unterberg A, Jung C, Forsting M, Rinkel G (2013) European stroke organization guidelines for the management of intracranial aneurysms and subarachnoid haemorrhage. Cerebrovasc Dis 35:93–112

6. Wiebers DO, Whisnant JP, Huston J 3rd, Meissner I, Brown RD Jr, Piepgras DG et al (2003) Unruptured intracranial aneurysms: natural history, clinical outcome, and risks of surgical and endovascular treatment. Lancet 362:103–110

7. Kosierkiewicz TA, Factor SM, Dickson DW (1994) Immunocytochemical studies of atherosclerotic lesions of cerebral berry aneurysms. J Neuropathol Exp Neurol 53:399–406

8. Chalouhi N, Ali MS, Jabbour PM, Tjoumakaris SI, Gonzalez LF, Rosenwasser RH et al (2012) Biology of intracranial aneurysms: role of inflammation. J Cerebr Blood Flow Metab 32:1659–1676

9. Aoki T, Kataoka H, Morimoto M, Nozaki K, Hashimoto N (2007) Macrophage-derived matrix metalloproteinase-2 and -9 promote the progression of cerebral aneurysms in rats. Stroke 38:162–169

10. Sakaki T, Kohmura E, Kishiguchi T, Yuguchi T, Yamashita T, Hayakawa T (1997) Loss and apoptosis of smooth muscle cells in intracranial aneurysms. Studies with in situ DNA end labeling and antibody against single-stranded DNA. Acta Neurochir 139:469–474. discussion 474–465

11. Chalouhi N, Hoh BL, Hasan D (2013) Review of cerebral aneurysm formation, growth, and rupture. Stroke 44:3613–3622

12. Sforza DM, Putman CM, Cebral JR (2012) Computational fluid dynamics in brain aneurysms. Int J Numer Meth Biomed Eng 28:801–808

13. Metaxa E, Tremmel M, Natarajan SK, Xiang J, Paluch RA, Mandelbaum M et al (2010) Characterization of critical hemodynamics contributing to aneurysmal remodeling at the basilar terminus in a rabbit model. Stroke 41:1774–1782

14. Meng H, Wang Z, Hoi Y, Gao L, Metaxa E, Swartz DD et al (2007) Complex hemodynamics at the apex of an arterial bifurcation induces vascular remodeling resembling cerebral aneurysm initiation. Stroke 38:1924–1931

15. Boussel L, Rayz V, McCulloch C, Martin A, Acevedo-Bolton G, Lawton M et al (2008) Aneurysm growth occurs at region of low wall shear stress: patient-specific correlation of hemodynamics and growth in a longitudinal study. Stroke 39:2997–3002

16. Ruigrok YM, Rinkel GJ, Wijmenga C (2005) Genetics of intracranial aneurysms. Lancet Neurol 4:179–189

17. Al Mufti F, Alkanaq A, Amuluru K, Nuoman R, Abdulrazzaq A, Sami T et al (2017) Genetic insights into cerebrovascular disorders: a comprehensive review. J Vasc Interv Neurol 9:21–32

18. Korja M, Silventoinen K, McCarron P, Zdravkovic S, Skytthe A, Haapanen A et al (2010) Genetic epidemiology of spontaneous subarachnoid hemorrhage: nordic twin study. Stroke 41:2458–2462

19. Onda H, Kasuya H, Yoneyama T, Takakura K, Hori T, Takeda J et al (2001) Genomewide-linkage and haplotype-association studies map intracranial aneurysm to chromosome 7q11. Am J Hum Genet 69:804–819

20. van der Voet M, Olson JM, Kuivaniemi H, Dudek DM, Skunca M, Ronkainen A et al (2004) Intracranial aneurysms in finnish families: confirmation of linkage and refinement of the interval to chromosome 19q13.3. Am J Hum Genet 74:564–571

21. Vlak MH, Algra A, Brandenburg R, Rinkel GJ (2011) Prevalence of unruptured intracranial aneurysms, with emphasis on sex, age, comorbidity, country, and time period: a systematic review and meta-analysis. Lancet Neurol 10:626–636

22. Vlak MH, Rinkel GJ, Greebe P, Algra A (2013) Independent risk factors for intracranial aneurysms and their joint effect: a case-control study. Stroke 44:984–987

23. Raps EC, Rogers JD, Galetta SL, Solomon RA, Lennihan L, Klebanoff LM et al (1993) The clinical spectrum of unruptured intracranial aneurysms. Arch Neurol 50:265–268

24. Juvela S, Poussa K, Porras M (2001) Factors affecting formation and growth of intracranial aneurysms: a long-term follow-up study. Stroke 32:485–491

25. Morita A, Kirino T, Hashi K, Aoki N, Fukuhara S, Hashimoto N et al (2012) The natural course of unruptured cerebral aneurysms in a Japanese cohort. N Engl J Med 366:2474–2482

26. International Study of Unruptured Intracranial Aneurysms Investigators (1998) Unruptured intracranial aneurysms—risk of rupture and risks of surgical intervention. N Engl J Med 339:1725–1733

27. Wermer MJ, van der Schaaf IC, Algra A, Rinkel GJ (2007) Risk of rupture of unruptured intracranial aneurysms in relation to patient and aneurysm characteristics: an updated meta-analysis. Stroke 38:1404–1410

28. Ujiie H, Tamano Y, Sasaki K, Hori T (2001) Is the aspect ratio a reliable index for predicting the rupture of a saccular aneurysm? Neurosurgery 48:495–502. discussion 502–493

29. Kashiwazaki D, Kuroda S (2013) Size ratio can highly predict rupture risk in intracranial small (<5 mm) aneurysms. Stroke 44:2169–2173

30. Raghavan ML, Ma B, Harbaugh RE (2005) Quantified aneurysm shape and rupture risk. J Neurosurg 102:355–362

31. Villablanca JP, Duckwiler GR, Jahan R, Tateshima S, Martin NA, Frazee J et al (2013) Natural history of asymptomatic unruptured cerebral aneurysms evaluated at ct angiography: growth and rupture incidence and correlation with epidemiologic risk factors. Radiology 269:258–265

32. Mehan WA Jr, Romero JM, Hirsch JA, Sabbag DJ, Gonzalez RG, Heit JJ et al (2014) Unruptured intracranial aneurysms conservatively followed with serial ct angiography: could morphology and growth predict rupture? J Neurointerv Surg 6:761–766

33. Backes D, Rinkel GJ, Laban KG, Algra A, Vergouwen MD (2016) Patient- and aneurysm-specific risk factors for intracranial aneurysm growth: a systematic review and meta-analysis. Stroke 47:951–957

34. Etminan N, Brown RD Jr, Beseoglu K, Juvela S, Raymond J, Morita A et al (2015) The unruptured intracranial aneurysm treatment score: a multidisciplinary consensus. Neurology 85:881–889

35. Williams LN, Brown RD Jr (2013) Management of unruptured intracranial aneurysms. Neurol Clin Pract 3:99–108

36. Hasan DM, Mahaney KB, Brown RD Jr, Meissner I, Piepgras DG, Huston J et al (2011) Aspirin as a promising agent for decreasing incidence of cerebral aneurysm rupture. Stroke 42:3156–3162

Chapter 2

Current Imaging Techniques of Aneurysms

Donald R. Cantrell, Sameer A. Ansari, and Alexander W. Korutz

Abstract

This chapter will review the role of noninvasive imaging, such as CT angiography and MR angiography, in the management of intracranial aneurysms. In addition, we will also review some of the new advanced imaging techniques that are being used to image intracranial aneurysms.

Key words CT angiography, MR angiography, Screening, Follow-up, Intracranial aneurysm

1 Introduction

It is estimated that 3% of the adult population harbors an intracranial aneurysm (IA), of which ~0.25% will rupture each year [1]. Approximately 26–36% of these patients will die, and an additional 6–7% will sustain severe neurological deficits, leaving many of them functionally dependent [2]. Although conventional cerebral angiography remains the gold standard for the identification and characterization of IAs, noninvasive imaging technologies, such as CT angiography (CTA) and MR angiography (MRA), provide accurate, low-risk alternatives for aneurysm diagnosis and surveillance. Technological advancements in CTA and MRA have increased the overall image quality and spatial resolution of these techniques, further establishing CTA/MRA vascular imaging as the primary noninvasive diagnostic imaging modalities in the evaluation and clinical management of patients with IAs. More recently, emerging advanced imaging technologies, including vessel wall imaging, permeability imaging, 4D flow imaging, and computational fluid dynamics, are providing new insights into the pathophysiology of IA formation and rupture. In this chapter, we will review the current role of noninvasive imaging in the clinical management of IAs and survey some of the new and promising advanced imaging techniques.

Fawaz Al-Mufti and Krishna Amuluru (eds.), *Cerebrovascular Disorders*, Neuromethods, vol. 170,
https://doi.org/10.1007/978-1-0716-1530-0_2, © Springer Science+Business Media, LLC, part of Springer Nature 2021

2   CT Angiography

Computed tomography (CT) is a fast and readily accessible imaging modality which is currently the primary diagnostic modality for the initial evaluation of patients presenting with headaches in the emergent setting. A noncontrast CT of the brain has a 98–99% sensitivity for the identification of acute subarachnoid hemorrhage (SAH) [3]. In cases that are negative for SAH, the noncontrast head CT may identify an alternative etiology for the patient's symptoms. In cases that are positive for SAH, the distribution of the hemorrhage may be informative (Fig. 1). For instance, the region of most abundant clot may indicate the location of an IA in a patient with aneurysmal SAH. The pattern of hemorrhage can also suggest a nonaneurysmal etiology, such as the "benign" perimesencephalic pattern of venous hemorrhage or high convexity cortical SAH associated with trauma or reversible cerebral vasoconstriction syndrome (RCVS) [4].

CTA can be acquired immediately following a noncontrast head CT without the need to move the patient, making it an expedient and convenient modality in the workup for intracranial hemorrhage. CTA pairs a thin-section, volumetric CT acquisition with peripheral intravenous contrast injection that is time-optimized for arterial opacification. At our institution, 50–60 mL iodinated contrast is administered using a power injector at 4 mL/s via a peripheral IV that is 18 G or larger. A bolus tracking method is utilized to synchronize the CT acquisition with the contrast injection. In the bolus tracking software, a region of interest is placed within the aortic arch (or on a cervical carotid artery at some institutions), and following the injection of contrast, the attenuation is repeatedly measured within the region of interest until a threshold of 70–80 Hounsfield units is reached, at which point the full spiral acquisition is triggered. An acquisition for CTA of the head will typically extend from the C1 level through the vertex. Inferior extension of imaging through the C1 level is important for visualization of the posterior inferior cerebellar artery (PICA) origins, as 18% of patients will have extracranial/extradural PICA origins [5]. A typical acquisition is performed at 120 kVp and 200 mAs, with a pitch ranging between 0.6 and 1. The volumetric acquisition is reconstructed into axial, sagittal, and coronal reformatted images with a slice thickness of ~1.25 mm and a field of view of 120 mm^2, yielding an in-plane resolution on the order of 0.23 mm^2. Maximum intensity projections along the course of the vessel and three-dimensional (3D) volumetric and surface rendered images can be obtained on a postprocessing workstation for adjunctive image interpretation or at the request of neurointerventional/neurovascular surgeons for treatment planning. However, it is the authors' opinion that multiplanar reformats are adequate for

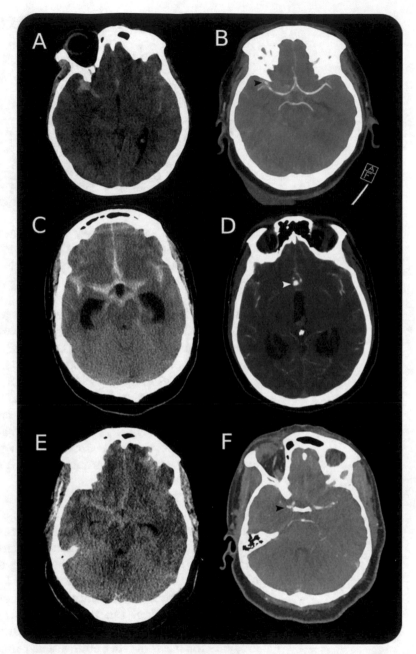

Fig. 1 The location of an aneurysm can often be inferred by the distribution of subarachnoid hemorrhage. Hemorrhage that is thickest in the right Sylvian fissure (**a**) is often the result of a right MCA bifurcation aneurysm (**b**). Anterior communicating artery aneurysms (**d**) will often result in hemorrhage that symmetrically fills the suprasellar cistern and tracks along the ACAs and bilateral MCAs (**c**). A posterior communicating artery aneurysm (**f**) may result in a pattern of hemorrhage (**e**) that is most pronounced in the right lateral aspect of the suprasellar cistern

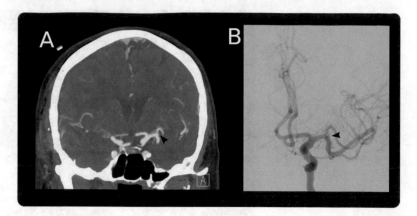

Fig. 2 Despite the tremendous utility of CTA, some of the smallest aneurysms remain below the limit of spatial resolution for the modality. In this case, a small left MCA bifurcation aneurysm (**b**) would be difficult to identify prospectively on CTA (**a**)

an accurate diagnosis in most cases. With proper implementation, CTA can achieve a spatial resolution of approximately 0.4–0.7 mm using a 64 MultiDetector CT (MDCT) and an overall sensitivity for detecting IAs between 96% and 98% when compared to Digital Subtraction Angiography (DSA) [3]. When stratified by size, CTA sensitivity for detecting IAs smaller than 3 mm is approximately 90–94%, while sensitivity for detecting an IA 4 mm or larger approaches 100% (Fig. 2) [3]. Continued advancements in multidetector CT technology are expected to yield further improvements in CTA spatial resolution, further increasing the sensitivity for detecting small IAs.

When imaging IAs after endovascular treatment, streak artifact related to both beam hardening and photon starvation limits evaluation of structures surrounding the coil mass. Since endovascular coils are constructed from metallic/platinum alloys that are designed to be visible under fluoroscopy, the required degree of radiopacity results in a very large amount of streak artifact on CT, significantly limiting the utility of this modality in the post–coil embolization setting. On the other hand, these alloys generate relatively little artifact on MRI, making MRA the preferred imaging modality for IAs following endovascular treatment (Fig. 3).

In contrast to endovascular coils, surgical aneurysm clips generate a disproportionately large amount of MR susceptibility artifact, which frequently precludes evaluation for recurrent or residual IAs with MRA (Fig. 3). As with aneurysm coils, surgical clips also result in CT streak artifact; however, this artifact is substantially less than what is typically seen from a dense coil mass and can be further mitigated with novel postprocessing metallic reduction algorithms and dual energy CT techniques (described below). Therefore, often following a conventional DSA study confirming stable post-surgical clipping findings, there is a transition to CTA for the noninvasive and routine follow-up of surgically clipped IAs [6, 7]. When utilizing CTA to evaluate for a residual or recurrent IA after surgical clipping, the CTA parameters can be optimized to reduce streak artifact by increasing the photon energy to 140 kVp and reducing the pitch to 0.6. Reducing pitch increases the number of photons sampled at the level of the clip, which improves image quality at the expense of increased radiation dose. Increasing the beam energy (kVp) improves photon penetration through the clip, but also decreases the conspicuity of vascular contrast. This negative effect might be counteracted by using a higher concentration iodine injection; however, in practice, this does not typically achieve the expected results [7]. Despite the reduced contrast conspicuity with a higher kVp, the improvement in signal-to-noise ratio that occurs with a higher kVp is preferable and typically used clinically.

Newer techniques derived from Dual Energy CT (DECT) can also improve IA visualization, particularly near high density structures such as bone or metal. Most CT scanners utilize a single polychromatic X-ray source which is paired with multiple rows of detectors that have the same broad spectral sensitivity. In such systems, a single X-ray attenuation value, reported in Hounsfield Units, can be generated for the full energy spectrum. Newer DECT scanners have begun to provide attenuation information as a function of photon energy. Vendors employ several different techniques to perform dual energy acquisitions. In the dual source platform, two separate X-ray sources are offset 90° with respect to each other and generate polychromatic photon spectra centered at two different energies. These two sources simultaneously acquire projectional data as both the sources and detectors rotate around the patient. In a rapid kilovolt switching platform, a single source is utilized, but two different photon spectral energy distributions are generated by rapidly alternating the X-ray source's voltage gradient. Finally, in a detector-based spectral CT, a single polychromatic X-ray source is driven at a single energy, but the CT detector is composed of two independent layers, having different (high or low) spectral energy sensitivities. Each of these commercially available platforms have their own unique set of advantages and disadvantages; however, all three can perform a dual energy acquisition [8].

The most common application of DECT is for material decomposition, a technique that can be performed in either the projection or image domains, depending on the vendor platform that is chosen. For two-material decomposition in the image domain, CT images are first constructed at both high and low polychromatic energies. Each voxel in these two images is then assumed to be composed of two basis materials of densities ρ_1 and ρ_2, each of which have known (precalculated) mass attenuation coefficients that are a function of photon energy $(\mu_1/\rho_1)(E)$ and $(\mu_2/\rho_2)(E)$.

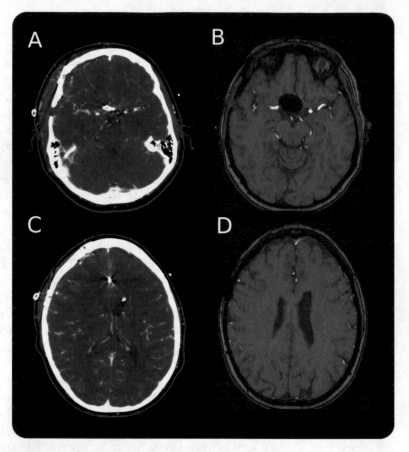

Fig. 3 Surgical clips often create substantial streak artifact (**a**), but the streak is much less limiting than the susceptibility artifact generated by surgical clips on MRA (**b**). In contrast, embolization coils generate extremely little susceptibility artifact (**d**) and a surprisingly large amount of streak (**c**). For these reasons, CTA is the preferred modality for aneurysms treated with surgical clipping, and MRA is the mainstay of follow-up for aneurysms treated by endovascular coiling

The recorded voxel attenuation is then assumed to be a linear combination of the two basis materials according to the equations:

$$\mu(E_L) = (\mu_1/\rho_1)(E_L) \times (\rho_1) + (\mu_2/\rho_2)(E_L) \times (\rho_2).$$
$$\mu(E_H) = (\mu_1/\rho_1)(E_H) \times (\rho_1) + (\mu_2/\rho_2)(E_H) \times (\rho_2).$$

Here, $(\mu_i/\rho_i)(E_j)$ are empirically precalculated values for each basis material at both the high and low polychromatic energy spectra, while $\mu(E_L)$ and $\mu(E_H)$ are the recorded CT voxel attenuation values. Thus, the system of two linear equations can be solved for the only unknown values (ρ_1 and ρ_2), which yields the concentration of both basis materials in each voxel. In neuroradiology, material decomposition using water and calcium basis materials can remove bone from an image to increase the conspicuity of skull base pathology such as IAs [9]. With the additional constraint of mass conservation, three material decomposition can also be performed [8].

DECT can also be used to generate virtual monochromatic images which can minimize beam hardening artifact, improving CTA of vessels at the skull base or adjacent to surgical clips (Fig. 4). Beam hardening artifact is created when a high-density material selectively attenuates low-energy photons from a polychromatic spectrum, leaving only high-energy photons in the X-ray beam. These high-energy photons more easily penetrate the surrounding tissues, causing them to appear darker on the reconstructed CT images. This artifact is often evident adjacent to surgical clips, embolization coils, and, to a lesser degree, adjacent to osseous structures. Beam hardening artifact arises from the polychromatic property of current X-ray sources and would be essentially nonexistent if a perfectly monochromatic source were utilized. Although DECT utilizes polychromatic X-ray sources, virtual monochromatic images can be reconstructed from basis material decomposition maps. For example, after the basis material densities ρ_1 and ρ_2 are computed in a two-material decomposition as described above, a virtual monochromatic attenuation $\mu(E_{\mathrm{mono}})$ at the energy E_{mono} can be generated for each voxel by multiplying each basis material density by its corresponding (known and precalculated) mass attenuation coefficient:

$$\mu(E_{\mathrm{mono}}) = (\mu_1/\rho_1)(E_{\mathrm{mono}}) \times (\rho_1) + (\mu_2/\rho_2)(E_{\mathrm{mono}}) \times (\rho_2).$$

Using this technique, high energy virtual monochromatic DECT can be helpful in reducing beam hardening artifact adjacent to high density structures [10]. The optimum virtual monochromatic energy ranges between 95 and 150 keV, depending on the composition of the material. While monochromatic imaging can mitigate beam hardening artifact, it cannot alleviate photon starvation, a phenomenon that is a large component of metallic streak artifact, as discussed above. Therefore, for very dense metallic materials, virtual monochromatic images are of only limited utility [10].

3 MR Angiography

In addition to being a widely available imaging modality, MRA carries the added benefit of using no ionizing radiation, and offers the ability to perform the examination without intravenous contrast. 3D time-of-flight MR angiography (TOF-MRA) is the most widely utilized MRA technique and can achieve a spatial resolution that approaches 1 mm at 1.5 T and 0.6 mm at 3 T [3]. Limitations of MRA include the increased sensitivity of MR imaging to patient

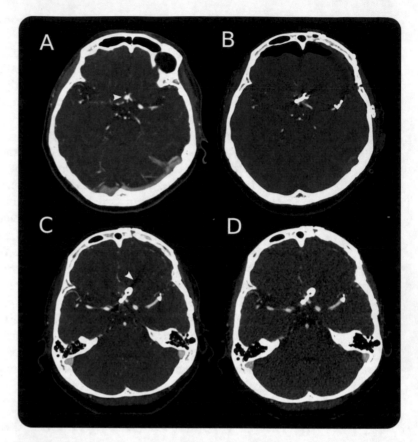

Fig. 4 Dual Energy CT can help to decrease streak artifact surrounding surgical clips by generating virtual monochromatic images. In this case, an anterior communicating artery aneurysm (**a**) was clipped and then imaged in the immediate postoperative setting with conventional CTA (**b**). Subsequently, a Dual Energy scanner was used for follow-up imaging. A large amount of streak artifact remains on the polychromatic reconstruction (**c**), but the streak is substantially reduced on the virtual monochromatic images (**d**)

motion with relatively longer scan times as well as artifacts related to susceptibility from adjacent osseous structures and slow or turbulent blood flow. In a meta-analysis of examinations performed at both 1.5 and 3 T, TOF-MRA was found to have a pooled sensitivity of 95%, with a 95% confidence interval between 89% and 98%, with studies performed on 3 T scanners trending toward higher performance [11]. This meta-analysis did not stratify sensitivity and specificity by IA size, but indicated that in the subset of studies reporting the sizes of missed IAs, 90% of false negatives were <5 mm in maximal dimension on MRA. In a single-institution study of 133 IAs using 1.5 T TOF-MRA, expert neuroradiologists recorded a sensitivity of 89% for IAs >3 mm, but this dropped to only 55% for IAs ≤3 mm, leading the authors to suggest 3 mm as a practical cutoff for TOF-MRA sensitivity (at least at 1.5 T) [12]. A more recent study which focused on detecting IAs <5 mm at 3 T found that experienced readers could identify these lesions with a sensitivity of 98.2–98.9% [13]. The mean maximal dimension of IAs in this study measured 3.17 ± 0.97 mm (Standard Error of Mean). Interestingly, this group reported that postprocessing with dedicated volume rendering software was necessary to identify 49 of the 276 aneurysms identified on MRA in their study, and that these 49 aneurysms were initially missed on review of the source images and standard maximum intensity projections (MIPS) alone. Thus, without volume rendering software support, the sensitivity of MRA for detecting small IAs at 3 T would have been approximately 81%.

TOF-MRA consistently reports a lower overall specificity than sensitivity for detecting IAs. This is primarily related to false positives that result from vascular infundibula where the small exiting vessels are below the resolution of MRA, or when a tortuous vessel origin cannot be fully resolved causing it to resemble a focal outpouching and be mistaken for an IA. For example, in the meta-analysis by Sailer et al., the pooled specificity for all IAs was 89%, with a 95% confidence interval between 80% and 95% (compared to a sensitivity of 95% detailed above) [11]. Li et al. reported a specificity of 93.2–94.9% (compared to a sensitivity of 98.2–98.9%) [13]. In situations where a vascular infundibulum or tortuous vessel origin is suspected but cannot be confirmed, it is the authors' opinion that the risk of a false positive aneurysm diagnosis may be mitigated by performing an additional 3 T TOF-MRA examination to optimize image resolution and signal-to-noise ratio, if the initial MRA was performed at 1.5 T. A separate and complementary CTA examination could also be considered to help exclude aneurysm mimics.

Ultrashort echo time (TE) techniques are used to minimize the impact of susceptibility when imaging vessels and aneurysms near metallic structures, most often endovascular coils. Most TOF-MRA sequences utilize an echo time ranging between 3 and 7 ms. A typical TOF-MRA sequence using a TE of 5 ms will create local distortion and signal dropout which extends approximately 1–2 mm beyond the outer margin of a platinum endovascular coil mass [6]. Even this small amount of artifact can obscure clinically significant residual and recurrent IAs. Ultrashort TE sequences which utilize echo times on the order of 2.5 ms help to minimize signal intensity loss around embolization coils by shortening the amount of time that the coil mass can interact with and alter adjacent nuclear spins. Contrast-enhanced imaging can also be used in conjunction with an ultra-short TE sequence to further improve the signal intensity in close vicinity to endovascular coils. Ultrashort TE and contrast enhanced MRA have become the standard of practice when imaging coiled IAs, especially in the setting of adjunctive intracranial stenting, as the combination of these techniques yields a sensitivity approaching 100% for residual IAs >3 mm [6]. Recent reports regarding gadolinium deposition may decrease the utilization of contrast enhanced MRA techniques for the serial follow-up of IAs status post-coil embolization, due to concerns of repeated patient exposure. However, studies also suggest that most residual IAs missed by an optimized MR angiography protocol are very small, typically measuring <2 mm, and in most cases do not warrant intervention [6].

To further reduce artifact related to a coil mass, some institutions double the receiver readout bandwidth, a technique which comes at the expense of decreased signal-to-noise ratio. Imaging these patients at 3 T can improve the signal, but increased susceptibility artifact would also be expected when doubling the field strength. There is relatively little data comparing the sensitivity of 1.5 and 3 T MRA in the evaluation of IAs after coil embolization; though small studies have suggested that 3 T imaging is not inferior [6, 14].

4 Advanced and Emerging Noninvasive Imaging Techniques

Digital Subtraction Angiography, CTA, and MRA are all anatomical imaging techniques that primarily evaluate the size and shape of the vascular lumen. However, recent histopathological studies are increasingly recognizing IAs as an inflammatory disease of the vessel wall which is characterized by the recruitment of macrophages, the disorganization of smooth muscle cells, the loss of the internal elastic lamina, and the disruption of the media [15]. This developing histopathological understanding has provided incentive for new extraluminal imaging techniques such as vessel wall imaging (VWI) and permeability imaging. Recent research has also established the role of hemodynamic stress in the initiation of IA formation, although its role in IA growth and rupture remains controversial [15]. Hemodynamic stress may be evaluated directly with 4D flow MRI or can be simulated using patient-specific anatomical images with computational fluid dynamics (CFD) modeling techniques.

VWI primarily consists of high resolution pre- and postcontrast T1 weighted black blood FSE/TSE MRI sequences which aim for maximal suppression of blood and CSF signal so as to optimally visualize the vessel wall. At our institution, we utilize T1- and T2-weighted 3D isotropic SPACE sequences (Sampling Perfection with Application Optimized Contrasts Using Different Flip Angle Evolution) for VWI. The clinical application of VWI remains in its infancy, but demand for this MR imaging technique is rapidly expanding. Increasing data suggests that VWI may play an important role in the assessment of IA rupture risk. Inflammatory changes in the IA wall are thought to promote or increase vasa vasorum and compromise wall permeability, which can be detected as enhancement on postcontrast VWI sequences [16]. As such, circumferential wall enhancement may be suggestive of IA inflammation and instability. A recent study of 108 aneurysms found that 87% of unstable aneurysms demonstrated circumferential aneurysm wall enhancement, while only 27% of stable aneurysms demonstrate this same finding [16]. Although further prospective data is required, future aneurysm risk stratification schemes may include circumferential aneurysm wall enhancement as an important predictive factor in addition to standard patient and aneurysm specific risk factors.

An additional promising application of VWI is in the clinical setting of a patient with subarachnoid hemorrhage and multiple IAs (Figs. 5 and 6). This is a relatively common scenario, as up to 30% of patients with aneurysmal subarachnoid hemorrhage are found to have more than one IA, making it difficult to determine which one should be targeted for urgent treatment [17]. Currently, the subarachnoid hemorrhage pattern is correlated with the IA size and morphology to identify the most likely culprit; however, in many cases the source of the hemorrhage may remain unclear. A recent pilot study of five patients with subarachnoid hemorrhage demonstrated enhancement of all ruptured IAs as well as the absence of enhancement in all unruptured IAs [18]. These early findings suggest that VWI may play a role in identifying both unstable and ruptured IAs in the future, although larger prospective research studies are needed.

Permeability imaging is another emerging technique for evaluating IA wall integrity (Fig. 7). This method is predicated on the assumption that a small, but measurable amount of gadolinium contrast leaks from the intravascular space through the aneurysm wall into the extravascular, extracellular space [19]. Permeability imaging is performed by mathematically modeling gadolinium leakage dynamics on a time-series of T1 weighted volumes which are acquired using standard Dynamic Contrast Enhanced (DCE) MRI protocols previously developed for imaging brain tumors. During postprocessing, the Modified Tofts Equation is applied to calculate the kinetic parameters governing the diffusion of the gadolinium tracer through the semipermeable aneurysm wall. K^{trans} is the primary derived parameter and is a direct measure of vessel wall permeability. A pilot study observed measurable gadolinium leakage through aneurysm walls and demonstrated that K^{trans} correlates to other known measures of rupture risk such as IA size, location, and symptomatology [19]. Studies are currently underway to evaluate aneurysm permeability as an independent predictor of IA rupture risk.

It is generally agreed that the complex processes of IA initiation, growth, and rupture are governed by an interplay of pathobiological changes in the vessel wall and mechanical stresses related

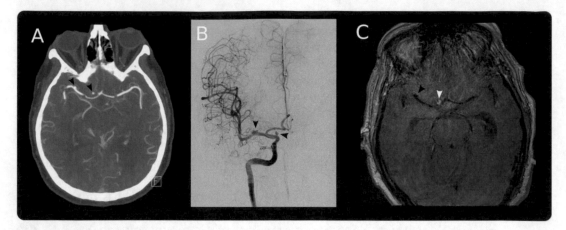

Fig. 5 Vessel Wall Imaging can help to identify the site of rupture in the setting of subarachnoid hemorrhage and multiple aneurysms. In this case, aneurysms were seen to arise from the right supraclinoid ICA and the M1 segment of the right MCA (**a**, **b**). The distribution of subarachnoid hemorrhage was not highly informative. Post-contrast T1 SPACE imaging demonstrated that only the right supraclinoid ICA aneurysm wall was avidly enhancing, identifying it as the likely source of hemorrhage

to altered hemodynamics [15], though the relative importance of these two processes at each stage of the aneurysm life cycle is unclear. CFD is a mathematical technique that allows for the evaluation of flow patterns and hemodynamic stresses within structures of any user-defined shape. This technique has been applied to analyze the flow within patient-specific intracranial arterial anatomy and has gained a great deal of traction within the literature [20]. CFD studies have demonstrated increased wall shear stress at the sites of vessel bifurcations in the Circle of Willis, providing evidence that wall shear stress may be an important factor in the initiation of aneurysm formation. However, the field has not reached a consensus regarding the specific hemodynamic parameters affecting IA growth and rupture. In fact, several high-profile papers seemingly report contradictory results regarding the role of wall shear stress in the growth or rupture of an IA [15], a scenario which has led to some frustration among clinicians [21].

Despite its current limitations in predicting aneurysm rupture, CFD continues to improve our understanding of flow dynamics within IAs. For instance, energy loss has emerged as a promising new hemodynamic parameter which reflects the difference in the hemodynamic energy at the inlet and outlet of an aneurysm. Preliminary work studying a small cohort of seven anterior communicating artery aneurysms suggested that this energy loss may be correlated with IA rupture risk [22].

While CFD simulations are inherently limited by mathematical simplifications and assumptions, new 4D Flow MRI methods have the capability to directly image the in vivo flow patterns and wall shear stresses within IAs (Fig. 8). Recent work has demonstrated

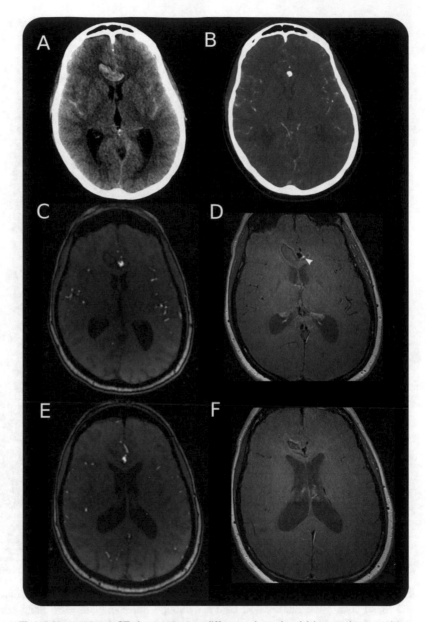

Fig. 6 Noncontrast CT demonstrates diffuse subarachnoid hemorrhage, with a focal hematoma along the course of the anterior cerebral arteries (**a**), and CTA reveals an aneurysm (**b**). TOF MRA (**c**) and post-contrast T1 weighted Vessel Wall Imaging (**d**) shows that the aneurysm wall is avidly enhancing (arrowhead). Both Vessel Wall Imaging and the distribution of blood identify this as the site of rupture. Slightly superior to the focal hematoma, a second ACA aneurysm is present (**e**), but its wall is non-enhancing (**f**)

that 4D Flow imaging can distinguish distinct patterns of hemodynamic flow and wall shear stress in IAs of variable morphology, some of which may predispose a specific IA to rupture or thrombus

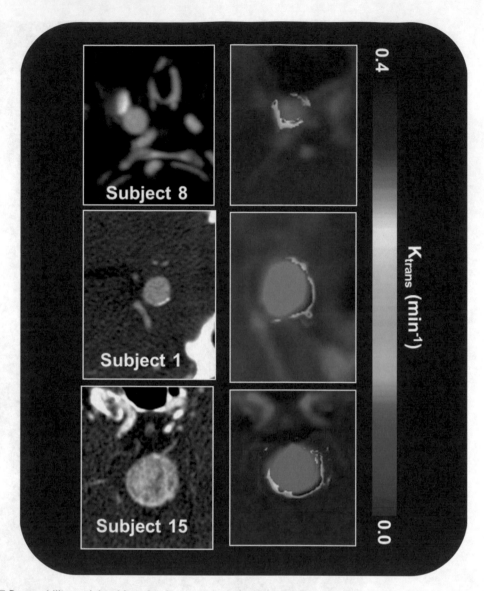

Fig. 7 Permeability-weighted imaging is a new technique that utilizes the Toft model of diffusion to quantify the leakage of gadolinium-based contrast through the aneurysm wall, most commonly described by the leakage coefficient K_{trans}. This method can identify weakened areas of the aneurysm wall that that may be at risk for future bleb formation or rupture. This figure is courtesy of Drs. Parmede Vakil, Charles Grady Cantrell, and Timothy Carroll

formation [23]. Since this technique directly measures in vivo blood flow, there is greater hemodynamic accuracy with elimination of the assumptions required for computational simulation. As this method continues to improve in spatial and flow/velocity resolution, it may yield greater insights into the mechanobiology of IA growth and rupture.

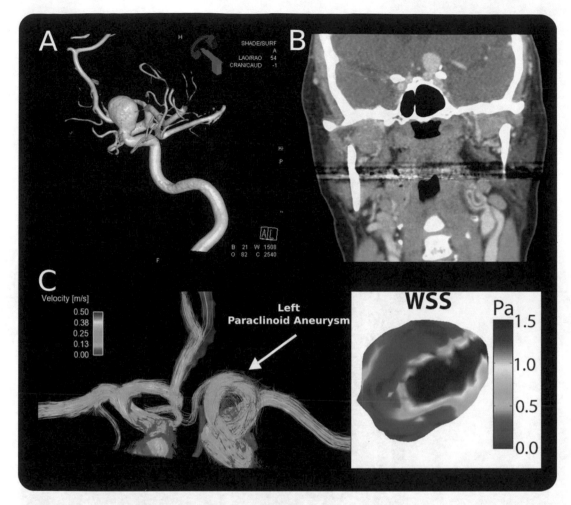

Fig. 8 Conventional cerebral angiography (**a**) and CTA (**b**) can visualize the morphology of this left ICA aneurysm, but fail to capture the underlying mechanobiology. 4D MR Flow imaging is a promising new technique that can directly visualize patterns of flow and wall shear stress within cerebral aneurysms. In the future, these distinguishable patterns may play an important role in risk stratification. This figure is courtesy of Dr. Susanne Schnell

5   The Role of Noninvasive Imaging in Clinical Management

IAs are frequently incidentally detected during high-risk screening or as the result of the diagnostic evaluation of headache or other neurological symptoms. Noninvasive imaging also plays a critical role in the surveillance of both treated and untreated IAs.

5.1   Incidental Aneurysms

As described above, it is estimated that 3% of the population harbors an IA [1] and these are often detected incidentally in clinical practice (Figs. 9 and 10).

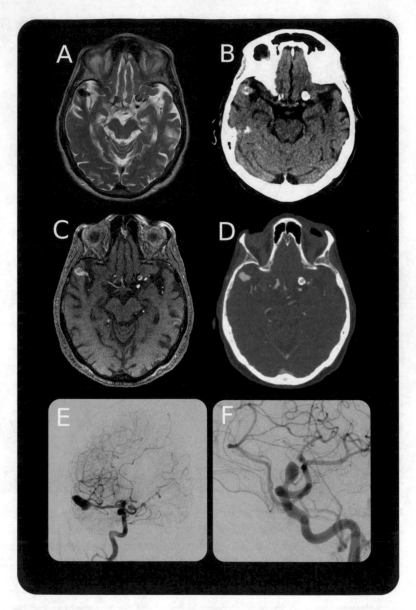

Fig. 9 Incidental aneurysms can be identified on T2-weighted MRI sequences (**a**) and noncontrast CT (**b**). Subsequent MRA (**c**) and CTA (**d**) demonstrate partially calcified right MCA bifurcation and left supraclinoid ICA aneurysms. Conventional cerebral angiography (**e**, **f**) better characterizes the aneurysm morphology

5.2 High-Risk Screening

It has been well established that certain patient populations have a genetic predisposition for the development of IAs and that screening of these patients is often warranted. These high-risk groups include patients with autosomal dominant polycystic kidney disease, Ehlers–Danlos syndrome, neurofibromatosis type 1, coarctation of the aorta, tuberous sclerosis, hereditary hemorrhagic

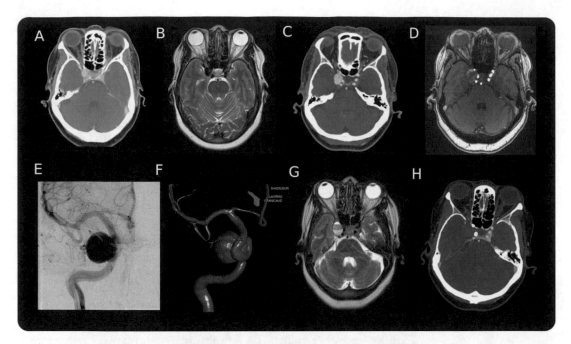

Fig. 10 This large right cavernous ICA aneurysm can be incidentally seen on the first slice of a CT Neck performed for submandibular pain (**a**) and an MR performed for headache (**b**). Subsequent CTA and MRA further characterize the aneurysm (**c, d**). Conventional cerebral angiography (**e**) with surface rendering (**f**) was performed prior to pipeline embolization. Shortly after embolization, stagnant blood and a hematocrit level can be seen within the excluded aneurysm on MR (**g**). There is complete exclusion and nonopacification of the aneurysm on post-embolization CTA (**h**)

telangiectasia, Klippel–Trenaunay–Weber syndrome, and Kawasaki disease [3].

Screening is also indicated in nonsyndromic patients with a family history of IAs. The relative prevalence of hereditary IAs is unknown, but estimates are as high as 6.7% of all IAs [3]. The genetics and mode of inheritance are not yet fully elucidated; however, some patients with a family history of IAs may have a higher risk for harboring an IA. In the Familial Intracranial Aneurysm Trial, aneurysms were found in 19.1% of patients who were greater than 30 years of age, had two or more relatives with an aneurysm, and also had a history of either smoking or hypertension [24].

5.3 Management of Asymptomatic, Unruptured Aneurysms

The International Study of Unruptured Intracranial Aneurysms prospectively evaluated 1692 patients and found the risk for IA rupture in patients with no prior history of SAH to be dependent on size. Although IAs that were <7 mm in size and located in the posterior circulation (or arising from the posterior communicating artery) harbored a 2.5% subarachnoid hemorrhage risk over 5 years (0.5% annually), no ruptures were observed for IAs <7 mm in size in the anterior circulation. Additional IA rupture risk factors revealed in the study included patient age >60 years, female sex, Japanese or Finnish descent, and an irregular morphology (presence of a daughter sac). A subsequent study in the United States followed 319 patients with IAs <7 mm for a mean of 2.4 years and showed no interval rupture or SAH complications [25]. The Small Unruptured Aneurysm Verification prospective study further assessed the rupture risk for IAs <5 mm and determined that the overall rupture risk is approximately 0.34%/year, with IA multiplicity and hypertension representing significant predictive factors for rupture [26]. Hence, for patients of American and European descent with an IA measuring <5 mm in size, it is generally considered reasonable to manage these patients conservatively with medical management/risk factor control and noninvasive imaging follow-up [27]. If conservative management is initiated, the 2015 American Heart Association guidelines recommend an initial follow-up examination at 6–12 months after initial discovery, with subsequent noninvasive imaging performed every 1–2 years once IA stability has been documented [1]. For patients without contraindication to MR imaging, MRA is recommended for long-term follow-up rather than CTA to minimize cumulative patient exposure to ionizing radiation [1]. The American Heart Association Guidelines further suggest that noninvasive imaging follow-up can be discontinued only when the patient's age or medical comorbidities make them ineligible for invasive interventions due to an inordinately high risk or lack of significant benefit.

Imaging surveillance is critical because recent studies have demonstrated a dramatically increased rupture risk associated with IAs that have enlarged on serial examinations. In a study of 1325 IAs, 18 demonstrated an increase in size on serial imaging, and these enlarging lesions demonstrated an 18.5% annual risk for rupture [28]. For this reason, patients with interval IA enlargement should be offered prompt treatment in the absence of prohibitive comorbidities.

Aside from interval IA progression on serial follow-up imaging, it remains uncertain as to which asymptomatic aneurysms should be treated and which may be managed conservatively. Shortcomings in our current knowledge of the pathophysiology of IA rupture limit the value of any clinical guidelines proposed for the management of these patients, and many authors seem reticent to offer such recommendations. Acknowledging this limitation, the Task Force Team for the Korean Society for Interventional Neuroradiology has provided useful guidelines suggesting that IA treatment should be offered to patients with a life expectancy of >10–15 years with one or more of the following risk factors.

1. IA size ≥5 mm.
2. Presence of neurological symptoms.

3. IA location in the posterior circulation or arising from the anterior or posterior communicating arteries.

4. history of prior subarachnoid hemorrhage.

5. documented enlargement or change in morphology during imaging surveillance.

6. hypertension and multiple aneurysms in patients <50 years of age.

7. IA with a large height to neck width (aspect) ratio, a lobule, or bleb [27].

The authors have also suggested that it is reasonable to offer treatment to patients who have severe anxiety or depression secondary to the presence of an untreated IA [27].

5.4 Diagnostic Workup of Headache or Neurological Symptoms

Unruptured aneurysms >7 mm may present with headache, seizures, thromboembolic events, or neurological symptoms secondary to local mass effect such as cranial nerve, motor, or sensory deficits [1]. Excluding lesions identified during the workup of headaches, symptomatic unruptured IAs have a fourfold increased risk of rupture [1]. For this reason, these symptomatic IAs are most often treated with either endovascular coiling or microsurgical clipping, depending on their size and location. Interestingly, although some small observational studies have suggested a reduction in headache frequency or severity after treatment, larger studies have failed to demonstrate a correlation between IA rupture risk and the symptomatic presentation with headache, suggesting that the majority of IAs identified during headache evaluation are, in fact, incidental.

5.5 Diagnosis and Management of Ruptured Aneurysms

Ruptured IAs will present with severe headache secondary to subarachnoid hemorrhage, classically with sudden acute onset and characterized as the worst headache of a patient's life. In the emergent evaluation of severe, acute-onset headache, CT is the most appropriate initial imaging study. Within the first 6 h, CT is reported to have a 100% sensitivity; however, this efficacy declines as the blood products age and redistribute during the late acute and subacute phases. For instance, CT has an 85% sensitivity for detecting SAH after 5 days and only a 50% sensitivity for detecting hemorrhage after 1 week. Thus, as the time after symptom onset increases, CT could be paired with additional diagnostic studies to increase the sensitivity for detection of SAH such as a CTA to assess for an associated vascular lesion, an MRI to assess for sulcal or cisternal FLAIR hyperintensity for subtle SAH, or a lumbar puncture to assess for CSF xanthochromia or occult SAH. For instance, McCormack et al. have determined that in the subacute setting, a negative noncontrast CT and a subsequent negative CTA yields a 99.43% posttest probability to exclude SAH [29].

In the event that a ruptured IA is identified as the etiology for SAH, urgent treatment is indicated. A recent study of 459 cases of aneurysmal SAH treated with either microsurgical clipping or endovascular coil embolization demonstrated a 44% relative risk reduction for death or functional dependence when treatment was performed within 24 h of the initial bleed. In this study, patients that were treated more than 24 h after their initial presentation had a 14.4% risk for death or functional dependence at 6 months, while those that were treated within the 24 h period had a risk of only 8.0% [30]. Endovascular versus microsurgical treatment options and decisions should be agreed upon by subspecialized neurovascular surgeons and neurointerventionalists based on the location, anatomy, and characteristics of the IA as well as patient specific factors such as age, comorbidities, and SAH grade.

5.6 Surveillance of Treated Aneurysms

As previously discussed, CTA is the preferred modality for noninvasive surveillance of microsurgically clipped IAs and is highly effective in the detection of even small residual and recurrent aneurysms. For instance, Dehdashti et al. have reported that high-quality CTA achieved a 100% sensitivity and specificity for the detection of two small (<2 mm) residual aneurysms in their study population consisting of 60 clipped IAs [31].

In contrast, MRA utilizing an ultrashort echo time is the noninvasive imaging methodology of choice for the follow-up of IAs treated with endovascular coil embolization. The reported sensitivity of MRA to detect residual IAs ranges from 90% to 100%, and in most studies, the size of the missed remnants was ≤3 mm [6]. In a study to assess the sensitivity of MRA following endovascular treatment, Farb et al. determined that no IA remnants >3 mm were missed when utilizing both noncontrast time of flight and contrast-enhanced MRA [32].

Irrespective of the type of treatment, DSA remains the gold standard for diagnosis of residual or recurrent IAs. However, noninvasive imaging modalities offer important safety benefits over conventional angiography. Even in skilled neurointerventional practices, diagnostic cerebral angiography carries an approximate risk of 0.1–0.3% for a clinically significant adverse event. Although this risk is small, annual or biannual DSA surveillance studies can impart a relatively substantial cumulative risk to be balanced against the 0.1–0.2% annual risk that a successfully embolized IA will rebleed (Cerebral Aneurysm Rerupture After Treatment and ISAT trials). Microsurgical clipping of IAs are associated with an even lower rate of rebleeding (~0.06% annual risk in the ISAT trial). For this reason, noninvasive imaging techniques should be strongly considered as the primary modality for IA surveillance following treatment, with DSA reserved for further investigation, problem-solving, and/or treatment planning of a recurrent or enlarging residual IA.

6   Conclusion

Less than 30 years ago, there were no noninvasive imaging alternatives to conventional cerebral angiography. With the introduction of helical CT scanners in the 1990s and computers powerful enough to manage the increased data burden, CTA has improved as a viable noninvasive imaging alternative [6]. Currently, the number of noninvasive neurovascular imaging studies far exceeds the number of catheter-based examinations. Noninvasive vascular imaging is a critical component in the management of IAs. CTA and MRA offer highly sensitive, accessible, and low-risk alternatives to conventional DSA for screening, diagnosis, and IA surveillance. Furthermore, emerging techniques such as vessel wall imaging, permeability imaging, patient-specific computational fluid dynamics simulation, and 4D Flow MRI aim to visualize aspects of the complex pathophysiology and mechanobiology of IA development, growth, and rupture for novel risk stratification paradigms. Rapid ongoing innovation in these new modalities ensures that the central role of noninvasive vascular imaging in IA diagnosis, characterization, and surveillance will continue.

References

1. Thompson BG, Brown RD Jr, Amin-Hanjani S, Broderick JP, Cockroft KM, Connolly ES Jr et al (2015) Guidelines for the management of patients with unruptured intracranial aneurysms: a guideline for healthcare professionals from the American Heart Association/American Stroke Association. Stroke 46(8):2368–2400

2. Connolly ES Jr, Rabinstein AA, Carhuapoma JR, Derdeyn CP, Dion J, Higashida RT et al (2012) Guidelines for the management of aneurysmal subarachnoid hemorrhage: a guideline for healthcare professionals from the American Heart Association/American Stroke Association. Stroke 43(6):1711–1737

3. Hacein-Bey L, Provenzale JM (2011) Current imaging assessment and treatment of intracranial aneurysms. AJR Am J Roentgenol 196(1):32–44

4. Ansari SA, Rath TJ, Gandhi D (2011) Reversible cerebral vasoconstriction syndromes presenting with subarachnoid hemorrhage: a case series. J Neurointerv Surg 3(3):272–278

5. Tomandl BF, Kostner NC, Schempershofe M, Huk WJ, Strauss C, Anker L et al (2004) CT angiography of intracranial aneurysms: a focus on postprocessing. Radiographics 24(3):637–655

6. Wallace RC, Karis JP, Partovi S, Fiorella D (2007) Noninvasive imaging of treated cerebral aneurysms, part I: MR angiographic follow-up of coiled aneurysms. AJNR Am J Neuroradiol 28(6):1001–1008

7. Wallace RC, Karis JP, Partovi S, Fiorella D (2007) Noninvasive imaging of treated cerebral aneurysms, part II: CT angiographic follow-up of surgically clipped aneurysms. AJNR Am J Neuroradiol 28(7):1207–1212

8. Patino M, Prochowski A, Agrawal MD, Simeone FJ, Gupta R, Hahn PF et al (2016) Material separation using dual-energy CT: current and emerging applications. Radiographics 36(4):1087–1105

9. Hegde A, Chan LL, Tan L, Illyyas M, Lim WE (2009) Dual energy CT and its use in neuroangiography. Ann Acad Med Singap 38(9):817–820

10. Yu L, Leng S, McCollough CH (2012) Dual-energy CT-based monochromatic imaging. AJR Am J Roentgenol 199(5 Suppl):S9–S15

11. Sailer AM, Wagemans BA, Nelemans PJ, de Graaf R, van Zwam WH (2014) Diagnosing intracranial aneurysms with MR angiography: systematic review and meta-analysis. Stroke 45(1):119–126

12. Okahara M, Kiyosue H, Yamashita M, Nagatomi H, Hata H, Saginoya T et al (2002)

Diagnostic accuracy of magnetic resonance angiography for cerebral aneurysms in correlation with 3D-digital subtraction angiographic images: a study of 133 aneurysms. Stroke 33 (7):1803–1808

13. Li MH, Li YD, Gu BX, Cheng YS, Wang W, Tan HQ et al (2014) Accurate diagnosis of small cerebral aneurysms </=5 mm in diameter with 3.0-T MR angiography. Radiology 271(2):553–560

14. Majoie CB, Sprengers ME, van Rooij WJ, Lavini C, Sluzewski M, van Rijn JC et al (2005) MR angiography at 3T versus digital subtraction angiography in the follow-up of intracranial aneurysms treated with detachable coils. AJNR Am J Neuroradiol 26 (6):1349–1356

15. Chalouhi N, Hoh BL, Hasan D (2013) Review of cerebral aneurysm formation, growth, and rupture. Stroke 44(12):3613–3622

16. Edjlali M, Gentric JC, Regent-Rodriguez C, Trystram D, Hassen WB, Lion S et al (2014) Does aneurysmal wall enhancement on vessel wall MRI help to distinguish stable from unstable intracranial aneurysms? Stroke 45 (12):3704–3706

17. Juvela S (2000) Risk factors for multiple intracranial aneurysms. Stroke 31(2):392–397

18. Matouk CC, Mandell DM, Gunel M, Bulsara KR, Malhotra A, Hebert R et al (2013) Vessel wall magnetic resonance imaging identifies the site of rupture in patients with multiple intracranial aneurysms: proof of principle. Neurosurgery 72(3):492–496. discussion 6

19. Vakil P, Ansari SA, Cantrell CG, Eddleman CS, Dehkordi FH, Vranic J et al (2015) Quantifying intracranial aneurysm wall permeability for risk assessment using dynamic contrast-enhanced MRI: a pilot study. AJNR Am J Neuroradiol 36(5):953–959

20. Robertson AM, Watton PN (2012) Computational fluid dynamics in aneurysm research: critical reflections, future directions. AJNR Am J Neuroradiol 33(6):992–995

21. Kallmes DF (2012) Point: CFD—computational fluid dynamics or confounding factor dissemination. AJNR Am J Neuroradiol 33 (3):395–396

22. Hu P, Qian Y, Lee CJ, Zhang HQ, Ling F (2015) The energy loss may predict rupture risks of anterior communicating aneurysms: a preliminary result. Int J Clin Exp Med 8 (3):4128–4133

23. Schnell S, Ansari SA, Vakil P, Hurley M, Carr J, Batjer H et al (2012) Characterization of cerebral aneurysms using 4D FLOW MRI. J Cardiovasc Magn Reson 14(Suppl 1):W2

24. Brown RD Jr, Huston J, Hornung R, Foroud T, Kallmes DF, Kleindorfer D et al (2008) Screening for brain aneurysm in the familial intracranial aneurysm study: frequency and predictors of lesion detection. J Neurosurg 108(6):1132–1138

25. Chien A, Liang F, Sayre J, Salamon N, Villablanca P, Vinuela F (2013) Enlargement of small, asymptomatic, unruptured intracranial aneurysms in patients with no history of subarachnoid hemorrhage: the different factors related to the growth of single and multiple aneurysms. J Neurosurg 119(1):190–197

26. Sonobe M, Yamazaki T, Yonekura M, Kikuchi H (2010) Small unruptured intracranial aneurysm verification study: SUAVe study, Japan. Stroke 41(9):1969–1977

27. Jeong HW, Seo JH, Kim ST, Jung CK, Suh SI (2014) Clinical practice guideline for the management of intracranial aneurysms. Neurointervention 9(2):63–71

28. Inoue T, Shimizu H, Fujimura M, Saito A, Tominaga T (2012) Annual rupture risk of growing unruptured cerebral aneurysms detected by magnetic resonance angiography. J Neurosurg 117(1):20–25

29. McCormack RF, Hutson A (2010) Can computed tomography angiography of the brain replace lumbar puncture in the evaluation of acute-onset headache after a negative noncontrast cranial computed tomography scan? Acad Emerg Med 17(4):444–451

30. Phillips TJ, Dowling RJ, Yan B, Laidlaw JD, Mitchell PJ (2011) Does treatment of ruptured intracranial aneurysms within 24 hours improve clinical outcome? Stroke 42 (7):1936–1945

31. Dehdashti AR, Binaghi S, Uske A, Regli L (2006) Comparison of multislice computerized tomography angiography and digital subtraction angiography in the postoperative evaluation of patients with clipped aneurysms. J Neurosurg 104(3):395–403

32. Farb RI, Nag S, Scott JN, Willinsky RA, Marotta TR, Montanera WJ et al (2005) Surveillance of intracranial aneurysms treated with detachable coils: a comparison of MRA techniques. Neuroradiology 47(7):507–515

Chapter 3

Surgical Management of Intracranial Aneurysms: General Principles

Neil Majmundar, Michael Cohen, and Chirag D. Gandhi

Abstract

The surgical management of intracranial aneurysms has evolved rapidly over the past few decades, and despite the increasingly common use of alternative endovascular treatment options, it still plays a critical role in treatment. This chapter briefly reviews the pathophysiology and clinical presentation of the disease but largely focuses on indications for treatment as well as key preoperative and intraoperative considerations. Such a systematic approach is designed to provide a guide for meticulous surgical planning *and* execution and ensure the highest quality outcomes.

Key words Aneurysm, Subarachnoid hemorrhage, Aneurysm clip, Indocyanine green angiography, Pterional craniotomy, Orbitozygomatic craniotomy

1   Introduction

Approximately six million people in the USA, or 1 in 50 people, harbor an unruptured intracranial aneurysm [1]. While the majority of these patients have aneurysms which will go undetected and are unlikely to be problematic, approximately 30,000 patients suffer annually from a ruptured aneurysm in the USA. Aneurysms are generally discovered incidentally or after they rupture. The decisions of whether to treat, and how to treat aneurysms, have evolved as outcomes-based research has altered the treatment paradigms.

Historically, the use of microsurgical clips in the treatment of aneurysms can be traced back to Walter Dandy in 1938, when he published his case report as "the first attempt to cure an aneurysm at the Circle of Willis by direct attack upon the aneurysm." [2] Clip application continued to be the primary treatment modality for aneurysms until the 1990s when Guglielmi coils were used to treat aneurysms via an endovascular approach [3]. Endovascular treatment has since been expanded, and has become the primary treatment modality in most instances at institutions across the USA and Europe. While there has been a shift over the past 20 years

Fawaz Al-Mufti and Krishna Amuluru (eds.), *Cerebrovascular Disorders*, Neuromethods, vol. 170,
https://doi.org/10.1007/978-1-0716-1530-0_3, © Springer Science+Business Media, LLC, part of Springer Nature 2021

toward endovascular treatment, open surgical treatment of aneurysms will remain a necessary treatment modality for complex aneurysms that cannot be safely treated with endovascular techniques.

This chapter explores the general principles of aneurysm surgery, including preoperative considerations and common surgical approaches.

2   Pathophysiology and Clinical Presentation

Aneurysmal subarachnoid hemorrhage (SAH) is often a devastating event with a high mortality and morbidity. After aneurysmal SAH, a patient is at substantial risk of rebleeding: 3–4% in the first 24 h and 1–2% each day in the first month [4]. Rerupture is associated with a mortality that is estimated to be 70%. Aneurysm repair is the only effective treatment to prevent this occurrence [4].

Unruptured cerebral aneurysms may manifest clinically by their mass effect on adjacent neurologic structures, or they may be discovered incidentally when a patient has a neuroimaging study for another indication. Such aneurysms have a future risk of rupture that depends in part on their size and location, and necessitate treatment in order to prevent impending SAH.

Surgical and endovascular techniques are available for aneurysm treatment. In many cases, anatomic considerations, such as size, location, along with other morphological features determine which treatment is most appropriate for the patient. Several trials have attempted to investigate and compare the outcomes between the two treatment modalities available for the treatment of ruptured and unruptured aneurysms. The International Subarachnoid Aneurysm Trial (ISAT), published in 2012, demonstrated that endovascular coil embolization had better survival free of disability, was more likely to result in independent survival at 1 year, had a significantly lower risk of death at 5 years, yet had a slightly increased risk of bleeding [4–6]. Criticisms of ISAT include the greater than 80% exclusion rate, and the fact that the majority of patients had an anterior circulation aneurysm less than 10 mm with favorable-grade SAH. The Barrow Ruptured Aneurysm Trial (BRAT) has published outcomes at 1, 3, and 6 years, with intent to publish outcomes at 10 years after enrollment [7–9]. The BRAT showed that while coil embolization and surgical clipping had similar outcomes in patients with anterior circulation aneurysms, coil embolization provided a sustained benefit for patients with posterior circulation aneurysms [9]. Several criticisms of BRAT also exist, including the fact that 68 patients crossed over from the coiling group to the clipping group, possibly reducing the theoretical benefit from coil embolization of anterior circulation aneurysms. As the authors of BRAT noted, the difference between endovascular treatment and surgical clipping are not due to differences between the treatment effectiveness, but due to the morbidity and mortality associated with open surgery [9]. Both the ISAT and BRAT trials demonstrate that despite significant advances in endovascular treatment, surgical management of aneurysms will always be a necessary treatment modality.

---

3 Management Strategies

3.1 Preoperative Consideration

3.1.1 Imaging

Imaging is of utmost importance in cases of surgical clipping. At our institution, along with a standard CTA with coronal and sagittal reconstruction, the senior author prefers to obtain multiple 3-dimensional views to appreciate the surgical orientation of the aneurysm, surrounding vessels, and adjacent bony landmarks. Preoperative digital subtraction angiography (DSA) provides further detail regarding the vascular anatomy, and is particularly useful to study the extracranial and intracranial circulation dynamically. Furthermore, angiography provides the ability to appreciate the precise orientation of vessels encountered during the approach to the aneurysm, including veins encountered during dissection of the Sylvain fissure. Occasionally, balloon test occlusion can be useful to plan for vessel sacrifice for giant proximal ICA aneurysms, but is not routinely performed.

3.1.2 Anesthesia

Important anesthetic considerations in cases of surgical clipping include stringent management of blood pressure to avoid intraoperative rupture, reduction of intracranial pressure, and maintaining adequate cerebral perfusion pressure. Propofol and fentanyl are the anesthetic and narcotic combination generally used throughout aneurysm cases. The effects of anesthetics upon blood pressure, ICP, and neuromonitoring must be strongly considered. Therefore, high concentrations of volatile anesthetics are avoided. At the onset of surgery, prophylactic antibiotics are administered and most neurosurgeons will request an antiepileptic drug, dexamethasone and/or mannitol, and euvolemia. Throughout the surgery, from pinning until closure, blood pressure must be carefully monitored with an arterial line and modified at the neurosurgeon's request to avoid intraoperative complications and provide adequate cerebral blood flow. Blood pressure control is of utmost importance in avoiding intraoperative rupture and providing adequate cerebral blood flow given retractor use and placement of temporary clips. Using an EVD placed preoperatively, ICP is routinely monitored while clipping ruptured aneurysms. An experienced neuroanesthesiologist with an understanding of cerebral autoregulation parameters is crucial while clipping ruptured aneurysms. In addition to mannitol, mild hyperventilation may also be transiently employed in the reduction of ICP.

In cases with intraoperative aneurysm rupture, the patient may become hemodynamically unstable and present the anesthesiologist with a more labile situation. The blood pressure, ICP, and CPP must be closely monitored, as well as any decline in cardiac function secondary to stress induced cardiomyopathy in these patients. In addition, patients may also develop early onset vasospasm, which presents the neurosurgeon with an additional challenge in providing a good outcome. The anesthesiologist must always have two large bore intravenous lines prepared preoperatively to provide adequate resuscitation in the event of intraoperative aneurysm rupture.

If a major intracranial artery must be temporarily clipped to allow for safe dissection of the aneurysm neck, gentle blood pressure augmentation ensures adequate cerebral perfusion via the PCOM, ACOM, and EC-IC collaterals. When temporary clip placement is not deemed safe or poses a risk of ischemia, adenosine has been administered to induce cardiac arrest, particularly in cases of giant aneurysms, to soften the aneurysm enough to dissect the dome and place a permanent clip. Adenosine is only used in very specific cases, as most cases will allow for safe placement of temporary clips.

3.2 Intraoperative Consideration

3.2.1 Temporary and Permanent Clips

Aneurysm clips have evolved since Dandy first applied a V-shaped malleable clip to treat an aneurysm. While clips have been developed in many shapes and sizes by a variety of manufacturers, they all serve two primary functions: (1) temporary occlusion of blood flow during temporary clipping, and (2) permanent occlusion of blood flow into the aneurysm restructuring the vessel anatomy.

Temporary clips serve two functions: occlude blood flow at the site of proximal control in cases where the aneurysm has ruptured, and decrease blood flow to the aneurysm to facilitate dome and neck dissection. Temporary clips exert less closing pressure upon the vessel walls than permanent clips. In cases of intraoperative aneurysm rupture, a temporary clip may be applied to the site of proximal control to slow the bleeding enough for the surgeon to coagulate the site of bleeding with bipolar cautery or to quickly dissect the neck and place a permanent clip. In cases where the aneurysm dome is rigid and perforators cannot be visualized behind the dome, temporary clipping can be used to soften the aneurysm dome allowing for visualization of hidden perforators or arterial branches.

Once the aneurysm has been dissected, the neck is fully visualized, and the surrounding afferent and efferent vessels are identified, permanent clipping of the aneurysm can be performed. The closing pressure of permanent clips varies by manufacturer, clip type, and clip configuration. Clips come in a variety or lengths, shapes, and configurations. Clips shapes include but are not limited to straight, angled, curved, bayonetted, L-shaped, fenestrated, and

Fig. 1 A variety of aneurysm clips which may be used during aneurysm clipping. The top row of clips is straight, curved, and angled. The bottom row of clips is fenestrated, allowing the clip to circumnavigate a vessel which may be overlying the aneurysm

nonfenestrated (Fig. 1). Multiple clip lengths and configurations allow the neurosurgeon to alter the aberrant anatomy of the vessel wall without negatively impacting blood flow. Multiple clips may be applied to aneurysms with wide necks or more complex anatomy where one clip is insufficient to eliminate blood flow to the aneurysm. Clips can be stacked in multiple configurations, and oriented to reconstruct the parent vessel lumen.

3.2.2 Intraoperative Imaging

Intraoperative imaging is intended to ensure patency of arterial branches and perforators adjacent to the aneurysm after clip placement. Intraoperative doppler ultrasound, Indocyanine green (ICG) video angiography, and intraoperative DSA can all be used to investigate the patency of vessels.

Intraoperative Doppler Ultrasound

Intraoperative Doppler ultrasound is used to evaluate the flow through vessels after temporary and permanent clip placement in order to avoid inadvertent obstruction of surrounding vessels. The Doppler probe is placed directly upon the vessels seen in the operative field and provides real time feedback regarding patency and flow. Vessels must be visible and accessible to the probe in order to be evaluated.

Indocyanine Green (ICG) Video Angiography

Intraoperative fluorescence with ICG video angiography allows for intraoperative visual assessment of flow through surrounding vessels and perforators. As ICG is used in combination with the operative microscope, it can be used directly after clip placement and allows for prompt removal of a clip with suboptimal placement. This is one advantage over intraoperative DSA, as DSA requires time to set up after clip placement and may increase the risk of ischemic injury. Furthermore, ICG allows for direct visualization of vessels as they appear in the operative field under the microscope, allowing for much easier interpretation. Blood products and anything else which may obstruct direct visualization of the vessel must be removed to avoid incorrect interpretation.

Intraoperative DSA

Intraoperative DSA allows for almost immediate confirmation of treatment after clip placement. The ability to return to the operative field without having to prepare both the operating room and operative field again allow for quicker resolution of poorly placed clips, and immediate treatment of post clipping complications. At many centers where intraoperative DSA cannot be performed, the surgeon may elect to obtain DSA immediately following skin closure, with intention to return to the OR if the clip placement is inadequate.

3.3 Management: General Treatment Approach of Intracranial Aneurysms

While most intracranial aneurysms are amenable to either endovascular coil embolization or surgical clipping, treatment largely depends upon the aneurysm characteristics and surgeon preferences.

- A variety of factors ultimately impact the decision for treatment choice.
  - Patient age.
  - Neurological status.
  - Medical comorbidities.
  - Ruptured vs. Unruptured status.
  - Presence of perianeurysmal hematoma.
  - Aneurysm location.
  - Aneurysm orientation.
  - Aneurysm shape.
  - Presence of mass effect.
- Aneurysms amenable to endovascular treatment.
  - Neck size less than 4 mm.
  - Dome to neck ratio of 2:1.
  - Challenging location access microsurgically.
- Factors favoring surgical clipping include the following.
  - Aneurysms with a surrounding hematoma after rupture requiring evacuation.
  - Surgically accessible aneurysm location (MCA bifurcation).
  - Wide neck or complex morphology.
  - Young patient age.

– Aneurysms exerting mass effect upon eloquent structures [10].

Aneurysm location plays a significant role in selecting surgical treatment. Aneurysms located in the posterior circulation are generally more likely to be treated with endovascular techniques due to the significantly higher surgical morbidity associated with clipping [5, 9, 11]. With the advent of balloon assisted coiling, stent assisted coiling, and flow diverting stents, the variety of aneurysms able to be treated with endovascular techniques, especially posterior circulation aneurysms, continues to grow [12, 13]. While this statement is generally true, microsurgical treatment still has a role in posterior circulation aneurysms and management of these aneurysms can vary by institution and surgeon [9, 14].

Aneurysm rupture status also influences the choice for treatment modality. Some studies have shown favorable outcomes in endovascular treatment of aneurysms amenable to coil embolization in patients with SAH [11, 15]. Furthermore, for patients with a high grade SAH, coil embolization can secure the ruptured aneurysm in the acute period with a lower relative procedural morbidity [16]. Both BRAT and ISAT, trials mentioned earlier in this chapter, demonstrated that endovascular treatment with coiling resulted in better outcomes in patients presenting with SAH [4–6, 9].

Aneurysm treatment modalities may also vary by institution, availability of technology, and surgeon training. Institutions with neurosurgeons experienced with microsurgical techniques will generally have better functional outcomes for patients with aneurysms located in deeper locations or challenging anatomy [17]. Microsurgical and endovascular outcomes can largely vary depending on centers and their patient volume [18]. One retrospective study investigated outcomes for patients with unruptured intracranial aneurysms, and noted the differences in outcomes even amongst large volume centers [18]. Institutional expertise can affect the modality of treatment chosen to treat equivocal aneurysms.

3.4 Management; Surgical Approaches

3.4.1 Pterional

The pterional or frontotemporal craniotomy is the most classically used approach for access to anterior circulation aneurysms. This surgical approach may be modified depending upon the aneurysm and surgeon preference. The patient is first placed in a three-pin Mayfield head holder, typically with two pins posterior and one anterior (Fig. 2). Head positioning depends upon the aneurysm and surgeon, but most commonly the head is rotated anywhere from 30° to 60° toward the contralateral shoulder and extended inferiorly 10–20° so that the malar eminence is the highest point in the vertical plane. The degree of head rotation and extension can position the Sylvian fissure at the highest point to facilitate temporal and frontal lobe relaxation for Sylvian fissure splitting. The extent

Fig. 2 This intraoperative image demonstrates the head position during the clipping of a left sided ruptured posterior communicating (PCOM) artery aneurysm. The red line denotes the incision line for a pterional craniotomy. The blue line demonstrates the incision line for an orbitozygomatic (OZ) craniotomy

of rotation is based upon the optimal working angle and determined by the location of the aneurysm.

• ACOM aneurysms at 60°
• ICA aneurysms at 45°
• MCA lesions at 30°

The head should be elevated above the body and neck compression avoided to facilitate venous return through the internal jugular vein. Head positioning should be selected in such a way that the frontal and temporal lobes fall out of the dissection plane thereby minimizing brain retractors.

Once the head is positioned, the incision begins 1 cm anterior to the tragus above the root of the zygoma, and is carried out in a curvilinear fashion medially and anteriorly toward the vertex of the patient's hairline at midline.

• Preservation of the superficial temporal artery during the exposure is vital in case a microvascular bypass is necessary.
• Frontal sinus size should be evaluated on preoperative imaging, as a vascularized pericranial graft may be required for closure in cases where the sinus must be violated.

Once the incision is made and dissection begins, the scalp flap can be reflected in multiple ways (subfascial, interfascial, myocutaneous) to preserve the frontalis branch of the facial nerve.

- Routinely, a myocutaneous flap is used along with a Leyla bar, which can retract the flap with "fishhooks."

- In cases where more sphenoid exposure is necessary, an interfascial or subfascial dissection can be used to release the temporalis inferiorly and keep the muscle bulk from obstructing the surgeon's view and angle of attack.

After the scalp flap is reflected away, burr hole placement is performed.

- Burr holes are placed at the pterion keyhole, above the root of the zygoma on the middle fossa floor, and posteriorly below the superior temporal line.

- Additional burr holes are useful in elderly patients with adherent dura.

- A craniotome with footplate is then used to perform the craniotomy.

Depending upon the aneurysm, the extent of bony removal following craniotomy can vary. The craniotomy can always be extended down to the middle fossa floor to create space for the temporal lobe to be released inferiorly. The sphenoid wing is routinely flattened with a high-speed drill, providing a direct view to the opticocarotid cistern once dura is opened. The meningoorbital band is an excellent landmark of the depth of sphenoid wing exposure necessary for a routine pterional craniotomy. Division of the meningoorbital band is only necessary when performing an anterior clinoidectomy or obtaining cavernous sinus exposure. Dural tack-up sutures are then placed to aid in epidural hemostasis. The middle meningeal artery is identified and coagulated prior to opening the dura. The dura is opened in a C-shaped fashion with the pedicle of the dural flap oriented toward the sphenoid wing. The dura can then be secured along the sphenoid wing with sutures, providing a direct view to the opticocarotid cistern and proximal Sylvian fissure.

Once the opening has been completed, hemostasis has been achieved from both the cutaneous and bony edges, and the surgical site is cleared of any debris, the retractor frame (if used) can be attached to the Mayfield and the microscope may be brought into the operative field.

3.4.2 Orbitozygomatic

The orbitozygomatic craniotomy (OZ) can be considered an extension of the pterional craniotomy, requiring additional bony removal of the zygoma (if necessary) and the lateral wall and roof of the orbit to provide a wider operative corridor.

- The OZ provides exposure of both anterior and middle cranial fossae, basilar apex, and the upper clivus [19].
- The OZ approach expands the working angle to the aneurysm, and can reduce the amount of brain retraction required during the case to visualize critical structures.
- The OZ approach is extremely versatile and provides access to aneurysms residing at the ICA complex, A1-A2-ACOM complex, MCA bifurcation, and Basilar tip.

  Multiple modifications of the OZ have been published and different variations provide different working angles.

- Removal of the lateral wall and roof of the orbit provides an improved working angle posteriorly and medially and is the most useful component of the OZ.
- Removal of the zygoma is rarely necessary, and provides a better working angle along the middle fossa floor and tentorial incisura and can be useful with high-riding basilar apex aneurysms being approached using a combined OZ-sub-temporal exposure.

3.4.3   Other Surgical Approaches

In addition to the traditional pterional and OZ, subtemporal, retrosigmoid, and far lateral approaches may be used in order to obtain adequate access for clipping [20–24]. The majority of aneurysms can be treated with the pterional and OZ craniotomies. For PICA aneurysms, a far lateral craniotomy can be required. While the full step-by-step approach is beyond the scope of this chapter due to the complex anatomy encountered, a brief review is provided.

The far lateral craniotomy allows access to the lower brainstem, particularly the cerebellomedullary and premedullary cisterns [25]. The patient may be positioned in a lateral park bench or three-quarter prone position which allows for optimal access to the posterior fossa. A variety of skin incisions can be made behind the ear in the posterior auricular and suboccipital regions depending on surgeon preference. Dissection of the muscular layers can be performed in a multilayer fashion, with the goal of avoiding vertebral artery injury and avoiding postoperative muscle atrophy. A lateral suboccipital craniotomy is then performed, with additional bony removal depending upon the location of the aneurysm and size of working corridor required for access. Once the opening has been completed, hemostasis has been achieved from both the cutaneous and bony edges, and the surgical site is cleared of any debris, the retractor frame (if used) can be attached to the Mayfield and the microscope may be brought into the operative field.

4   Conclusion

While many centers continue to shift toward endovascular treatment of aneurysms, surgical clipping will always remain an important treatment modality. This chapter summarizes the most basic considerations and approaches for surgical treatment. Both patient and aneurysm characteristics impact the decision to surgically clip an aneurysm. Furthermore, a surgical team consisting of the neurosurgeon, neuroanethesiologist, neuromonitoring technician, and experienced operating room staff is essential to optimize patient outcomes.

References

1. Brain aneurysms: detection and treatment (2017) Accessed 19 Nov 2017
2. Dandy WE (1938) Intracranial aneurysm of the internal carotid artery: cured by operation. Ann Surg 107(5):654–659
3. Guglielmi G, Vinuela F, Dion J, Duckwiler G (1991) Electrothrombosis of saccular aneurysms via endovascular approach. Part 2: preliminary clinical experience. J Neurosurg 75 (1):8–14
4. Molyneux A, Kerr R, Stratton I et al (2002) International Subarachnoid Aneurysm Trial (ISAT) of neurosurgical clipping versus endovascular coiling in 2143 patients with ruptured intracranial aneurysms: a randomised trial. Lancet 360(9342):1267–1274
5. Molyneux AJ, Kerr RS, Birks J et al (2009) Risk of recurrent subarachnoid haemorrhage, death, or dependence and standardised mortality ratios after clipping or coiling of an intracranial aneurysm in the International Subarachnoid Aneurysm Trial (ISAT): long-term follow-up. Lancet Neurol 8(5):427–433
6. Molyneux AJ, Kerr RS, Yu LM et al (2005) International subarachnoid aneurysm trial (ISAT) of neurosurgical clipping versus endovascular coiling in 2143 patients with ruptured intracranial aneurysms: a randomised comparison of effects on survival, dependency, seizures, rebleeding, subgroups, and aneurysm occlusion. Lancet 366(9488):809–817
7. McDougall CG, Spetzler RF, Zabramski JM et al (2012) The barrow ruptured aneurysm trial. J Neurosurg 116(1):135–144
8. Spetzler RF, McDougall CG, Albuquerque FC et al (2013) The barrow ruptured aneurysm trial: 3-year results. J Neurosurg 119 (1):146–157
9. Spetzler RF, McDougall CG, Zabramski JM et al (2015) The barrow ruptured aneurysm trial: 6-year results. J Neurosurg 123 (3):609–617
10. Guresir E, Schuss P, Setzer M, Platz J, Seifert V, Vatter H (2011) Posterior communicating artery aneurysm-related oculomotor nerve palsy: influence of surgical and endovascular treatment on recovery: single-center series and systematic review. Neurosurgery 68 (6):1527–1533. discussion 1533–1524
11. Natarajan SK, Sekhar LN, Ghodke B, Britz GW, Bhagawati D, Temkin N (2008) Outcomes of ruptured intracranial aneurysms treated by microsurgical clipping and endovascular coiling in a high-volume center. AJNR Am J Neuroradiol 29(4):753–759
12. Kallmes DF, Brinjikji W, Cekirge S et al (2017) Safety and efficacy of the Pipeline embolization device for treatment of intracranial aneurysms: a pooled analysis of 3 large studies. J Neurosurg 127(4):775–780
13. Becske T, Potts MB, Shapiro M et al (2017) Pipeline for uncoilable or failed aneurysms: 3-year follow-up results. J Neurosurg 127 (1):81–88
14. Sanai N, Tarapore P, Lee AC, Lawton MT (2008) The current role of microsurgery for posterior circulation aneurysms: a selective approach in the endovascular era. Neurosurgery 62(6):1236–1249. discussion 1249-1253
15. Bracard S, Lebedinsky A, Anxionnat R et al (2002) Endovascular treatment of Hunt and Hess grade IV and V aneurysms. AJNR Am J Neuroradiol 23(6):953–957
16. Jain R, Deveikis J, Thompson BG (2004) Endovascular management of poor-grade aneurysmal subarachnoid hemorrhage in the geriatric population. AJNR Am J Neuroradiol 25(4):596–600

17. Chyatte D, Porterfield R (2001) Functional outcome after repair of unruptured intracranial aneurysms. J Neurosurg 94(3):417–421

18. Zacharia BE, Bruce SS, Carpenter AM et al (2014) Variability in outcome after elective cerebral aneurysm repair in high-volume academic medical centers. Stroke 45 (5):1447–1452

19. Lemole GM Jr, Henn JS, Zabramski JM, Spetzler RF (2003) Modifications to the orbitozygomatic approach. Technical note. J Neurosurg 99(5):924–930

20. Al-Mefty O (1987) Supraorbital-pterional approach to skull base lesions. Neurosurgery 21(4):474–477

21. Hakuba A, Liu S, Nishimura S (1986) The orbitozygomatic infratemporal approach: a new surgical technique. Surg Neurol 26 (3):271–276

22. Hakuba A, Tanaka K, Suzuki T, Nishimura S (1989) A combined orbitozygomatic infratemporal epidural and subdural approach for lesions involving the entire cavernous sinus. J Neurosurg 71(5 Pt 1):699–704

23. Al-Mefty O, Anand VK (1990) Zygomatic approach to skull-base lesions. J Neurosurg 73(5):668–673

24. Pellerin P, Lesoin F, Dhellemmes P, Donazzan M, Jomin M (1984) Usefulness of the orbitofrontomalar approach associated with bone reconstruction for frontotemporosphenoid meningiomas. Neurosurgery 15 (5):715–718

25. Benet A, Lawton MT, Kliot M, Berger MS (2016) Surgical anatomy of the skull base. In: Winn HR (ed) Youmans and Winn neurological surgery, vol 1, 7rh edn. Elsevier, Amsterdam, pp 76–106

Chapter 4

Surgical Management of Intracranial Aneurysms: Technique and Pitfalls

Michael Cohen, Neil Majmundar, and Chirag D. Gandhi

Abstract

Intracranial aneurysms represent a heterogenous disease and each anatomical location has a unique set of surgical considerations that must be methodically understood. This chapter summarizes the most high-yield operative principles as well as potential pitfalls for each of the most common types of intracranial aneurysms.

Key words Aneurysm, Subarachnoid hemorrhage, Aneurysm clip, PCOM artery, MCA, ACOM, Basilar, PICA

1 Introduction

Microsurgical clipping remains an important treatment modality for the management of both ruptured and unruptured intracranial aneurysms. The rapidly evolving sophistication of endovascular techniques such as flow diversion and intrasaccular embolization for routine aneurysms has increasingly relegated surgical clipping techniques for the more complex lesions [1–6]. Aneurysms treated with microsurgery tend to be larger, wide-necked, or incorporate parent arteries. This chapter explores the surgical techniques and complication avoidance strategies that are relevant to this heterogeneous group of aneurysms arising from the most common anatomical locations: the posterior communicating (PCOM) segment of the internal carotid artery (ICA), middle cerebral artery (MCA), anterior communicating artery (ACOM), basilar artery apex (BA), and posterior inferior cerebellar artery (PICA).

Fawaz Al-Mufti and Krishna Amuluru (eds.), *Cerebrovascular Disorders*, Neuromethods, vol. 170,
https://doi.org/10.1007/978-1-0716-1530-0_4, © Springer Science+Business Media, LLC, part of Springer Nature 2021

2   Pathophysiology and Clinical Presentation: Relevant Anatomy

Microsurgical treatment of cerebral aneurysms requires a detailed understanding of the relevant vascular and skull base anatomy and microsurgical techniques to ensure good clinical outcomes.

2.1   PCOM Segment ICA Aneurysms

- The ICA has cervical (C1), petrous (C2), lacerum (C3), cavernous (C4), clinoid (C5), ophthalmic (C6), and posterior communicating (C7) segments [7].

- The PCOM artery divides the supraclinoid ICA into C6 (from distal dural ring to PCOM artery) and C7 (from PCOM artery to ICA terminus) segments.

- Major branches from the C7 segment of the ICA are the PCOM artery and anterior choroidal (ACh) artery.

- The PCOM artery arises from the posteromedial surface of the ICA approximately midway between the ophthalmic artery and the ICA terminus and travels posteromedially, to join the PCA (Fig. 1) [8].

- The PCOM artery is superior and medial to the occulomotor nerve (CN 3) and parallels the course of this cranial nerve.

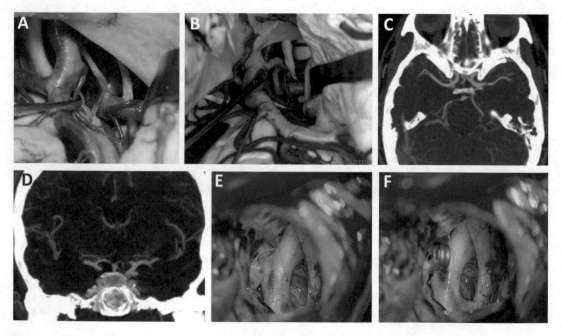

Fig. 1 (a) Illustrated model showing the anterior choroidal artery retracted to show the PCOM artery traveling to the PCA. Both the optic and oculomotor nerves can be seen in relationship to the ICA and PCOM artery. **(b)** The PCOM artery is again visualized with several perforators. **(c, d)** CTA in axial and coronal views showing left PCOM artery aneurysm. **(e)** Intraoperative images showing fully visualized aneurysm neck prior to clipping and **(f)** after clip placement with exclusion of neck (Intraoperative images courtesy of Charles Kulwin, MD)

- A fetal PCOM artery is an important anatomic variant where a large PCOM directly supplies the PCA without significant contribution from the basilar artery.

- The superior and lateral surfaces of the PCOM artery contain numerous perforators that supply various eloquent structures including the optic tract and chiasm, internal capsule, pituitary stalk, thalamus, and hypothalamus [8].

- The ACh artery arises distal to the PCOM artery and is typically closer to the PCOM artery than the ICA terminus.

- Usually a PCOM segment ICA aneurysm arises just distal the PCOM artery and projects posteriorly and laterally and can compress the occulomotor nerve.

2.2 Middle Cerebral Artery (MCA) Bifurcation Aneurysms

- The M1 segment of the MCA arises from the ICA terminus and parallels the sphenoid ridge as it gives off medial and lateral lenticulostriate perforators within the proximal aspect of the sylvian fissure.

- The anterior temporal artery arises from the M1 segment proximal to the bifurcation.

- The M1 segment bifurcates or trifurcates just proximal to the genu (90° turn), where the M1 segments bend around the limen insula (Fig. 2) [8].

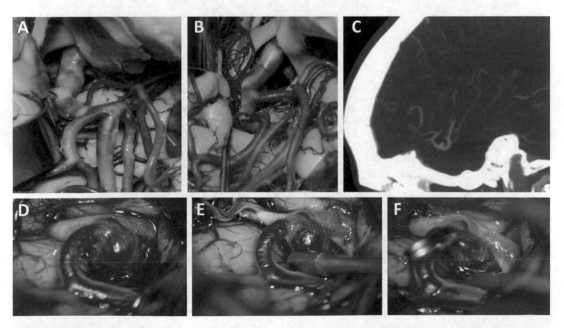

Fig. 2 (a) Illustrated model showing the short M1 segment of the middle cerebral artery ends in an MCA trifurcation. (b) The M1 trifurcates into three M2 vessels where pass over the insula. (c) CTA MIP in sagittal view showing right MCA aneurysm. (d) Intraoperative images showing fully visualized laterally projecting MCA bifurcation aneurysm with (e) temporary clip placed on M1 segment, and (f) final clip placement with exclusion of neck (Intraoperative images courtesy of Charles Kulwin, MD)

- The majority of MCA aneurysms occur at the bifurcation/ trifurcation [9].

- The M2 segments of the MCA begins at the genu, pass over the insular cortex, and end at the circular sulcus of the insula.

- The M3 segments travels from the depth of the sylvian fissure toward its distal surface, starting at the circular sulcus of the insula, where the M3 segments make a 180° turn, and ending at the cortical surface.

- The M4 segments begin at the cortical surface and travel along the frontal and parietal lobes.

2.3 Anterior Communicating Artery (ACOM) Aneurysms

- The A1 segment of the ACA begins at the ICA terminus within the carotid cistern, travels above the optic nerve and chiasm within the lamina terminalis cistern, and terminates at the ACOM (Fig. 3).

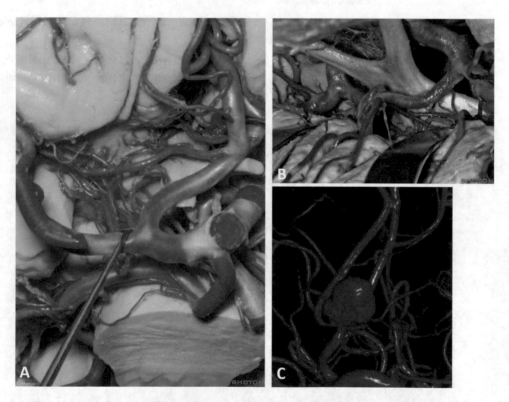

Fig. 3 (**a**) The A1 travels from the ICA terminus toward the ACOM artery. The recurrent artery of Heubner is the first branch seen, traveling retrograde parallel to the A1 toward the sylvian fissure. The orbitofrontal artery is the second branch traveling toward the gyrus rectus. The dissector is just above the ACOM artery, and between both A2 branches. (**b**) In surgical orientation, the frontal lobe is retracted and the A1 is followed to the ACOM artery above the optic chiasm. (**c**) A 3D reconstruction of a digital subtraction angiogram showing a large ACOM aneurysm

- While the chiasmatic cistern contains the optic nerve and chiasm, the lamina terminalis cistern contains both A1 segments, the ACOM (and aneurysm), A2 segments, and both recurrent arteries of Heubner.

- 10% of A1 segments are unilaterally hypoplastic; A1 hypoplasia is present in 85% of ACOM aneurysms [8].

- 60% of ACOM arteries are singular, 30% are duplicated, and 10% are triplicated [10].

- The size of the ACOM is linearly related to the difference between each A1 segment; the larger the difference between A1 segments, the larger the ACOM.

- The A2 segment arises at the ACOM and follows the curvature of the rostrum of the corpus callosum, ending at the junction between rostrum and genu.

- The A1, ACOM, and A2 contain a variable number of perforating branches, that typically arise from the superior and posterior surfaces of the artery.

- A1 and ACOM perforators primarily supply the optic chiasm, anterior third ventricle, and hypothalamus.

- The recurrent artery of Heubner (recurrent artery) is the largest perforating branch arising from the distal A1 or proximal A2 (within 5 mm of the ACOM) and travels retrograde along the A1, above the ICA bifurcation and toward the anterior perforated substance to supply the caudate, putamen, globus pallidus, and anterior limb of the internal capsule [8].

- The orbitofrontal artery is the second named branch arising from the A2 segment, which travels over the gyrus rectus toward the olfactory tract.

- The frontopolar artery is the third major branch arising from the A2 segment and travels along the medial surface of the frontal lobe.

2.4 Basilar Apex Aneurysms

- The basilar artery gives rise to bilateral superior cerebellar arteries (SCA), which are often duplicated, then terminates as it gives rise to bilateral posterior cerebral arteries (PCA) (Fig. 4).

- The oculomotor nerve (CN III) arises from the midbrain, medial to the cerebral peduncle, and travels through the interpeduncular cistern, superior to the SCA and inferior to the PCA [11].

- Basilar artery aneurysms most commonly arise at the terminal aspect of the basilar artery as it branches into the PCAs.

- The basilar apex most commonly lies within the interpeduncular cistern, contained by the two cerebral peduncles posteriorly, the

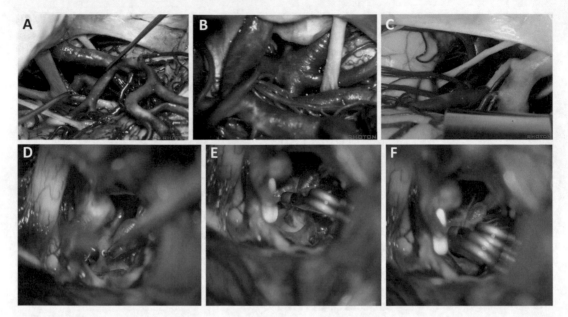

Fig. 4 (**a**) The opticocarotid triangle is demonstrated by retracting the optic nerve. The basilar apex is visualized along with both P1 segments of the PCA. (**b**) The oculomotor-carotid triangle is demonstrated by retracting the ICA, showing the basilar apex. (**c**) The tentorial incisura is exposed using a subtemporal approach. The PCA and SCA are exposed with the oculomotor nerve traveling between these two arteries. The upper basilar trunk is exposed as a point for proximal control. (**d**) Intraoperative images showing fully visualized basilar terminus aneurysm with (**e**) first, and (**f**) second clip placement with exclusion of neck (Intraoperative images courtesy of Charles Kulwin, MD)

dorsum sellae and clivus anteriorly, the uncus and tentorium laterally, and the mammillary bodies superiorly [12].

- Significant variation in basilar apex location exists, and the basilar artery can terminate as high as the mammillary bodies, or as low as the mid-section of the clivus [13].

- The interpeduncular cistern is separated from the chiasmatic cistern by Liliequist's membrane, a sheet of arachnoid that originates from the dorsum sellae and inserts on the mammillary bodies [12].

- Thalamoperforating arteries arise from the posterior aspect of the basilar artery, the PCAs, and the PCOMs [13].

2.5 Posterior Inferior Cerebellar Artery (PICA) Aneurysms

- The PICA arises from the vertebral artery (VA) anterior to the medulla approximately at the level of the rootlets of CN XII.

- The VA pierces the dura lateral to the C1 nerve roots and travels superiorly, passing under the dentate ligament, positioned anterolateral to the medulla.

- The anteromedullary segment of PICA typically arises from the VA distal to the dentate ligament and travels posteriorly and superiorly adjacent to the rootlets of CN XII until the inferior olive.

- The lateral medullary segment of PICA continues posteriorly from the inferior olive until the rootlets of CN IX, X, and XI.

- The tonsillomedullary segment of PICA turns inferiorly after passing the rootlets of CN IX, X, and XI, descends inferiorly into the foramen magnum and forms a "caudal loop" around the cerebellar tonsil.

- The telovelotonsillar segment of PICA begins after the caudal loop as the artery ascends toward the fourth ventricle, where it forms a "cranial loop" before terminating in cortical branches that supply the cerebellar hemispheres.

- Most PICA aneurysms arise in the proximal anteromedullary and lateral medullary segments, although more distal aneurysms can also occur.

3   Management Strategies; Surgical Steps

3.1   PCOM Segment ICA Aneurysms (Fig. 5)

- A standard pterional craniotomy, with the patient's head rotated 45 degrees to the contralateral side and extended, is performed and the lesser sphenoid wing flattened with a drill.

- The opticocarotid cistern is opened and the optic nerve and ophthalmic segment of the ICA are identified.

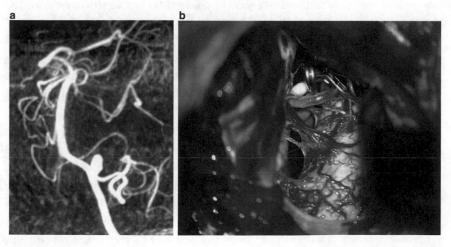

Fig. 5 (**a**) 3D reconstruction of a digital subtraction angiogram showing a PICA aneurysm. (**b**) Intraoperative images showing fully visualized PICA aneurysm with clip exclusion of neck (Intraoperative images courtesy of John Steele, MD)

- The proximal sylvian fissure is split to identify the ICA terminus; additional sylvian dissection may be necessary to relax the frontal and temporal lobes.

- The ophthalmic segment of the ICA is dissected away from the optic nerve and a site for temporary clipping is identified for proximal control.

- Dissection along the ICA is performed to identify the PCOM artery, the ACh artery, and the aneurysm neck.

- Typically a simple straight or slightly curved clip is used to occlude the aneurysm, ensuring that the PCOM artery, the ACh artery, and perforators arising from the ICA or PCOM artery are not occluded by the clip.

    Pitfalls

- Gentle frontal lobe retraction is typically well tolerated, but the temporal lobe should not be retracted as a laterally projecting PCOM aneurysm can be adherent to the uncus and retraction can cause intraoperative rupture.

- While every effort is made to maintain patency of the PCOM artery, PCOM artery occlusion is typically well tolerated when there is an ipsilateral P1 segment of the PCA.

- The ACh artery is a small but important artery that supplies the optic tract, thalamus, and the posterior limb of the internal capsule; occlusion can have varying neurological manifestations but will almost always result in a clinically significant stroke and should be avoided at all costs.

3.2   Middle Cerebral Artery (MCA) Bifurcation Aneurysms

- A pterional craniotomy with the head rotated 30 degrees to the contralateral side is tailored to the location of the MCA aneurysm within the sylvian fissure; more distal MCA aneurysms may require posterior extension of the skin incision and craniotomy.

- The opticocarotid cistern and proximal sylvian fissure is opened using microscissors to cut the sphenoidal arachnoid covering the ICA terminus and M1 segment of the MCA; a site along the M1 for proximal control is dissected.

- The distal sylvian fissure is opened using the operative microscope by identifying and incising cortical arachnoid on the frontal side of the sylvian venous complex.

- Microscissors are used to widen the cortical opening and dissection is carried deep through the sylvian arachnoid until the M2 or M3 segment of the MCA is identified.

- Dissection follows the M2 or M3 segment from distal to proximal until the aneurysm is reached; the MCA creates a natural separation and plane of dissection between the frontal and temporal lobes.

- In some circumstances, particularly ruptured aneurysms, it is safer to approach the aneurysm from proximal to distal, where the M1 segment is exposed and followed distally to the aneurysm.

- Dissection is continued from the point of proximal control toward the aneurysm neck; post-bifurcation branches are identified.

- A straight or curved clip can be used and is applied parallel with the parent vessels if possible, avoiding perforators and all post-bifurcation branches in the clip construct.

- Because of significant variety of branch vessels in this location, complex clip reconstructions may be required in this location.

  Pitfalls

- Blunt dissection should be avoided when splitting the sylvian fissure. Sharp dissection is preferred, especially when dissecting around the aneurysm where traction can cause an intraoperative rupture.

- Subpial dissection plane should be avoided while splitting the sylvian fissure.

- Lenticulostriates must be identified and protected to avoid a postoperative stroke.

- While small veins crossing the sylvian fissure can be sacrificed, the large sylvian vein(s) should be protected.

- MCA branches within the sylvian fissure always supply either frontal or temporal lobes, but never both.

- The sylvian vein(s) should always remain on the temporal side of the sylvian fissure dissection.

3.3 Anterior Communicating Artery (ACOM) Aneurysms

- A standard pterional craniotomy with the head rotation at 60 degrees is performed on the side of the dominant A1, where the neck of the aneurysm typically arises; the right side is utilized when the A1 segments are symmetric.

- The sphenoid wing must be aggressively flattened to provide an adequate angle of exposure toward the ACOM; an orbitozygomatic or modified orbitozygomatic craniotomy can be useful to increase the working angle for the operative microscope.

- The dura is opened and the sylvian fissure can be split as needed to relax the fontal and temporal lobes.

- The chiasmatic cistern is opened and the arachnoid on either side of the optic nerve is dissected, mobilizing the ICA and frontal lobe.

- The ICA is dissected within the carotid cistern toward its termination and the ipsilateral A1 segment is exposed.

- Dissection of the ipsilateral A1 is continued, past the ACOM, to expose the ipsilateral A2 and recurrent artery of Heubner.

- The contralateral A1 is exposed by following the optic chiasm to identify the contralateral optic nerve and contralateral A1.

- The interhemispheric fissure is opened as widely as possible to identify the contralateral A2 and recurrent artery, distal to the aneurysm dome.

- Gyrus rectus resection may be necessary if interhemispheric fissure opening does not expose the contralateral A2 and recurrent artery.

- Temporary clipping of the contralateral A1 is useful to isolate the ipsilateral A1 as the sole point of proximal control, should an intraoperative rupture occur.

- Bilateral recurrent arteries and all ACOM and A1 perforators must be dissected away from the aneurysm neck prior to clip application.

- Posterior and superior projecting aneurysms tend to reside within the interhemispheric fissure, making for an easier plane of dissection.

- Anterior projecting aneurysms tend to reside partially within the interhemispheric fissure, while inferior projecting aneurysms tend to adhere to the optic chiasm and surrounding dura.

- Final clip selection depends largely on the direction of aneurysm dome projection and the presence of adherent arteries to the aneurysm dome; a straight simple or fenestrated clip is typically used.

  Pitfalls

- ACOM aneurysms are among the most challenging to treat microsurgically due to the myriad of arteries associated with the aneurysm, the bilateral arterial supply to the aneurysm, and the numerous eloquent perforators associated with the aneurysm.

- The recurrent artery of Heubner and the orbitofrontal artery arise several millimeters from each other; these arteries can only be differentiated based on their trajectories: the recurrent artery parallels A1 throughout its course, while the orbitofrontal travels toward the gyrus rectus.

- Elevation of the frontal lobes may avulse inferiorly projecting ACOM aneurysms, which tend to be adherent to the dura surrounding the optic chiasm.

- ACOM aneurysms, in particular, require extensive aneurysm neck dissection and exposure of the arteries and perforators that may be adherent to the aneurysm neck.

3.4   Basilar Apex Aneurysms

- The basilar apex can be approached through either the transsylvian (pterional) or sub-temporal corridors depending on the orientation and height of the aneurysm relative to the posterior clinoid.

- Right sided approaches are typically performed to avoid dominant temporal lobe injury, unless there is preexisting right hemiparesis, left CN 3 palsy, or a second left sided aneurysm that needs to be treated simultaneously.

- Pterional transsylvian approach.

  - A standard pterional craniotomy is performed, typically on the right side.

  - The sphenoid wing must be flattened to provide an adequate angle of exposure toward the basilar apex; an orbitozygomatic or modified orbitozygomatic craniotomy can increase the working angle for the operative microscope.

  - The dura is opened and the sylvian fissure is split widely to free the temporal lobe.

  - The anterior bridging vein to the sphenoparietal sinus is typically sacrificed to further release the temporal lobe.

  - The optic-carotid triangle and oculomotor-carotid triangle are the two main working corridors through the membrane of Liliequist to the interpeduncular cistern.

  - The PCA is identified by following the PCOM from the ICA; if the oculomotor-carotid triangle must be widened, the PCOM can be sacrificed if there is no fetal configuration.

  - The PCA is followed through the membrane of Liliequist into the interpeduncular cistern and the basilar artery apex is identified.

  - Dissection along the PCOM and PCA is on the inferior surface of the vessels to avoid perforators.

  - Bilateral PCA and SCA are identified and a perforator-free zone on the basilar trunk is identified for proximal control.

  - The aneurysm neck is explored and perforators are dissected away from the neck.

  - Posterior perforators are explored just before clip application; a temporary clip can be applied to the basilar trunk to soften the aneurysm.

  - The clip is applied, ensuring all anterior and posterior perforators are outside the clip and that the clip blades are as close as possible to the PCAs without occluding their flow.

- Subtemporal approach.
  - Osmotic diuresis and a lumbar drain or external ventricular drain provides brain relaxation necessary to minimize temporal lobe retraction.

- A temporal craniotomy is performed using a linear or reverse-horseshoe shaped incision centered on the posterior 2/3 of the zygomatic arch.

- The craniotomy must extend down to the middle fossa floor to provide an adequate working angle to the tentorial incisura.

- After the dura is opened, the temporal lobe is gently retracted and the ambient and crural cisterns can be opened to drain CSF and relax the brain, revealing CN III and the SCA.

- If exposure of the tentorial incisura is inadequate due to elevated intracranial pressure or recent subarachnoid hemorrhage, resection of the inferior temporal gyrus increases the working space.

- The SCA is followed proximally toward its origin and the basilar artery is exposed, and a site for proximal control is identified without perforators.

- Bilateral SCA and PCA are identified, the PCOM is identified at the P1-P2 junction, and CN III can be seen traveling inferior to the PCA and superior to the SCA.

- The aneurysm neck is dissected, first anteriorly along the superior surface of bilateral P1 segments; perforators are then carefully dissected off the posterior neck of the aneurysm, creating space for clip application.

- A fenestrated clip is useful to apply the clip low on the neck of the aneurysm, keeping the PCA inside the fenestration to prevent occlusion.

Pitfalls

- Temporal lobe release and relaxation is paramount when treating basilar apex aneurysms from the pterional trans-sylvian and subtemporal approaches.

- More than any other aneurysm type, visualization of all perforators anterior and posterior to the aneurysm neck is essential before clip application; strokes arising from inadvertent thalamoperforator occlusion often cause devastating neurologic sequelae.

- Perforators can often be hidden behind the aneurysm neck; temporary clipping can soften the aneurysm and allow for final dissection in these blind spots prior to final clip application.

- Proximal control is difficult to obtain in low-riding basilar apex aneurysms using a pterional approach; posterior clinoidectomy or a transcavernous approach [14–17] can increase exposure in these cases; alternatively, a subtemporal approach provides better proximal control in low-riding basilar apex aneurysms.

3.5 Posterior Inferior Cerebellar Artery (PICA) Aneurysms

- The far lateral approach or one of its many variations is most often utilized for PICA aneurysms; a retrosigmoid approach or midline suboccipital approach can be sufficient for distal PICA aneurysms or high-riding PICA origins.

- Partial removal of the occipital condyle increases the lateral exposure of the cerebellomedullary fissure and can be useful if the standard far lateral approach provides inadequate visualization.

- Exposure of the extradural VA is not routinely necessary, as proximal control can be obtained from the intradural VA.

- Once the dura is opened, CSF can be drained from cisterna magna to relax the cerebellum.

- The dentate ligament is cut, exposing the intradural VA and a site for proximal control.

- Dissection along the vertebral artery is continued until the PICA origin is exposed.

- The tonsillomedullary segment of PICA is exposed in the cerebellomedullary fissure and dissection is continued proximally until the aneurysm neck is exposed.

- Depending on the site of origin of the PICA and the location of the aneurysm, the working corridor to dissect the neck of the aneurysm may be above or below CN XII; often a combination of working corridors is required to effectively dissect the neck.

- Small medullary perforators arise from the medial VA and must be protected when clipping aneurysms that arise from the PICA origin.

- Complex anatomic relationships often exist between the lower cranial nerves and the aneurysm neck; fenestrated clips can be used to protect adjacent cranial nerves or perforators.

  Pitfalls

- Aneurysm neck dissection is often less complicated than other aneurysm types due to fewer arachnoid adhesions and perforators.

- The challenge with PICA aneurysms is the limited maneuverability and small working corridors created by the multiple lower cranial nerves in the vicinity of the PICA origin; this is less of an issue with distal PICA aneurysms that reside within the cerebellomedullary fissure.

4 Conclusion

This chapter summarizes the relevant anatomy, key surgical steps, and pitfalls in the surgical treatment of the most common cerebral aneurysms.

References

1. Becske T, Brinjikji W, Potts MB et al (2017) Long-term clinical and angiographic outcomes following pipeline embolization device treatment of complex internal carotid artery aneurysms: five-year results of the pipeline for uncoilable or failed aneurysms trial. Neurosurgery 80(1):40–48. https://doi.org/10.1093/neuros/nyw014.

2. Becske T, Kallmes DF, Saatci I et al (2013) Pipeline for uncoilable or failed aneurysms: results from a multicenter clinical trial. Radiology 267(3):858–868. https://doi.org/10.1148/radiol.13120099.

3. Lylyk P, Miranda C, Ceratto R et al (2009) Curative endovascular reconstruction of cerebral aneurysms with the pipeline embolization device: the Buenos Aires experience. Neurosurgery 64(4):632–642.; discussion 642–643; quiz N6. https://doi.org/10.1227/01.NEU.0000339109.98070.65

4. Nelson PK, Lylyk P, Szikora I, Wetzel SG, Wanke I, Fiorella D (2011) The pipeline embolization device for the intracranial treatment of aneurysms trial. AJNR Am J Neuroradiol 32 (1):34–40. https://doi.org/10.3174/ajnr.A2421

5. Spiotta AM, Derdeyn CP, Tateshima S et al (2017) Results of the ANSWER trial using the PulseRider for the treatment of broad-necked, bifurcation aneurysms. Neurosurgery 81(1):56–65. https://doi.org/10.1093/neuros/nyx085.

6. Turjman F, Levrier O, Combaz X et al (2015) EVIDENCE trial: design of a phase 2, randomized, controlled, multicenter study comparing flow diversion and traditional endovascular strategy in unruptured saccular wide-necked intracranial aneurysms. Neuroradiology 57 (1):49–54. https://doi.org/10.1007/s00234-014-1439-7.

7. DePowell JJ, Froelich SC, Zimmer LA et al (2014) Segments of the internal carotid artery during endoscopic transnasal and open cranial approaches: can a uniform nomenclature apply to both? World Neurosurg 82(6 Suppl): S66–S71. https://doi.org/10.1016/j.wneu.2014.07.028.

8. Rhoton AL (2002) The supratentorial arteries. Neurosurgery 51(Suppl_1):S1-S53–S1-S120. https://doi.org/10.1097/00006123-200210001-00003

9. Rinne J, Hernesniemi J, Niskanen M, Vapalahti M (1996) Analysis of 561 patients with 690 middle cerebral artery aneurysms: anatomic and clinical features as correlated to management outcome. Neurosurgery 38(1):2–9. https://doi.org/10.1097/00006123-199601000-00002

10. Perlmutter D, Rhoton AL (1976) Microsurgical anatomy of the anterior cerebral-anterior communicating-recurrent artery complex. J Neurosurg 45(3):259–272. https://doi.org/10.3171/jns.1976.45.3.0259

11. Saeki N, Rhoton AL (1977) Microsurgical anatomy of the upper basilar artery and the posterior circle of Willis. J Neurosurg 46 (5):563–578. https://doi.org/10.3171/jns.1977.46.5.0563

12. Matsuno H, Rhoton AL, Peace D (1988) Microsurgical anatomy of the posterior fossa cisterns. Neurosurgery 23(1):58–80. https://doi.org/10.1227/00006123-198807000-00012

13. Rhoton AL (2002) Aneurysms. Neurosurgery 51(4 Suppl):S121–S158

14. Chanda A, Nanda A (2002) Anatomical study of the orbitozygomatic transsellar—transcavernous—transclinoidal approach to the basilar artery bifurcation. J Neurosurg 97 (1):151–160. https://doi.org/10.3171/jns.2002.97.1.0151

15. Dolenc VV, Skrap M, Sustersic J, Skrbec M, Morina A (1987) A transcavernous-transsellar approach to the basilar tip aneurysms. Br J Neurosurg 1(2):251–259

16. Figueiredo EG, Zabramski JM, Deshmukh P, Crawford NR, Preul MC, Spetzler RF (2006) Anatomical and quantitative description of the transcavernous approach to interpeduncular and prepontine cisterns. Technical note. J Neurosurg 104(6):957–964. https://doi.org/10.3171/jns.2006.104.6.957.

17. Krisht AF (2005) Transcavernous approach to diseases of the anterior upper third of the posterior fossa. Neurosurg Focus 19(2):1–10. https://doi.org/10.3171/foc.2005.19.2.3

Chapter 5

Endosaccular Approaches to Intracranial Aneurysms

Elie Dancour and Phillip M. Meyer

Abstract

Intracranial aneurysms are local dilatations or "weaknesses" in cerebral artery walls, which are prone to growing and rupturing, causing intracranial bleeding. The endovascular treatment of intracranial aneurysms encompasses many techniques, all of which had origins of development with direct endosaccular approaches. Endosaccular treatment involves direct catheterization of the aneurysm dome. The scope of this chapter will focus on endovascular treatment modalities including primary coil embolization coiling, balloon-assisted coil embolization and stent-assisted coil embolization. Basic endosaccular approaches to cerebral aneurysms will continue to be important as viable treatment options, as well as benchmarks to compare newer technology.

Key words Cerebral aneurysm, Subarachnoid hemorrhage, Dissecting aneurysms, Blood blister aneurysms, Fusiform aneurysms, Serpentine aneurysms, Coil embolization, Balloon-assisted coil embolization, Stent-assisted coil embolization, Woven Endo Bridge Device, Onyx

1 Introduction

Intracranial aneurysms (IAs) are local dilatations or "weaknesses" in cerebral artery walls, which are prone to growing and rupturing, causing intracranial bleeding. The overall prevalence of unruptured IAs is between 2% and 3.2% in the general population with a male to female ratio of 1:2 [1]. Cerebral aneurysmal rupture is responsible for 85% of subarachnoid hemorrhages (SAHs) [2]. Typically 10–15% of patients with a SAH die prior to reaching the hospital, and of those who survive, 42% will be dependent, 46% will have some form of disability, and 12% will be left severely impaired. The treatment techniques and management guidelines for IAs have been continually developing since the 1990s, and these rapid developments have caused difficulty for clinicians in the field to respond to the changes and operate within the ever-changing guidelines.

Fawaz Al-Mufti and Krishna Amuluru (eds.), *Cerebrovascular Disorders*, Neuromethods, vol. 170,
https://doi.org/10.1007/978-1-0716-1530-0_5, © Springer Science+Business Media, LLC, part of Springer Nature 2021

2   Pathophysiology and Clinical Manifestations

2.1   Types of Aneurysms

Although there are many types of aneurysms, they are typically classified into four different groups. Saccular aneurysms are the most common types. They resemble a round outpouching with a well-defined dome and neck connecting to the parenting vessel. These aneurysms tend to favor bifurcation locations like between the middle cerebral artery (MCA) and posterior cerebral artery (PCA), between the two anterior cerebral arteries (ACA), and the ACA and the bifurcation of the MCA branches [3].

Microaneurysms are IAs with diameters smaller than 2 mm. Most of these are related to hypertension and are called Charcot–Bouchard aneurysms. They often occur in blood vessels smaller than 0.3 mm and are prone to develop in basal ganglia vessels. The rest tend to be infectious IAs such as mycotic aneurysms and account for less than 1% of all IAs [4]. The infectious IAs seen in the distal MCAs are related to septic emboli from infective endocarditis, while proximal branches are more likely to be affected via infection spread from cavernous sinus thrombosis or meningitis. These aneurysms tend to resemble more blisters and patients tend to present with infectious symptoms. Despite low incidence, the morbidity and mortality in such cases has been reported to be as high as 80% in ruptured IAs [5].

Giant IAs (GIAs) have a diameter of over 25 mm. While they account for only 5% of all IAs, their prognosis tends to be dismal [6]. Untreated GIAs have over 50% risk of rupture and 88–100% mortality at 2 year follow-up postrupture. This is theorized to be due to the mass effect causing increased intracranial hypertension for a prolonged period of time [6, 7].

Fusiform aneurysms refer to a widened segment of artery affecting at least 270° degrees of the lumen's circumference. They are significantly less common than saccular aneurysms but are more difficult to treat and have a devastating natural history with up to 80% mortality in 5 years if left untreated [8]. Treatment of these aneurysms tends to pose a challenge from both an endovascular and surgical viewpoint given the presence of vital perforators, which may be located within the dilated segment of vessel. In the past, endovascular approaches were limited to the "deconstructive" techniques of partial coil embolization, glue embolization, or parent artery occlusion after passing a balloon test occlusion.

2.2   Special Types of Aneurysms

Dissecting aneurysms are arterial dissections that typically start with a minor tear to the inner wall of the vessel followed by further dissection by the shearing force of blood flow causing a pseudoaneurysm. Most dissections are caused by either trauma or as a complication of endovascular interventions. Spontaneous dissections usually occur between the V3 and V4 segments of the vertebral artery, due to the tortuous nature of the artery promoting turbulence and high shearing forces. The V3 segment is anchored by ligaments surrounding it and therefore has little mobility. However, those anchoring ligaments do not exist as the vessel enters the dura and becomes the V4 segment; it is at this location that the vessel is at greatest shearing risk [9].

Blood blister-like aneurysms are small aneurysms that originate from nonbranching sites of the terminal internal carotid artery (ICA) with a broad base [10]. These aneurysms are rare and account for anywhere from 0.9% to 6.5% of all IAs. They have an extremely fragile wall and are highly prone to spontaneous rupture. While little is known about their pathophysiology, they are theorized to arise from dissections [11].

Giant serpentine aneurysms are a subtype of giant aneurysms first described in 1977 as partially thrombosed aneurysms [12]. Slow blood flow through these aneurysms leads to intraluminal thrombus formation, which give the appearance of a serpent under digital subtraction angiography. These aneurysms are chronic manifestations and the thrombus tends to be highly fibrosed, predisposing them to stiff and rubber-like textures. These aneurysms tend to cause seizures or ischemic strokes secondary to their mass effect, rather than hemorrhagic complications [13].

"De novo" aneurysms are phenomena first described by Graf and Hamby in 1964 and refer to IA formation in previously normal locations that are remote from the original lesion. These are reported to be extremely rare and account for less than 2% of all IAs; however, they may be underreported due to lack of follow-ups [14]. Their pathogenesis is most likely related to changes induced by treatment of preexisting aneurysms. Studies have suggested that the risk of de novo formation is higher in patients with multiple IAs as well as in patients who have undergone surgical clipping as opposed to endovascular embolization [14].

3   Management Strategies

Conventional treatment options for IAs are either open-surgical or endovascular. Open-surgical techniques include but are not limited to simple clipping, contralateral MCA clipping techniques, wrapping and clipping of aneurysms, and bypass techniques, either extracranial-to-intracranial or intracranial-to-intracranial. However, the scope of this chapter will focus on endovascular treatment modalities.

3.1   Primary Coil Embolization

Detachable coils were invented by Guglielmi in the 1990s and transluminal embolization techniques have been in development since then. Simple coiling refers to the process of navigating a

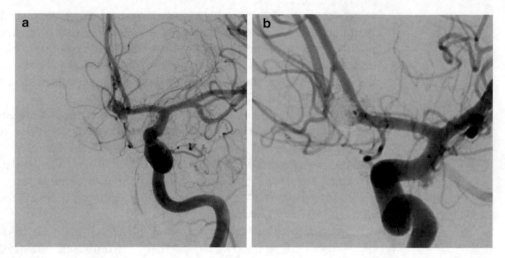

Fig. 1 Angiogram of a 65-year-old woman presenting with a subarachnoid hemorrhage found to have a (**a**) ruptured anterior communicating artery aneurysm, whch was treated with (**b**) primary coil embolization

microcatheter in the dome of the aneurysm with the help of micro-guidewires with subsequent delivery and packing of detachable coils within the aneurysmal sac. The goal of coil embolization is to achieve dense packing and induce rapid blood clot formation within the aneurysm so as to isolate it from active blood circulation (Fig. 1). Simple coiling can be employed for all IAs that have a desirable dome-to-neck ratio of >2.0. Blood blister-like aneurysms are not amenable to coiling secondary to their fragile walls, which can be easily perforated [15].

3.2 Double Catheter Technique

The double catheter technique is employed for IAs with a slightly unfavorable dome-to-neck ratio (1.5–2.0). Before coiling, two microcatheters are positioned in the proximal and distal aspects of the aneurysm. The first coil is deployed from the proximal catheter to create a supporting frame and then the rest of the coils are deposited from the distal catheter (Fig. 2). The first framing coil is not detached until satisfactory packing is obtained. This technique is safe and effective for elongated IAs, especially those at MCA bifurcations. However, this technique comes with the risk of coils shifting at the withdrawal of the distal microcatheter [16].

3.3 Balloon-Assisted Coiling

Balloon-assisted coiling (BAC) was initially described by Moret et al. in 1997, while describing the treatment of IAs with a wide neck [17]. In this technique, one or more nondetachable balloons are temporarily inflated to block the aneurysm neck while the coils are packed (Fig. 3). Depending on the complexity of the aneurysm, multiple balloons can be used simultaneously.

Balloon-assisted coiling was used frequently in IAs with an unfavorable dome-to-neck ratio (1–1.5). Analysis of the trial Treatment by Endovascular Approach of Non-Ruptured Aneurysms

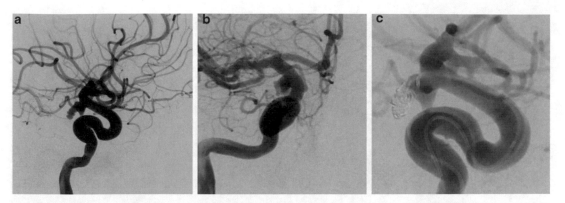

Fig. 2 Angiogram of a 70-year-old woman presenting with subarachnoid hemorrhage found to have a right-sided posterior communicating artery ruptured aneurysm on (**a**) lateral view and (**b**) AP view. (**c**) The double-catheter technique was employed to coil her aneurysm

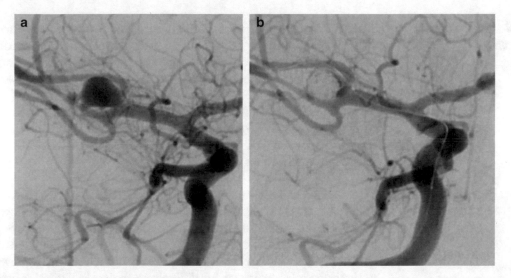

Fig. 3 Angiogram of a 62-year-old woman who presented with a subarachnoid hemorrhage and found to have a (**a**) ruptured right middle cerebral artery bifurcation aneurysm which was (**b**) coiled with the assistance of a balloon positioned distal to the aneurysm to prevent coil migration

(ATENA) revealed that intraoperative aneurysmal rupture rates were higher in the BAC group than simple coiling (3.2% vs. 2.2%) and BAC was associated with higher permanent morbidity and mortality; however, this study was underpowered and the difference could not be proven to be significant [18]. The trial Clinical and Anatomical Results in the Treatment of Ruptured Intracranial Aneurysms (CLARITY) also suggested higher thromboembolic rates (12.7% vs. 11.3%), morbidity (3.9% vs. 2.5%), and mortality (1.3% vs. 1.2%) in the BAC group when compared to simple coiling [18].

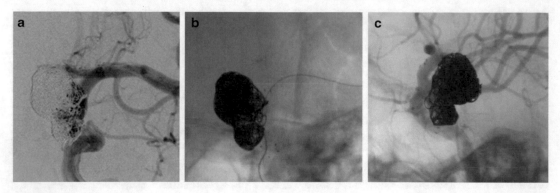

Fig. 4 Angiogram of a 55-year-old woman who presented with a subarachnoid hemorrhage and was found to have (**a**) a ruptured paraclinoid left ICA aneurysm which was treated with coil embolization. (**b–c**) A micro-guidewire was then navigated past the coiled aneurysm and a stent was deployed across the neck of the aneurysm

3.4 Stent-Assisted Coiling

The first report of stent-assisted coiling (SAC) for IAs was also in 1997 by Higashida et al. [19]. The SAC technique can overcome the limitations of wide-necked, gigantic, and fusiform aneurysms. This technique is similar to a BAC technique; however, here a stent is deployed at the neck of the aneurysm prior to, or after, coil packing (Fig. 4). The IAs with an extremely unfavorable dome-to-neck ratio (<1.0) require SAC due to the need for permanent support to prevent coil prolapse and migration [20]. There are four major SAC techniques:

3.4.1 Simple Stent-Assisted Coiling

This is also called the mesh technique. The first step involves deployment of a stent across the neck of the aneurysm followed by navigating a microcatheter into the aneurysmal lumen through the interstices of the stent. Coils are then delivered through the microcatheter. This technique requires a high level of guidewire navigating skills, as it can be difficult to maintain the microcatheter's position during coil deployment [21].

3.4.2 Stent-Jail Technique

In this technique, the microcatheter is positioned inside the aneurysmal lumen prior to stent deployment over the aneurysmal neck. This allows the microcatheter to remain within the aneurysmal dome as it is essentially trapped by the stent. In the early days of this technique there were reports of stent migration during retrieval of the coil microcatheter [22]. Today this is less of an issue due to the developments of retrievable stents such as the Enterprise (Cordis, Florida) and the Solitaire (ev3, Irvine, California). With the advent of these retrievable devices, the stent is semideployed during coiling and only fully deployed after retrieval of the microcatheter. This enables readjustments of the stent position in case of migration. It has been shown to be safe and effective and is one of the most used SAC techniques employed today [23].

3.4.3 Stent-Jack Technique

In this technique, one coil is first deposited into the aneurysmal lumen prior to deployment of the stent. The coil typically will form a large loop, which is then pushed back into the aneurysmal lumen by the deployment of the stent. This results in better coil wall positioning [24]. There is also a semijacking technique using retrievable stents where the stent is only partially deployed in the jacking movement, allowing for multiple movements to position the coil against the wall of the aneurysm. However, this technique is not recommended due to increased risk of rupture and the possibility of compromising the structural integrity and stability of the stent [22].

3.4.4 Y-Stenting Technique

The Y-stenting technique was developed for treating IAs located at bifurcation points. In this technique, typically one microcatheter remains within the aneurysmal dome as two stents are deployed in each limb of the bifurcation to block the aneurysmal neck (Fig. 5). This is one optimal method of treating basilar artery aneurysms, as their anatomic positions typically make them difficult to treat any other way. Aneurysms at the top of the basilar artery are usually in close proximity to the PCAs, and one cannot protect both limbs with one stent as the contralateral side would remain vulnerable to coil prolapse and migration [25]. Normally, nonretrievable stents such as Neuroform (Boston Scientific Neurovascular, Fremont, California) or Wingspan (Striker/Boston Scientific SMART, Fremont, California) are used. Operators typically attempt to deploy the stent slowly so that the stent structures overlay close to each other and create a better wall effect.

3.4.5 Other SAC Techniques

There are many other stenting techniques based on similar principles, one of which is the waffle cone/ice-cone technique [26]. It is performed using the Neuroform or Enterprise stent, where the stent is deployed into the proximal neck of the aneurysm and the coils are packed in a waffle-cone formation. This technique is more suitable for large wide-neck bifurcation IAs. It requires the flare ends of the stent (4.5 mm) be wider than the aneurysmal neck. Compared to the Y-stent, this approach uses a single stent and thus reduces risks of thromboembolic events and the probability of in-stent stenosis. While it is more flexible and can be used in a wide range of vessel configurations, it is technically demanding on operators. It is currently recommended for IAs with elongated domes, wide necks, and on bifurcating vessels where Y-stenting or surgical clipping are unsuitable. In IAs with shorter dome heights, the forward tension on the microcatheter could result in the backward migration of the stent during coiling [26].

3.5 Simple Stenting for Intracranial Dissecting Aneurysms

In cases of intracranial dissecting aneurysms, simple stenting is used as the most effective approach to treat the flap and close the tear in order to restore wall integrity. It is, however, essential to identify

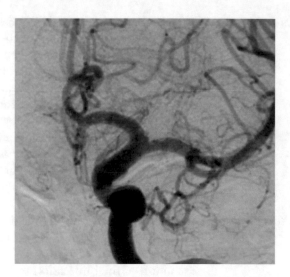

Fig. 5 Y-stent and coiling of a left internal carotid artery terminus aneurysm

the true lumen from the pseudolumen before stenting, which can often be a challenge [27].

3.6 Salvation Techniques

Salvation techniques are used when dealing with ruptured IAs or vascular trauma during interventional procedures [21]. The main goal of these techniques is to restore vessel wall integrity and to stop bleeding. The most commonly used technique is to deploy a covered stent to block the perforation in the vessel wall, which often sacrifices in situ perforators in that region. Overlapping stents as well as flow-diverting stents are also used to close the defect by trapping tissues with dense stent structure and is considered safer for preserving perforators.

3.7 Intrasaccular Flow Disruptions

Besides detachable coils, other types of intrasaccular flow disruption devices are being developed, one of which is the Woven Endo Bridge Device (WEB; Sequent Medical, Aliso Viejo, California). WEB is deployed inside the aneurysmal sac to induce fast thrombosis [28]. It is suitable for most saccular IAs and even ruptured IAs, as it facilitates acute aneurysmal occlusion. WEB does not place adjacent perforating arteries at risk and there is no need for antiplatelet therapy immediately after the procedure [29]. So far, two single-center series demonstrated high technical success of treatments with no mortality and morbidity less than 5%. However, some experts point out its limitation in treating irregular dome-shaped IAs as good wall apposition may be difficult [30].

3.8 Liquid Embolic Material

There have been attempts at using liquid embolic agents as aneurysmal intrasaccular embolization agents. Onyx (Covidien/EV3, Irvine, California) is a liquid embolic agent containing ethylene vinyl alcohol (EVOH) copolymer and dimethyl sulfoxide

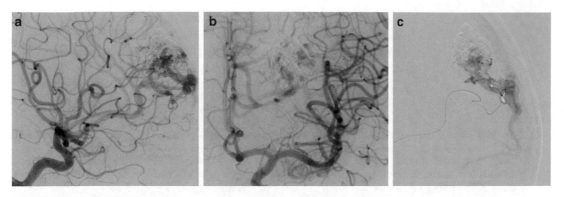

Fig. 6 Angiogram of an 11-year-old girl performed for evaluation of headaches and found to have a left frontal unruptured AVM treated with Onyx. (**a**) Lateral view (**b**) AP view (**c**) Super-selective run using a micro-catheter

(DMSO) in a volume ratio of 3:2 and tantalum powder (28%, wt/wt) as radiopaque marker [31]. The lower the concentration of the copolymer, the less viscous the agent and the more distal penetration can be achieved. The viscosity of Onyx 18, Onyx 20, and Onyx 34 is a factor of the concentration of EVOH (6%, 6.5%, 8% respectively). Onyx 18, 20, and 34 are typically used for embolization of arteriovenous malformation (AVM) nidus (Fig. 6).

High density Onyx 500 (containing 20% of EVOH) may be used for aneurysm occlusions. In the USA, Onyx is approved for treatment of intracranial saccular, sidewall aneurysms that present with a wide neck (neck $\geq$ 4 mm or dome-to-neck ratio < 2) and are not amenable to surgical clipping. In a prospective, multicenter, European registry of Onyx aneurysm embolizations performed in 97 patients, Molyneaux et al. reported a 79% complete aneurysm occlusion rate and a 10% recanalization rate in the 71 patients for whom 1-year follow-up angiography was available. Procedure-related morbidity and mortality rates were 8.2% and 2%, respectively [32]. Although Onyx may be suitable for complex and irregularly shaped IAs, operators must be aware of concerns regarding DMSo toxicity, the need for concurrent balloon-assistance and the possibility of nontarget liquid embolization [33].

4   Conclusion

Since the first use of the Guglielmi detachable coil system for cerebral aneurysm embolization in 1990, various endovascular methods have been developed to treat aneurysms. The main strategic techniques to date include primary coiling, balloon-assisted coil embolization, and stent-assisted coil embolization. The development and introduction of newer devices such as flow diverting devices and flow disruption devices, offers tailored treatment for individual patients based on presentation, anatomy, and the physician's familiarity with the techniques. However, basic endo-saccular approaches to cerebral aneurysms will continue to be important as viable treatment options, as well as benchmarks to compare newer technology.

References

1. Vlak MH et al (2011) Prevalence of unruptured intracranial aneurysms, with emphasis on sex, age, comorbidity, country, and time period: a systematic review and meta-analysis. Lancet Neurol 10(7):626–636

2. van Gijn J, Kerr RS, Rinkel GJ (2007) Subarachnoid haemorrhage. Lancet 369 (9558):306–318

3. Bharatha A et al (2010) Comparison of computed tomography angiography with digital subtraction angiography in the assessment of clipped intracranial aneurysms. J Comput Assist Tomogr 34(3):440–445

4. Kannoth S et al (2007) Intracranial infectious aneurysm: presentation, management and outcome. J Neurol Sci 256(1–2):3–9

5. Bohmfalk GL et al (1978) Bacterial intracranial aneurysm. J Neurosurg 48(3):369–382

6. Gobble RM et al (2012) Extracranial-intracranial bypass: resurrection of a nearly extinct operation. J Vasc Surg 56 (5):1303–1307

7. Barrow DL, Alleyne C (1995) Natural history of giant intracranial aneurysms and indications for intervention. Clin Neurosurg 42:214–244

8. Yahia AM et al (2008) Complications of Neuroform stent in endovascular treatment of intracranial aneurysms. Neurocrit Care 8 (1):19–30

9. Ducruet AF et al (2013) Reconstructive endovascular treatment of a ruptured vertebral artery dissecting aneurysm using the Pipeline embolization device. J Neurointerv Surg 5(4): e20

10. Çinar C et al (2013) Endovascular treatment of ruptured blister-like aneurysms with special reference to the flow-diverting strategy. Neuroradiology 55(4):441–447

11. Ishikawa T et al (1997) Pathological consideration of a "blister-like" aneurysm at the superior wall of the internal carotid artery: case report. Neurosurgery 40(2):403–405. discussion 405–406

12. Segal HD, McLaurin RL (1977) Giant serpentine aneurysm. Report of two cases. J Neurosurg 46(1):115–120

13. Christiano LD et al (2009) Giant serpentine aneurysms. Neurosurg Focus 26(5):E5

14. Zali A, Khoshnood RJ, Zarghi A (2014) De novo aneurysms in long-term follow-up computed tomographic angiography of patients with clipped intracranial aneurysms. World Neurosurg 82(5):722–725

15. Gross BA, Du R (2013) Microsurgical treatment of ophthalmic segment aneurysms. J Clin Neurosci 20(8):1145–1148

16. Kwon SC et al (2006) A double catheter technique for elongated middle cerebral artery bifurcation aneurysm. A case report. Interv Neuroradiol 12(1):41–44

17. Moret J et al (1997) Reconstruction technic in the treatment of wide-neck intracranial aneurysms. Long-term angiographic and clinical results. Apropos of 56 cases. J Neuroradiol 24 (1):30–44

18. Pierot L et al (2009) Endovascular treatment of unruptured intracranial aneurysms: comparison of safety of remodeling technique and standard treatment with coils. Radiology 251 (3):846–855

19. Higashida RT et al (1997) Intravascular stent and endovascular coil placement for a ruptured fusiform aneurysm of the basilar artery. Case report and review of the literature. J Neurosurg 87(6):944–949

20. Wakhloo AK et al (2008) Stent-assisted reconstructive endovascular repair of cranial fusiform atherosclerotic and dissecting aneurysms: long-term clinical and angiographic follow-up. Stroke 39(12):3288–3296

21. Yang PF et al (2010) Preliminary experience and short-term follow-up results of treatment of wide-necked or fusiform cerebral aneurysms with a self-expanding, closed-cell, retractable stent. J Clin Neurosci 17(7):837–841

22. Hong B et al (2009) Semi-jailing technique for coil embolization of complex, wide-necked intracranial aneurysms. Neurosurgery 65 (6):1131–1138. discussion 1138–1139

23. Lopes DK, Wells K (2009) Stent remodeling technique for coiling of ruptured wide-neck cerebral aneurysms: case report. Neurosurgery 65(5):E1007–E1008. discussion E1008

24. de Paula Lucas C et al (2008) Stent-jack technique in stent-assisted coiling of wide-neck aneurysms. Neurosurgery 62(5 Suppl 2):

ONS414–ONS416.   discussion
ONS416–ONS417

25. Jankowitz BT et al (2012) Y stenting using kissing stents for the treatment of bifurcation aneurysms. J Neurointerv Surg 4(1):16–21

26. Padalino DJ et al (2013) Enterprise stent for waffle-cone stent-assisted coil embolization of large wide-necked arterial bifurcation aneurysms. Surg Neurol Int 4:9

27. Lee DH et al (2013) The difference of each angiographic finding after multiple stent according to stent type in bilateral vertebral artery dissection. J Cerebrovasc Endovasc Neurosurg 15(3):229–234

28. Zhao J et al (2015) The influence of porosity on the hemocompatibility of polyhedral oligomeric silsesquioxane poly (caprolactone-urea) urethane. Int J Biochem Cell Biol 68:176–186

29. Pierot L et al (2011) Remodeling technique for endovascular treatment of ruptured intracranial aneurysms had a higher rate of adequate postoperative occlusion than did conventional coil embolization with comparable safety. Radiology 258(2):546–553

30. Lubicz B et al (2014) WEB-DL endovascular treatment of wide-neck bifurcation aneurysms: short- and midterm results in a European study. AJNR Am J Neuroradiol 35(3):432–438

31. Loh Y, Duckwiler GR (2010) A prospective, multicenter, randomized trial of the Onyx liquid embolic system and N-butyl cyanoacrylate embolization of cerebral arteriovenous malformations. J Neurosurg 113(4):733–741

32. Molyneux AJ et al (2004) Cerebral Aneurysm Multicenter European Onyx (CAMEO) trial: results of a prospective observational study in 20 European centers. AJNR Am J Neuroradiol 25(1):39–51

33. Carlson AP et al (2013) Continued concern about parent vessel steno-occlusive progression with Onyx HD-500 and the utility of quantitative magnetic resonance imaging in serial assessment. Neurosurgery 72(3):341–352. discussion 352

Chapter 6

Endoluminal Approaches to Intracranial Aneurysms

Eric R. Cohen, Fawaz Al-Mufti, Gaurav Gupta, and Sudipta Roychowdhury

Abstract

Treatment of intracranial aneurysms with complex morphology such as giant wide-necked or fusiform aneurysms is challenging. Stent-assisted coiling and balloon-assisted coiling are alternative treatment techniques, but studies have shown less than expected efficacy given their high rate of recanalization. Over the past few years, endoluminal techniques and flow-diverting devices (FDDs) have revolutionized the treatment of such complex aneurysms, for which the natural history of microsurgical or conventional neuroendovascular management has traditionally been poor. FDDs are implanted within the parent artery rather than the aneurysm sac. By modifying intra-aneurysmal and parent-vessel flow dynamics at the aneurysm/parent vessel interface, flow diversion triggers a cascade of gradual intra-aneurysmal thrombosis. As endothelialization of the FDD progresses, the parent vessel reconstructs while preserving the patency of normal perforators and side branch vessels. Herein we review FDDs, mechanisms of actions, current data and technical updates, complications and their management, and ongoing studies in relevance to the practicing neurointerventionalist.

Key words Intracranial aneurysms, Endovascular, Flow diversion, Subarachnoid hemorrhage, Pipeline embolization device, Pipeline flex embolization device, Silk flow diverter, Flow-redirection endoluminal device (FRED), Surpass flow diverting stent

1 Introduction

Endoluminal methods have been recognized as a definitive therapeutic alternative in the endovascular treatment of intracranial arterial aneurysms with the recent advent of flow diverting devices (FDDs), which will be the focus of this chapter. The approach and underlying physiological mechanism of FDDs actually embody a new conceptual paradigm that is a departure from existing endovascular therapeutic methods. Whereas coil embolization only occludes the aneurysm, and parent vessel sacrifice deconstructs the diseased artery, flow diversion reconstructs the diseased

---

The original version of this chapter was revised. The correction to this chapter is available at https://doi.org/10.1007/978-1-0716-1530-0_23

Fawaz Al-Mufti and Krishna Amuluru (eds.), *Cerebrovascular Disorders*, Neuromethods, vol. 170,
https://doi.org/10.1007/978-1-0716-1530-0_6, © Springer Science+Business Media, LLC, part of Springer Nature 2021,
Corrected Publication 2021

segment of the parent artery while concomitantly excluding, thrombosing, and often shrinking the aneurysm [1–4]. These characteristics are particularly advantageous for treating giant, wide-necked aneurysms causing symptomatic mass effect, fusiform aneurysms, and other aneurysms arising from segmentally dysplastic vessels. Although flow diversion can be performed in conjunction with endosaccular coil embolization, flow diversion itself is a durable treatment option without the associated risk of coil compaction and endosaccular recanalization [3]. Overall complication rates of flow diversion are comparable to existing endovascular methods and may have declined over the past several years as operators have become more familiar with using FDDs [5]. In this chapter, we will discuss the mechanism and procedural details for using FDDs, namely the Pipeline Embolization Device(PED) [6].

There are certain considerations associated with the use of FDDs. First, the primary indication for use of the PED, and the one for which it is FDA-approved, is the treatment of large or giant wide-necked intracranial aneurysms in the internal carotid artery from the petrous to the superior hypophyseal segments [1]. This role is expanding, however, and FDDs have been applied to all types of intracranial aneurysms, including small aneurysms, fusiform aneurysms, posterior circulation aneurysms, and vascular dissections [7, 8]. Second, although stagnation of intra-aneurysmal flow is typically noticeable immediately after FDD deployment, aneurysm occlusion progresses gradually over a course of several weeks to months. Complete occlusion at the end of a flow-diverting procedure is typically only seen with small aneurysms [3]. For large aneurysms, complete occlusion may take up to 12 months. Third, because flow diversion involves permanent placement of a device within the parent vessel, the risk of thromboembolic complications is higher than that of traditional or balloon-assisted coil embolization without permanent stent placement) [9]. Pre- and postprocedural dual antiplatelet treatment is generally recommended to reduce the risk of thromboembolic events, but this preventative measure itself carries a risk of hemorrhagic complications. As a result, the use of flow diversion may be limited in the setting of ruptured aneurysms, although this is an area of active investigation [10].

2   Pathophysiology

A hallmark of flow-diverting treatment is the exclusion of the aneurysm with maintenance of normal flow through the parent vessel and arterial side branches covered by the device [6, 11, 12]. Following deployment of an FDD, the aneurysm begins to thrombose and subsequently shrink and collapse around the device construct [1]. Over the ensuing 6–12 months, aneurysmal thrombosis continues, and the parent vessel is reconstructed with eventual aneurysmal occlusion [3, 4, 13].

Two underlying mechanisms work in concert to achieve this result: (1) an immediate mechanical disruption of blood flow into the aneurysm and redirection of flow along the parent artery, initiating thrombosis in the aneurysm, and (2) endothelialization of the walls of the flow-diverting construct [7]. To disrupt flow across the aneurysm neck, FDDs are designed with greater metallic surface coverage and correspondingly less porosity than a conventional stent. This optimal balance of surface coverage and porosity is what allows FDDs to exclude the aneurysm from circulation but preserve the patency of covered branch vessels (e.g., anterior choroidal or ophthalmic arteries) [14–16]. The flow demand for normal tissue perfusion generates a sufficient pressure gradient to maintain antegrade flow across the stent wall and into arterial side branches [17]. Table 1 shows the surface coverage and composition of the PED as well as those of other FDDs approved for use in Europe. Note that the Pipeline Flex embolization device (PFED) is identical to the PED except in its updated deployment mechanism.

The second mechanism by which FDDs achieve their goal is by providing an endoluminal scaffold for endothelialization, which occurs over the 6–12 months after FDD deployment, definitively isolating the aneurysm from the parent artery [18]. Ultimately, the entire parent artery-aneurysm complex is reconstructed such that the interface between the parent artery and the aneurysm is occluded [19].

3 Management Strategies

3.1 Preprocedure Preparation

Placement of an endoluminal device can cause activation and adherence of platelets, resulting in in situ thrombosis and possible distal embolization. Dual antiplatelet therapy (DAT) has been demonstrated to effectively reduce the risk of thromboembolic complications associated with coil embolization and stent-assisted coiling of intracranial aneurysms [20–22]. DAT is generally advised, and commonly practiced, in the setting of flow diversion, as well. DAT generally consists of daily doses of aspirin (i.e., acetylsalicylic acid), a cyclo-oxygenase-1 (COX-1) inhibitor, and clopidogrel (Plavix), a thienopyridine (i.e., a platelet P2Y12 receptor antagonist). Although regimens vary across institutions, the use of DAT is supported by retrospective analyses, which favor high-dose aspirin (>160 mg) rather than low-dose and that DAT be continued for *at least* 6 months postprocedure [23]. Although currently there is no standardized pre- or postprocedure protocol for DAT, nor are there any prospective, randomized controlled studies to help determine an optimal therapeutic strategy, the dose and timing of clopidogrel administration are known to be important factors in achieving therapeutic levels of platelet inhibition [7, 23]. At a daily clopidogrel dose of 75 mg, it can take 3–7 days to achieve a therapeutic

Table 1

Comparison of the basic characteristics of existing flow-diverting stents

Device	Manufacturer	Composition	Surface coverage (%)	Approval
Pipeline embolization device (PED) and Pipeline Flex embolization device (PFED)	ev3/Covidien (Irvine, CA, USA)	25% platinum, 75% nickel-cobalt chromium alloy	35	FDA, CE
Silk flow diverter	Balt Extrusion (Montmorency, France)	Nitinol	35–55 (Sfyroeras et al., JVS 2012) *Porosity 45-60* (Alderazi et al.)	CE
Flow redirection endoluminal device (FRED)	MicroVention, Inc. (Tustin, CA, USA	Nitinol	22–44 (Microvention FRED Brochure)	CE
Surpass flow diverter	Stryker Neurovascular (Fremont, CA, USA)	Cobalt-chromium	30	CE
P64 flow modulation device	Phenox (Bochum, Germany)	Nitinol	35–40 (Briganti et al., JNIS 2015)	CE

FDA Food and Drug Administration, *CE* Conformité Européenne

Metal surface coverage (%) = metal covered surface area/total surface area × 100 = 100 − porosity (%)

level of platelet inhibition, whereas a single loading dose can achieve maximum platelet inhibition within hours [7, 23]. Therefore, DAT regimens for FDD placement commonly consist of loading doses of aspirin and clopidogrel days before the procedure followed by clopidogrel 75 mg daily and aspirin 81–325 mg daily.

The use and timing of preprocedural loading doses of aspirin and clopidogrel is variable across institutions. In a retrospective meta-analysis of patients undergoing aneurysmal PED treatment, preprocedural loading doses of clopidogrel were associated with fewer symptomatic permanent thrombotic events [23]. Preprocedure loading doses of both aspirin and clopidogrel were also associated with a decreased incidence of symptomatic permanent hemorrhagic events. The latter finding is paradoxical in light of the increased risk of spontaneous intracranial hemorrhage associated with DAT, a risk which is even higher in patients undergoing treatment with the PED, and requires further investigation [23]. Finally, it has been shown that a 600-mg bolus of clopidogrel achieves a faster onset of action (i.e., 2–4 h to peak effect) without a change in the rate of hemorrhagic complications relative to a

300-mg loading dose, but this finding was demonstrated in patients undergoing coronary intervention [7].

Use of point-of-care (POC) platelet function testing (PFT) varies among neurointerventional practitioners, ranging from no use at all to routine testing before every case. VerifyNow (Accriva Diagnostics, San Diego, California, USA) is a popularly used POC PFT system. There is insufficient evidence that routine preprocedural PFT decreases morbidity from FDD-associated thromboembolic or hemorrhagic complications. However, there are few practical options for PFT, and the results do offer some needed guidance in identifying and managing patients who are resistant or hyporesponsive to clopidogrel. Therefore, we advocate its routine use. Of note, the timing of PFT with respect to the timing and doses of clopidogrel may be important due to the pharmacokinetics of the drug [7, 23, 24].

At our institution, the following preprocedure protocol is used.

- Fourteen days prior to the procedure, baseline PFT is performed using the VerifyNow P2Y12 Test (also known as the PRU Test) and Aspirin Test. DAT is initiated with loading doses of clopidogrel 300 mg and aspirin 650 mg.

- Preprocedural DAT is continued with aspirin 325 mg daily and clopidogrel 75 mg daily.

- Seven days prior to the procedure, PFT is repeated. If platelet function is less than or equal to 200 P2Y12 reaction units (PRU), our threshold for adequate platelet inhibition, then aspirin and clopidogrel are continued at the same doses. If platelet function is greater than 200 PRU, then the daily clopidogrel dose is increased.

- On the day of the procedure, PFT is repeated. If platelet function measures less than 200 PRU, no clopidogrel dosing changes are made. If platelet function measures greater than or equal to 200 PRU, then a loading dose of clopidogrel is administered and PFT is repeated.

3.2 Procedure Steps

The procedure is typically performed under general anesthesia. The main steps are then as follows:

- If cerebral angiography was not previously performed, DSA is performed on the bilateral internal carotid arteries (ICA's) and vertebral arteries. This provides a baseline examination of the cerebral circulation and collateral flow. When studying the artery to be treated, rotational DSA may be used to find the obliquity that optimizes visualization of the aneurysm and arterial segment for stent deployment.

- Vascular measurements are made to determine the diameter, length, and number of PED's necessary to adequately cover the aneurysm and the diseased parent vessel segment. Accurate measurements and correct sizing of the device are of considerable importance. The diameter of the PED should approximate the diameter of the target vessel as closely as possible and should not be deliberately oversized. The PED, like other FDDs and unlike conventional stents, is at braided stent, composed of a woven mesh of thin wires. As the device is unsheathed during deployment, it expands and foreshortens towards its nominal diameter and length, respectively. The metallic surface coverage of the device, which is the inverse of its porosity, varies with the device's constrained diameter, length, and curvature. As the device is bent, stretched, or shortened relative to its nominal dimensions, porosity, and therefore its metallic surface coverage, is altered. This changes the hydrodynamic properties of the device, which may have detrimental effects if deliberatley oversized. Deliberately oversizing is therefore potentially detrimental to the treatment of the aneurysm and generally not recommended. Maximum surface coverage is achieved when the device is maximally expanded to 0.25 mm greater than its nominal diameter; so accurately matching the diameter of the selected device to that of the parent artery is advised.

- Constraining the device along its long axis will also change its porosity. This may become apparent at the transition zones between the parent artery and a wide-necked or fusiform aneurysm, particularly if the arterial landing zone for the device is short, or if there is considerable difference in diameter between the proximal and distal landing. Stretching the PED, which may occur in angulated vessel segments, or in attempting to span the width of an aneurysm, is undesirable because it changes the device porosity and predisposes to subsequent device migration. Additionally, any deformation of the PED that does not allow the device to adequately appose the vessel wall can potentially lead to an "endoleak."

- One possible way to avoid detrimental deformation of the device is to telescope multiple shorter PED's of appropriate diameter to more accurately approximate the length and variable diameter of the diseased vessel segment. Also important is choosing a distal landing zone of sufficient length; the manufacturer's instructions state that this is least 2–3 mm beyond the distal edge of the aneurysm, but in reality, a longer landing zone than this may be needed, especially in an angled vessel segment.

- Prior to the therapeutic phase of the procedure, heparin is administered as an intravenous (IV) bolus of 50–100 U/kg body weight (approximately 3000–5000 U). Additional intraprocedural heparin maintenance doses are administered at the operator's discretion. A generally accepted ACT target range is 250–300 s (2.5–3 times the baseline ACT) [25]. Protamine sulfate should be readily available for emergent administration in the event of acute intra-procedural intracranial hemorrhage. The typical dose of protamine sulfate is 1 mg per 100 U of heparin administered.

- A long sheath, such as the 8-French AXS Infinity LS (Stryker Neurovascular, Fremont, California, USA), is used to access the proximal vessel of interest. For maximum intravascular support and stability, a triaxial system is typically used for PED deployment. A distal access catheter, such as the 5-French CAT5 (Penumbra, Alameda, California, USA), is used as a guide catheter and is advanced through the Infinity sheath. A Marksman microcatheter (Covidien/Medtronic, Minneapolis, Minnesota, USA) is the preferred microcatheter for delivery of the PED. It has the required inner diameter of 0.027 in. necessary to accommodate the PED.

- The Marksman is advanced over a wire, typically of 0.014-in. diameter, until its tip is distal to the aneurysm. For an ICA aneurysm, the microcatheter tip is usually advanced into a branch of the middle cerebral artery (MCA). The length of the distal end of the PED delivery wire is 15 mm; so the tip of the Marksman should be advanced beyond the distal landing zone by at least this distance.

- The PED is loaded from its introducer sheath into the Marksman microcatheter. The PED is then advanced into the microcatheter under fluorsoscopy until the distal tip of the delivery wire is aligned with the distal tip of the microcatheter.

- Deploying the PED is different than that of a conventional stent and requires a combination of pushing and unsheathing in order to cause the device, which is stretched and elongated in its sheathed configuration, to properly expand and foreshorten. After confirming fluoroscopically that the PED is in the desired location with respect to the aneurysm and target vessel, the microcatheter is slowly retracted while the PED position is maintained with the delivery wire, exposing the distal end of the delivery wire. Retraction of the microcatheter is continued just until the tip of the microcatheter is slightly proximal to the distal tip of the PED. The PED is then deployed by predominantly pushing the delivery wire to expose the device. After about 10 mm of the PED is exposed, its distal tip is expected to release from the distal capture coil of the delivery wire. When the distal end of the PED releases from the delivery wire, the device will simultaneously expand and slightly retract. Note that one may wish to anticipate this foreshortening if it is crucial to the positioning of the PED, for instance with respect to the origin of a perforating vessel or the edge of the aneurysm, and position the distal tip of the unsheathed device slightly more distally than its expected final position. If the PED does not release after ten full turns of the delivery wire, the entire delivery system should be removed (ev3, PED Instructions for Use). Once the PED is fully deployed, the Marksman microcatheter is advanced through the PED until the distal end of the delivery wire is recaptured, and the Marksman and the delivery wire are then withdrawn and removed together.

- Control cerebral DSA is performed upon completing the procedure in order to assess the immediate efficacy of the treatment, to evaluate for iatrogenic occlusion of side branches or distal embolization, and to ensure the absence of vascular dissection.

- The catheters and femoral sheath are removed. Closing the arteriotomy site with an arterial closure device, such as Perclose (Abbott Vascular, Abbott Park, Illinois, USA) or Angio-Seal (St. Jude Medical, Minnetonka, Minnesota, USA), is recommended when an antiplatelet or a thrombolytic rescue drug has been administered.

3.3 Managing Periprocedure Adverse Technical Events

As suggested above, the physical properties of FDDs and the mechanisms of their deployment, particularly in the setting of tortuous vascular anatomy, make these devices more prone to suboptimal deployment and overt technical failure than conventional stents [26]. Situations are therefore bound to arise that require quick and sometimes creative remediation to correct a malpositioned device and prevent thromboembolic complications [27]. Incomplete or suboptimal device expansion and proximal device migration are two of the relatively more common adverse technical events reported with the PED [1, 25, 28–34]. Prolapse of a flow-diverting device into the aneurysm sac has also been discussed with some degree of frequency [30, 34–36].

As detailed above, the PED is unsheathed from distal to proximal. Upon deployment, if the distal end of the device is in an undesirable location or configuration, the device can be removed, but the catheter and entire delivery system must be removed with the FDD as a unit. Once the entire device has been unsheathed, however, it cannot be easily extracted. This presents a dilemma when standard deployment maneuvers fail to effectuate full device expansion to match the size of the parent vessel. "Wagging" the PED during deployment or bumping the proximal end of the device with the microcatheter after deployment may be sufficient to induce the proximal end of the stent to fully expand [37]. However, these auxiliary techniques may increase the risk of vascular injury, and bumping the stent may cause the entire device to foreshorten and migrate [5, 26, 27, 32, 33].

If adequate apposition of the device with the parent vessel wall has not been achieved with gentle manipulation, balloon angioplasty should be considered. In the PUFS Trial, Bescke et al. noted "a few instances" in which a suboptimally expanded PED was dilated by balloon angioplasty [1]. Burrows et al., in a series of 100 consecutive procedures, reported incomplete expansion upon initial deployment of the PED in 9% of cases, which was treated with balloon angioplasty in two cases [25]. Despite the reported instances of success with this approach, we stress the cautionary note that this maneuver carries a risk of vessel rupture [38]. It is recommended that the angioplasty balloon be completely maintained within the device when inflated rather than using the balloon to push the proximal device open, as the latter can result in arterial injury [33, 38].

Lin et al. described their technique of using a distal intracranial catheter (DIC) to successfully complete deployment of nine incompletely expanded PED's and two PFED's [26]. While this adjunct maneuver may be sufficient in some cases, angioplasty may still be required for complete device expansion, as it was in six out of the 11 cases in which the DIC was used. When antegrade access of a constricted FDD in the ICA is unsuccessful, balloon expansion may still be possible via a retrograde approach, as demonstrated by Navarro et al., in which a proximally unopened PED was ultimately accessed by traversing the anterior communicating artery. It is worth noting, as a word of caution, that a carotid-cavernous fistula (CCF) was caused during repeated attempts to access the PED via an antegrade approach, although the CCF remained clinically silent [33].

Proximal device migration was seen in 12 out of the 100 cases reviewed by Burrows et al., making it the most frequent technical event observed in this study [25]. Migration is often accompanied by foreshortening and may also result in prolapse of the stent into the aneurysm sac. If this occurs, realignment of the stent may be possible. Martinez-Galdamez et al. described a unique approach in which an angioplasty balloon was advanced along the curvature of the aneurysm wall in order to direct it into the proximal, prolapsed end of the stent and realign the stent with the parent vessel [31]. Crowley et al. successfully realigned a prolapsed PED and contemporaneously deployed a second PED across the neck of the aneurysm [30]. If an FDD cannot be adequately repositioned or realigned, it may need to be retrieved with a microsnare so as not to obstruct flow in the parent artery. Hauck et al. presented a case in which this was carried out from a retrograde approach through the posterior communicating artery [35].

3.4 Managing Periprocedure Thromboembolic and Ischemic Complications

Thromboembolic complications in the setting of flow diversion include in-stent thrombosis, large-vessel occlusion, and perforator or side-branch occlusion. Symptomatic periprocedural thrombo-embolic complications associated with use of the PED have been reported in various studies with a frequency in the range of 2–8%, similar to that reported with coiling [1, 25, 39–45]. Interestingly, the frequency of *asymptomatic* ischemic events associated with use of FDDs, as detected by diffusion-weighted imaging (DWI), was actually found to be about an order of magnitude higher, 63% according to a recent study by Brasiliense et al. and 50.9% according to a study by Tan et al. [39, 44]. Without the benefit of a procedure control group for comparison, these results should be interpreted cautiously, as even diagnostic cerebral angiography can cause silent infarcts [46]. Nevertheless, as Tan et al. noted, DWI changes occurred at a much higher rate than that of new neurological symptoms, which were seen in 8.1% of patients [44].

Adequate preprocedure platelet inhibition is an important step in minimizing the frequency of thromboembolic complications [24, 47]. However, there are other contributory factors, and periprocedural ischemic events can occur despite this prophylactic measure. Tan et al. found that symptomatic thromboembolic changes were associated with longer procedure time (>116 min) and deployment of more than one Pipeline device [44]. Unfortunately, it is often the inherent complexity of the case that dictates these parameters, and one must simply be prepared with a contingency plan and the necessary materials with which to handle a thromboembolic crisis. Intra-procedural angiograms should be performed frequently to assess for filling defects in the parent artery and FDD and for disappearance of vascular side branches.

There are a number of possible endovascular methods to recanalize an acutely occluded FDD or parent artery, including both mechanical and chemical means. Because acute thrombi are platelet-rich (the so-called white thrombus), antiplatelet medications are a first-line approach and theoretically better suited than fibrinolytics for the management of intra-procedural thrombotic complications [48, 49]. The three FDA-approved glycoprotein (GP) IIb/IIIa receptor antagonists—abciximab (ReoPro), tirofiban (Aggrastat), and epitifibatide (Integrilin)—are potent inhibitors of platelet cross-linking and aggregation and are well-suited for the task of acute periprocedural clot disruption [50].

Of the three GP IIb/IIIa antagonists, Abciximab (ReoPro, Ely Lily, Indianapolis, Indiana, USA) has the most widely reported use as a thromboembolic rescue therapy during neuroendovascular procedures [25, 51]. It is administered either intra-arterially (IA) through the guide catheter or intravenously (IV), beginning with a rapid bolus at a weight-based dose of 0.25 mg/kg followed by a continuous maintenance infusion of 125μg/kg/min (to a maximum of 10 mg/min) for 12 h [52]. For greatest efficacy, we believe, as others have suggested, that it is best to deliver abciximab as an IA bolus, with the catheter tip directed at or within the obstructing thrombus, rather than as an IV bolus. In a small series of retrospectively reviewed cases, Song et al. noted that immediate thrombolysis was achieved with a much lower-dose bolus of abciximab when it was delivered IA, directly at the thrombus, than when it was administered IV [48].

There are several instances of successful trans-catheter IA abciximab use reported in the FDD literature spanning the gamut of periprocedural thromboembolic events. For example, De Vries et al. used IA abciximab to remove clots on the surface of a Surpass stent [51]. Aggressively removing endoluminal thrombi in a FDD before significant stenosis or complete occlusion occurs should be considered, as patients with luminal narrowing following deployment of an FDD are at higher risk of complete in-stent thrombosis [51].

There are caveats to the use of abciximab. One potential complication is that of paradoxical drug-induced platelet activation, which has been observed when lower levels of platelet inhibition were achieved. This phenomenon carries a concomitantly increased risk of thromboembolic complications. To minimize this risk, it is important to avoid partial dosing of abciximab, unless POC PFT indicates that adequate platelet inhibition has already been achieved. Another potential but rare complication of abciximab use is thrombocytopenia. As a precaution, platelet levels should be measured before administration and monitored daily throughout its use [53].

Eptifibatide is also administered as either an intra-arterial or intravenous bolus at a dose of 180μg/kg [50, 54]. This may be followed with a maintenance infusion at a rate of 2μg/kg/min for 20–24 h at the discretion of the operator [54]. Tirofiban is infused at an initial rate of 0.4μg/kg over 30 min or more followed by infusion at a rate of 0.1μg/kg over several hours.

Intracranial hemorrhage (ICH) is one of the risks associated with the use of all GP IIb/IIIa inhibitors, and the rate of this complication is higher in patients undergoing endovascular treatment of ruptured aneurysms (REF). Having a basic understanding of the pharmacologic properties of the GP IIb/IIIa inhibitors is beneficial in making appropriate management decisions. Abciximab has a high affinity for the GP IIb/IIIa receptor, which translates into a slow rate of dissociation and hence a prolonged inhibitory effect on the bound platelets, which can persist for days after infusion of the drug has been discontinued [55]. The unbound drug is rapidly cleared from the plasma. This prolonged platelet inhibition is disadvantageous in the event of a sudden bleed, but the antiplatelet effect can be overcome with platelet transfusion, which is why many operators prefer this agent over the others [55].

Eptifibatide and tirofiban, on the other hand, have low affinity for the GP IIb/IIIa receptor and exist in a relatively higher unbound fraction in the plasma. They have relatively short durations of action, with biological half-lives of 2–4 h for epitifibatide and 1.5–2 h for tirofiban [50]. Clearance of eptifibatide and tirofiban is predominantly renal, and as long as renal clearance mechanisms are intact, the platelet inhibitory effect of these drugs will be minimal several hours after discontinuing infusion [55]. In cases of renal insufficiency, their doses should be [53]. While some authors mention platelet transfusion as a potential reversal mechanism for eptifibatide and tirofiban, in reality, a platelet transfusion is ineffective against these drugs until they have been cleared from the plasma; otherwise, the circulating drug will simply inactivate the transfused platelets [50, 53].

Despite great concern over the possibility of side-branch occlusion when the PED was initially being investigated, this complication seems to occur fairly infrequently, with one study reporting a rate as low as 1.4% (i.e., two out of 140 cases) at 6-month follow-up and tends to be clinically silent [56, 57]. Higher rates of occlusion were reported in a smaller case series and have been discussed in case reports and reviews [2, 42, 51]. In Yu et al.'s study, both of the occlusions occurred in posterior communicating artery branches covered by one PED, although side-branch occlusion is generally believed to occur more frequently when two or more overlapping stents cover the origin of a branch artery [2, 57]. This complication is also seems to occur more commonly in the posterior circulation than the anterior circulation [56, 58]. At 6-month follow-up, for instance, De Vries et al. found absence of antegrade flow following Surpass implantation in two out of 15 cases of covered ophthalmic arteries (15%) and four out of 15 cases of covered posterior communicating arteries (31%), fortunately without neurological symptoms [51]. As one could surmise, side-branch occlusion has been associated with placement of multiple overlapping flow diverters, since the porosity of the barrier between the parent artery and the branch vessel is greatly reduced in areas where the stents are overlapping [2]. In addition to adequate antiplatelet preparation, parsimonious use of individual flow-diverting devices in building the total construct may be the best preventative measure one could take. Nevertheless, in the event of an acutely occluded side-branch after deploying an FDD, IA abciximab has been shown to be effective in recanalization [42].

IA fibrinolytics, such as recombinant tissue plasminogen activator (rtPA) and urokinase, have been used to treat periprocedural clot formation during endovascular aneurysm treatment, and their short half-life makes them well-suited for this application [59, 60]. However, in principle, these drugs may not be as effective as antiplatelet agents such as the GP IIb/IIIa inhibitors in disrupting an acute, platelet-rich clot. Like the GP IIb/IIIa inhibitors, thrombolytics also carry a risk of intracranial hemorrhage, which may particularly limit their use in the setting of ruptured aneurysm treatment. If the thrombus is occluding the parent artery or stent, mechanical disruption of the clot may be helpful in loosening the thrombus material and creating more surface area on which the antiplatelet or thrombolytic drug may act, similar to the technique advocated by Cronqvist et al. [59]

Mechanical thrombectomy can, in principle, be used to recanalize an acutely thrombosed FDD. De Vries et al. noted success with this approach in one case; however, no mention was made of the specific endovascular device used for this purpose nor any other technical details regarding the rescue procedure [51]. More studies are needed to demonstrate the safety of mechanical thrombectomy with regard to the theoretical risk of disrupting the flow-diverting construct, the risk of downstream embolization, and appropriate choice of mechanical thrombectomy device. Endoluminal stenosis of a flow-diverting device may also be treated with angioplasty [61]. Note that conservative measures, such as intravascular volume expansion and blood pressure elevation, have been a mainstay for dealing with periprocedural thromboembolic complications and may be implemented concurrently with chemical and mechanical clot disruption [48, 50].

3.5 Managing Periprocedure Vascular Injury

Iatrogenic vascular injury in the setting of flow diversion has been reported in two general procedural scenarios, during initial deployment of a FDD and during attempted balloon remodeling of an incompletely expanded FDD. The spectrum of clinical consequences from this type of complication ranges from minor to catastrophic. However, based on the limited number of cases reported in the literature, arterial wire perforations that occurred during FDD deployment or attempted deployment ultimately resulted in more favorable clinical outcomes than the vascular ruptures that occurred during attempted angioplasty of incompletely expanded FDDs, which caused a major stroke and death [4, 25, 36, 51, 62]. Careful monitoring of the distal tip of the delivery wire to avoid placement in small and angulated vessels and gentle manipulation of the FDD may be the best defenses against this type of complication [38, 62]. Balloon test occlusion (BTO) prior to attempting flow diversion may be worthwhile if procedural challenges are anticipated based on difficult vascular anatomy [33].

It is crucial to promptly recognize arterial perforation so that expedient measures can be taken to mitigate injury to the patient. An abrupt increase in systemic blood pressure and concomitant acute bradycardia (i.e., Cushing reflex), or a sudden rise in intracranial pressure (ICP), are telltale signs of acute ICH. Extravasation of contrast should be immediately confirmed by angiography and the site of perforation identified. However, multiple angiographic runs should be avoided, as they waste critical time and deposit additional contrast in the subarachnoid space.

A key step in management of iatrogenic ICH is immediate reversal of anticoagulation [25, 47, 63]. The effect of heparin is reversed with protamine sulfate, 10 mg IV per 1000 U of heparin administered, or 5 mg IV per 1000 U of heparin if it has been more than 30 min since heparin administration. Platelet transfusion should be initiated to reverse the effect of antiplatelet agents. Tight blood pressure control is also recommended. Placement of a ventricular drain may be necessary to control progressively increasing ICP.

If the vascular perforation is minor and occurs prior to deployment of the FDD, or at a point when the FDD can still be resheathed, conservative measures such as those described above may be sufficient. In a series of 37 patients treated with the Surpass flow diverter, De Vries et al. reported one case of a tiny wire perforation of the middle cerebral artery (MCA) resulting in an acute subarachnoid hemorrhage (SAH) prior to deployment of the FDD [51]. Heparin and antiplatelet medications were promptly reversed, and there were no clinical sequelae. The patient's aneurysm was successfully treated with the Surpass device 6 months later.

Some neurointerventionalists advocate leaving the perforating device in place until the source of bleeding has been controlled by other means, as the device itself may actually limit blood loss through the perforation, and withdrawing it may cause further damage to the vessel. This suggestion is particularly applicable if perforation occurs in the aneurysm itself, in which case a second microcatheter may be placed by which to treat the rupture [64].

Sacrifice of the parent vessel is generally considered a last resort, and clinical outcome primarily depends on the specific vessel involved and the collateral blood supply to the affected region. Pistocchi et al. reported one arterial perforation of the distal MCA in their series of 30 treated aneurysms caused by the distal wire tip during deployment of a Silk flow diverter [62]. The pierced parent artery was immediately occluded by coil embolization without adverse clinical consequences.

In some cases, open surgical intervention may be necessary. Nelson et al. reported one case of vascular rupture out of 31 patients treated in the PITA trial [4]. The event occurred when the study investigators attempted to use angioplasty to correct the diminished flow observed in the parent internal carotid artery (ICA) distal to an unsuccessfully deployed PED. The ruptured ICA required surgical ligation, and the patient suffered a large left-hemispheric infarct. Gentric et al. reported a case in which a partially opened PED, which was inadvertently deployed through a cell of a previously implanted Solitaire AB stent, ruptured upon attempted balloon expansion, resulting in a fatal hemorrhage [36]. The authors retrospectively suggested that leaving the collapsed FDD in place and considering a surgical bypass may have been a better alternative.

Iatrogenic dissection of the ICA is not specific to the implantation of flow diverters, and their management has been discussed elsewhere. However, it is interesting to note that one of the two iatrogenic cervical ICA dissections encountered by De Vries et al. in a study of the Surpass flow diverter was actually treated with a Surpass device [51]. The other ICA dissection went undetected at the time of the procedure and resulted in occlusion of the vessel [51]. Fischer et al. and Brzezicki et al. have demonstrated successful treatment of intracranial and cervical vascular dissections with the PED, as well [8, 65]. Although these two studies did not address treatment of iatrogenic dissections, the results, nevertheless, support the notion of using FDDs for this purpose [5, 58, 66].

3.6 Postprocedure Management: Antiplatelet Therapy

As previously noted, it is advisable to continue DAT for at least 6 months following flow-diverting treatment. By 12 months after flow-diverting therapy, endothelialization of the FDD is expected to be complete and the aneurysm completely occluded. Therefore, our practice is to continue DAT for one year with clopidogrel 75 mg daily and aspirin 81 mg daily. After one year, clopidogrel is discontinued and aspirin is continued indefinitely at 81 mg daily.

3.7 Postprocedure Management: Follow-Up Imaging and Delayed Complications

Protocols for postprocedure follow-up imaging vary according to individual practices and may need to be tailored to different clinical scenarios, but generally some combination of catheter angiography and magnetic resonance imaging (MRI) and magnetic resonance angiography (MRA) without and with intravenous gadolinium contrast are performed [1].

At our institution, the post-Pipeline embolization imaging protocol is as follows.

At 1 month: MRI and MRA.

At 6 months: Catheter angiography.

At 1 year: MRI and MRA.

At 2 years: Catheter angiography.

At 3 years: MRI and MRA.

At 4 years: MRI and MRA.

At 5 years: Catheter angiography.

MRA is then repeated annually until the aneurysm is deemed stable over several years.

One of the main indications for follow-up imaging is to assess the degree of aneurysm occlusion. Various classification schemes have been devised to categorize the completeness of aneurysm occlusion on follow-up imaging, such as the Raymond-Roy Occlusion Classification and the O'Kelly Marotta Scale. Ultimately, one must decide if there is sufficient exposed aneurysm to warrant further treatment, such as placement of an additional PED, or if serial imaging observation is sufficient, at least temporarily to determine if complete aneurysm occlusion occurs.

The patency and position of the flow-diverting construct are also assessed on follow-up imaging. Delayed parent vessel occlusion was observed in five cases by Bescke et al. through 1-year follow-up, and one was "definitively related to noncompliance with antiplatelet medication." The authors suggest that antiplatelet monitoring could be useful to reduce this complication.

Delayed subarachnoid hemorrhage (SAH) has been reported in 4% of PED cases and implies aneurysmal rupture [67]. One possible mechanism for delayed aneurysmal rupture include enzymatic degradation of the aneurysm wall during intra-aneurysmal thrombosis. Another hypothesized mechanism is the altered hemodynamics and change in arterial compliance in the parent vessel after FDD placement. This complication tends to be observed primarily in very large and giant symptomatic aneurysms [1, 25, 68]. Despite successful PED placement, aneurysm rupture occurred in 3.2% of cases. There were three cases of progressive aneurysm enlargement despite PED placement, which corresponded to adverse neurological outcomes.

Delayed intraparenchymal hemorrhage has been reported in 3% of cases [67]. The causes of such delayed distant hemorrhages remain unclear. Cerebral infarct with secondary hemorrhagic conversion due to dual antiplatelet therapy has been cited as a potential etiology. To date, there is no consensus on the management of such delayed hemorrhages. Risks should be weighed between discontinuing antiplatelet therapy to prevent hematoma expansion and the potential to develop in-stent thrombosis [63].

Vasogenic edema in the brain parenchyma surrounding aneurysms has been observed on MRI in the setting of both treated and untreated aneurysms. Untreated aneurysms demonstrating this phenomenon tend to be large, partially thrombosed, or both [69–71]. Postprocedural perianeurysmal edema has been reported following most types of endovascular aneurysmal therapy, including embolization with either bare-platinum coils or modified coils, parent vessel occlusion, and flow diversion [69, 70, 72–77].

Mural enhancement, which occurs in a minority of treated aneurysms, appears to be a normal inflammatory healing response that is confined to the aneurysm wall and is asymptomatic [70, 78]. Perianeurysmal edema, however, may represent an exaggeration of the normal inflammatory response that occurs in relation to some healing, thrombosing aneurysms [70]. The reasons behind this heightened response are not entirely clear, but the finding appears to be associated with large aneurysm size and close contact with adjacent brain parenchyma [72, 73]. Aneurysms such as those in the cavernous carotid artery, which are surrounded by dura mater or cerebral spinal fluid (CSF), may be less susceptible to parenchymal extension of the inflammation [73].

The relevance from a management standpoint is that perianeurysmal inflammation and edema may be symptomatic, presenting as an exacerbation of preexisting symptoms related to aneurysm mass effect or with new clinical findings, including persistent headache, cranial neuralgia, and visual symptoms [3, 72, 73]. Symptomatic improvement has been noted following administration of corticosteroids, but without control data, it is difficult to ascertain the effectiveness of this treatment [72, 73]. Optimal preventative and therapeutic measures for this complication require further investigation. In addition, diagnostic and treatment measures may need to address other potentially associated pathology, such as hydrocephalus, aseptic or chemical meningitis, and seizures, which have been reported in patients who developed perianeurysmal edema following coil embolization [70, 77, 79–81].

References

1. Becske T, Kallmes DF, Saatci I et al (2013) Pipeline for uncoilable or failed aneurysms: results from a multicenter clinical trial. Radiology 267(3):858–868

2. Szikora I, Berentei Z, Kulcsar Z et al (2010) Treatment of intracranial aneurysms by functional reconstruction of the parent artery: the Budapest experience with the pipeline embolization device. AJNR Am J Neuroradiol 31 (6):1139–1147

3. Lylyk P, Miranda C, Ceratto R et al (2009) Curative endovascular reconstruction of cerebral aneurysms with the pipeline embolization device: the Buenos Aires experience. Neurosurgery 64(4):632–642. discussion 642–633; quiz N636

4. Nelson PK, Lylyk P, Szikora I, Wetzel SG, Wanke I, Fiorella D (2011) The pipeline embolization device for the intracranial treatment of aneurysms trial. AJNR Am J Neuroradiol 32 (1):34–40

5. Jabbour P, Chalouhi N, Tjoumakaris S et al (2013) The pipeline embolization device: learning curve and predictors of complications and aneurysm obliteration. Neurosurgery 73 (1):113–120. discussion 120

6. Liou TM, Li YC (2008) Effects of stent porosity on hemodynamics in a sidewall aneurysm model. J Biomech 41(6):1174–1183

7. Fiorella D (2010) Anti-thrombotic medications for the neurointerventionist: aspirin and clopidogrel. J Neurointerv Surg 2(1):44–49

8. Fischer S, Vajda Z, Aguilar Perez M et al (2012) Pipeline embolization device (PED) for neurovascular reconstruction: initial experience in the treatment of 101 intracranial aneurysms and dissections. Neuroradiology 54 (4):369–382

9. Pierot L, Wakhloo AK (2013) Endovascular treatment of intracranial aneurysms: current status. Stroke 44(7):2046–2054

10. Phillips TJ, Wenderoth JD, Phatouros CC et al (2012) Safety of the pipeline embolization device in treatment of posterior circulation aneurysms. AJNR Am J Neuroradiol 33 (7):1225–1231

11. Augsburger L, Farhat M, Reymond P et al (2009) Effect of flow diverter porosity on intraaneurysmal blood flow. Klin Neuroradiol 19(3):204–214

12. Lieber BB, Stancampiano AP, Wakhloo AK (1997) Alteration of hemodynamics in aneurysm models by stenting: influence of stent porosity. Ann Biomed Eng 25(3):460–469

13. Kadirvel R, Ding YH, Dai D, Rezek I, Lewis DA, Kallmes DF (2014) Cellular mechanisms of aneurysm occlusion after treatment with a flow diverter. Radiology 270(2):394–399

14. Kallmes DF, Ding YH, Dai D, Kadirvel R, Lewis DA, Cloft HJ (2007) A new endoluminal, flow-disrupting device for treatment of saccular aneurysms. Stroke 38(8):2346–2352

15. Puffer RC, Kallmes DF, Cloft HJ, Lanzino G (2012) Patency of the ophthalmic artery after flow diversion treatment of paraclinoid aneurysms. J Neurosurg 116(4):892–896

16. Yavuz K, Geyik S, Saatci I, Cekirge HS (2014) Endovascular treatment of middle cerebral artery aneurysms with flow modification with the use of the pipeline embolization device. AJNR Am J Neuroradiol 35(3):529–535

17. Sadasivan C, Cesar L, Seong J et al (2009) An original flow diversion device for the treatment of intracranial aneurysms: evaluation in the rabbit elastase-induced model. Stroke 40 (3):952–958

18. Krishna C, Sonig A, Natarajan SK, Siddiqui AH (2014) The expanding realm of endovascular neurosurgery: flow diversion for cerebral aneurysm management. Methodist Debakey Cardiovasc J 10(4):214–219

19. Cantón G, Levy DI, Lasheras JC, Nelson PK (2005) Flow changes caused by the sequential placement of stents across the neck of sidewall cerebral aneurysms. J Neurosurg 103 (5):891–902

20. Hwang G, Jung C, Park SQ et al (2010) Thromboembolic complications of elective coil embolization of unruptured aneurysms: the effect of oral antiplatelet preparation on periprocedural thromboembolic complication. Neurosurgery 67(3):743–748. discussion 748

21. Yamada NK, Cross DT 3rd, Pilgram TK, Moran CJ, Derdeyn CP, Dacey RG Jr (2007) Effect of antiplatelet therapy on thromboembolic complications of elective coil embolization of cerebral aneurysms. AJNR Am J Neuroradiol 28(9):1778–1782

22. Brooks NP, Turk AS, Niemann DB, Aagaard-Kienitz B, Pulfer K, Cook T (2008) Frequency of thromboembolic events associated with endovascular aneurysm treatment: retrospective case series. J Neurosurg 108 (6):1095–1100

23. Skukalek SL, Winkler AM, Kang J et al (2016) Effect of antiplatelet therapy and platelet function testing on hemorrhagic and thrombotic complications in patients with cerebral aneurysms treated with the pipeline embolization device: a review and meta-analysis. J Neurointerv Surg 8(1):58–65

24. Delgado Almandoz JE, Crandall BM, Scholz JM et al (2013) Pre-procedure P2Y12 reaction units value predicts perioperative thromboembolic and hemorrhagic complications in patients with cerebral aneurysms treated with the pipeline embolization device. J Neurointerv Surg 5(Suppl 3):iii3–ii10

25. Burrows AM, Cloft H, Kallmes DF, Lanzino G (2015) Periprocedural and mid-term technical and clinical events after flow diversion for intracranial aneurysms. J Neurointerv Surg 7 (9):646–651

26. Lin LM, Colby GP, Jiang B et al (2016) Intra-DIC (distal intracranial catheter) deployment of the Pipeline embolization device: a novel rescue strategy for failed device expansion. J Neurointerv Surg 8(8):840–846

27. Park MS, Albuquerque FC, Nanaszko M et al (2015) Critical assessment of complications associated with use of the pipeline embolization device. J Neurointerv Surg 7(9):652–659

28. Lin LM, Colby GP, Jiang B et al (2015) Classification of cavernous internal carotid artery tortuosity: a predictor of procedural complexity in pipeline embolization. J Neurointerv Surg 7(9):628–633

29. Colby GP, Lin LM, Coon AL (2012) Revisiting the risk of intraparenchymal hemorrhage following aneurysm treatment by flow diversion. AJNR Am J Neuroradiol 33(7):E107. author reply E108

30. Crowley RW, Abla AA, Ducruet AF, McDougall CG, Albuquerque FC (2014) Novel application of a balloon-anchoring technique for the realignment of a prolapsed pipeline embolization device: a technical report. J Neurointerv Surg 6(6):439–444

31. Martínez-Galdámez M, Ortega-Quintanilla J, Hermosín A, Crespo-Vallejo E, Ailagas JJ, Pérez S (2017) Novel balloon application for rescue and realignment of a proximal end migrated pipeline flex embolization device into the aneurysmal sac: complication management. J Neurointerv Surg 9(1):e4

32. Miller TR, Jindal G, Gandhi D (2015) Focal, transient mechanical narrowing of a pipeline embolization device following treatment of an internal carotid artery aneurysm. J Neurointerv Surg 7(10):e35

33. Navarro R, Yoon J, Dixon T, Miller DA, Hanel RA, Tawk RG (2014) Retrograde trans-anterior communicating artery rescue of unopened pipeline embolization device with balloon dilation: complication management. BMJ Case Rep 2014:bcr2013011009

34. Pereira VM, Kelly M, Vega P et al (2015) New pipeline flex device: initial experience and technical nuances. J Neurointerv Surg 7 (12):920–925

35. Hauck EF, Natarajan SK, Langer DJ, Hopkins LN, Siddiqui AH, Levy EI (2010) Retrograde trans-posterior communicating artery snare-assisted rescue of lost access to a foreshortened pipeline embolization device: complication management. Neurosurgery 67(2 Suppl Operative):495–502

36. Gentric JC, Fahed R, Darsaut TE, Salazkin I, Roy D, Raymond J (2016) Fatal arterial rupture during angioplasty of a flow diverter in a recurrent, previously Y-stented giant MCA bifurcation aneurysm. Interv Neuroradiol 22 (3):278–286

37. Le EJ, Miller T, Serulle Y, Shivashankar R, Jindal G, Gandhi D (2017) Use of pipeline flex is associated with reduced fluoroscopy time, procedure time, and technical failure compared with the first-generation pipeline embolization device. J Neurointerv Surg 9 (2):188–191

38. Alderazi YJ, Shastri D, Kass-Hout T, Prestigia-como CJ, Gandhi CD (2014) Flow diverters for intracranial aneurysms. Stroke Res Treat 2014:415653

39. Brasiliense LB, Stanley MA, Grewal SS et al (2016) Silent ischemic events after pipeline embolization device: a prospective evaluation with MR diffusion-weighted imaging. J Neurointerv Surg 8(11):1136–1139

40. Briganti F, Napoli M, Tortora F et al (2012) Italian multicenter experience with flow-diverter devices for intracranial unruptured aneurysm treatment with periprocedural complications—a retrospective data analysis. Neuroradiology 54(10):1145–1152

41. Brilstra EH, Rinkel GJ, van der Graaf Y, van Rooij WJ, Algra A (1999) Treatment of intracranial aneurysms by embolization with coils: a systematic review. Stroke 30(2):470–476

42. Lall RR, Crobeddu E, Lanzino G, Cloft HJ, Kallmes DF (2014) Acute branch occlusion after pipeline embolization of intracranial aneurysms. J Clin Neurosci 21(4):668–672

43. Standhardt H, Boecher-Schwarz H, Gruber A, Benesch T, Knosp E, Bavinzski G (2008) Endovascular treatment of unruptured intracranial aneurysms with Guglielmi detachable coils: short- and long-term results of a single-centre series. Stroke 39(3):899–904

44. Tan LA, Keigher KM, Munich SA, Moftakhar R, Lopes DK (2015) Thromboembolic complications with pipeline embolization device placement: impact of procedure time, number of stents and pre-procedure P2Y12 reaction unit (PRU) value. J Neurointerv Surg 7(3):217–221

45. van Rooij WJ, Sluzewski M, Beute GN, Nijssen PC (2006) Procedural complications of coiling of ruptured intracranial aneurysms: incidence and risk factors in a consecutive series of 681 patients. AJNR Am J Neuroradiol 27 (7):1498–1501

46. Kato K, Tomura N, Takahashi S, Sakuma I, Watarai J (2003) Ischemic lesions related to cerebral angiography: evaluation by diffusion weighted MR imaging. Neuroradiology 45 (1):39–43

47. Delgado Almandoz JE, Crandall BM, Scholz JM et al (2014) Last-recorded P2Y12 reaction units value is strongly associated with thromboembolic and hemorrhagic complications occurring up to 6 months after treatment in patients with cerebral aneurysms treated with the pipeline embolization device. AJNR Am J Neuroradiol 35(1):128–135

48. Song JK, Niimi Y, Fernandez PM et al (2004) Thrombus formation during intracranial aneurysm coil placement: treatment with intra-arterial abciximab. AJNR Am J Neuroradiol 25(7):1147–1153

49. Ng PP, Phatouros CC, Khangure MS (2001) Use of glycoprotein IIb-IIIa inhibitor for a thromboembolic complication during Guglielmi detachable coil treatment of an acutely ruptured aneurysm. AJNR Am J Neuroradiol 22(9):1761–1763

50. Yi HJ, Gupta R, Jovin TG et al (2006) Initial experience with the use of intravenous eptifibatide bolus during endovascular treatment of intracranial aneurysms. AJNR Am J Neuroradiol 27(9):1856–1860

51. De Vries J, Boogaarts J, Van Norden A, Wakhloo AK (2013) New generation of Flow Diverter (surpass) for unruptured intracranial aneurysms: a prospective single-center study in 37 patients. Stroke 44(6):1567–1577

52. Cloft HJ, Samuels OB, Tong FC, Dion JE (2001) Use of abciximab for mediation of thromboembolic complications of endovascular therapy. AJNR Am J Neuroradiol 22 (9):1764–1767

53. Altenburg A, Haage P (2012) Antiplatelet and anticoagulant drugs in interventional radiology. Cardiovasc Intervent Radiol 35(1):30–42

54. Dumont TM, Kan P, Snyder KV, Hopkins LN, Siddiqui AH, Levy EI (2013) Adjunctive use of eptifibatide for complication management during elective neuroendovascular procedures. J Neurointerv Surg 5(3):226–230

55. Scarborough RM, Kleiman NS, Phillips DR (1999) Platelet glycoprotein IIb/IIIa antagonists. What are the relevant issues concerning their pharmacology and clinical use? Circulation 100(4):437–444

56. D'Urso PI, Lanzino G, Cloft HJ, Kallmes DF (2011) Flow diversion for intracranial aneurysms: a review. Stroke 42(8):2363–2368

57. Yu SC, Kwok CK, Cheng PW et al (2012) Intracranial aneurysms: midterm outcome of pipeline embolization device—a prospective study in 143 patients with 178 aneurysms. Radiology 265(3):893–901

58. Starke RM, Turk A, Ding D et al (2016) Technology developments in endovascular treatment of intracranial aneurysms. J Neurointerv Surg 8(2):135–144

59. Cronqvist M, Pierot L, Boulin A, Cognard C, Castaings L, Moret J (1998) Local intraarterial fibrinolysis of thromboemboli occurring during endovascular treatment of intracerebral aneurysm: a comparison of anatomic results and clinical outcome. AJNR Am J Neuroradiol 19(1):157–165

60. Hähnel S, Schellinger PD, Gutschalk A et al (2003) Local intra-arterial fibrinolysis of thromboemboli occurring during neuroendovascular procedures with recombinant tissue plasminogen activator. Stroke 34 (7):1723–1728

61. Estrade L, Makoyeva A, Darsaut TE et al (2013) In vitro reproduction of device deformation leading to thrombotic complications and failure of flow diversion. Interv Neuroradiol 19(4):432–437

62. Pistocchi S, Blanc R, Bartolini B, Piotin M (2012) Flow diverters at and beyond the level of the circle of willis for the treatment of intracranial aneurysms. Stroke 43(4):1032–1038

63. Cruz JP, Chow M, O'Kelly C et al (2012) Delayed ipsilateral parenchymal hemorrhage following flow diversion for the treatment of anterior circulation aneurysms. AJNR Am J Neuroradiol 33(4):603–608

64. Willinsky R, terBrugge K (2000) Use of a second microcatheter in the management of a perforation during endovascular treatment of a cerebral aneurysm. AJNR Am J Neuroradiol 21(8):1537–1539

65. Brzezicki G, Rivet DJ, Reavey-Cantwell J (2016) Pipeline embolization device for treatment of high cervical and skull base carotid artery dissections: clinical case series. J Neurointerv Surg 8(7):722–728

66. Amenta PS, Starke RM, Jabbour PM et al (2012) Successful treatment of a traumatic carotid pseudoaneurysm with the pipeline stent: case report and review of the literature. Surg Neurol Int 3:160

67. Brinjikji W, Murad MH, Lanzino G, Cloft HJ, Kallmes DF (2013) Endovascular treatment of intracranial aneurysms with flow diverters: a meta-analysis. Stroke 44(2):442–447

68. Kulcsár Z, Houdart E, Bonafé A et al (2011) Intra-aneurysmal thrombosis as a possible cause of delayed aneurysm rupture after flow-diversion treatment. AJNR Am J Neuroradiol 32(1):20–25

69. Craven I, Patel UJ, Gibson A, Coley SC (2009) Symptomatic perianeurysmal edema following bare platinum embolization of a small unruptured cerebral aneurysm. AJNR Am J Neuroradiol 30(10):1998–2000

70. Fanning NF, Willinsky RA, ter Brugge KG (2008) Wall enhancement, edema, and hydrocephalus after endovascular coil occlusion of intradural cerebral aneurysms. J Neurosurg 108(6):1074–1086

71. Jeon P, Kim BM, Kim DI et al (2012) Reconstructive endovascular treatment of fusiform or ultrawide-neck circumferential aneurysms with multiple overlapping enterprise stents and coiling. AJNR Am J Neuroradiol 33(5):965–971

72. Berge J, Biondi A, Machi P et al (2012) Flow-diverter silk stent for the treatment of intracranial aneurysms: 1-year follow-up in a multicenter study. AJNR Am J Neuroradiol 33 (6):1150–1155

73. Berge J, Tourdias T, Moreau JF, Barreau X, Dousset V (2011) Perianeurysmal brain inflammation after flow-diversion treatment. AJNR Am J Neuroradiol 32(10):1930–1934

74. Hammoud D, Gailloud P, Olivi A, Murphy KJ (2003) Acute vasogenic edema induced by thrombosis of a giant intracranial aneurysm: a cause of pseudostroke after therapeutic occlusion of the parent vessel. AJNR Am J Neuroradiol 24(6):1237–1239

75. Hampton T, Walsh D, Tolias C, Fiorella D (2011) Mural destabilization after aneurysm treatment with a flow-diverting device: a report of two cases. J Neurointerv Surg 3(2):167–171

76. Vu Dang L, Aggour M, Thiriaux A, Kadziolka K, Pierot L (2009) Post-embolization perianeurysmal edema revealed by temporal lobe epilepsy in a case of unruptured internal carotid artery aneurysm treated with bare platinum coils. J Neuroradiol 36 (5):298–300

77. White JB, Cloft HJ, Kallmes DF (2008) But did you use HydroCoil? Perianeurysmal edema and hydrocephalus with bare platinum coils. AJNR Am J Neuroradiol 29(2):299–300

78. Tulamo R, Frösen J, Hernesniemi J, Niemelä M (2010) Inflammatory changes in the aneurysm wall: a review. J Neurointerv Surg 2 (2):120–130

79. Cohen JE, Itshayek E, Attia M, Moscovici S (2012) Postembolization perianeurysmal edema as a cause of uncinate seizures. J Clin Neurosci 19(3):474–476

80. Horie N, Kitagawa N, Morikawa M, Tsutsumi K, Kaminogo M, Nagata I (2007) Progressive perianeurysmal edema induced after endovascular coil embolization. Report of three cases and review of the literature. J Neurosurg 106(5):916–920

81. Meyers PM, Lavine SD, Fitzsimmons BF et al (2004) Chemical meningitis after cerebral aneurysm treatment using two second-generation aneurysm coils: report of two cases. Neurosurgery 55(5):1222

Part II

Cerebral Aneurysms: Critical Care of Aneurysmal
Subarachnoid Hemorrhage

Chapter 7

Neurocritical Care Management of Aneurysmal Subarachnoid Hemorrhage, Early Brain Injury, and Cerebral Vasospasm

Neha S. Dangayach, Salman Assad, Christopher Kellner, and Stephan A. Mayer

Abstract

Aneurysmal subarachnoid hemorrhage (SAH) is a life-threatening emergency. Mortality and morbidity have improved with advances in neurocritical care and endovascular management. Key areas of focus in neurocritical care remain prevention of secondary neurological injury after SAH include early brain injury, cerebral edema, hydrocephalus, vasospasm, seizures, and systemic complications. This chapter will review the epidemiology, clinical presentation, and neurocritical care management of SAH.

Key words Aneurysmal subarachnoid hemorrhage, Delayed cerebral ischemia, Delayed ischemic neurologic deficits, Endovascular, Intra-arterial vasodilator, Transluminal balloon angioplasty, Vasospasm

1 Introduction - Subarachnoid Hemorrhage

Five to ten percent of all stroke subtypes are attributable to aneurysmal subarachnoid hemorrhage (SAH). Eighty percentage of all spontaneous SAH occurs due to a ruptured intracranial aneurysm. Other causes include vasculitis, arteriovenous malformations, and reversible cerebral vasoconstriction syndrome. The estimated incidence of SAH is 8–10 cases per 100,000 adults per year. SAH occurs most frequently in individuals between the ages of 55 and 60 years. With advances in neurocritical care and endovascular therapies, overall mortality from SAH has declined. However, neurocritical care management for SAH continues to remain challenging. Half the survivors suffer from long-term functional, cognitive, and behavioral decline [1]. Even survivors without neurological deficits suffer from behavioral consequences like anxiety, depression, and PTSD [2] The first goal of endovascular or open surgical treatment is prevention of re-rupture [3]. The focus of neurocritical

Fawaz Al-Mufti and Krishna Amuluru (eds.), *Cerebrovascular Disorders*, Neuromethods, vol. 170, https://doi.org/10.1007/978-1-0716-1530-0_7, © Springer Science+Business Media, LLC, part of Springer Nature 2021 care management in SAH is prevention of secondary neurological injury due to vasospasm, cerebral edema, hydrocephalus, seizures, and systemic complications. In this chapter we will review epidemiology, clinical presentation, and contemporary neurocritical care management for SAH.

2   Epidemiology and Etiology

The incidence of SAH varies globally from 2.0 cases per 100,000 persons in China to 22.5 cases per 100,000 persons in Finland [4]. The incidence of SAH in the USA is reported to be 9.7 per 100,000 while the nation-wide inpatient sample from 2003 estimated 14.5 discharges for SAH per 100,000 adults [5]. Case fatality ranges from 25% to 50% most likely due to the original aneurysmal bleeding or re-rupture [6]. This estimate does not include patients who die prior to seeking medical attention.

Factors associated with higher incidence of aneurysms include family history of first degree relative with SAH, with an even higher incidence with two or more first degree relatives have suffered SAH, among patients with Ehlers–Danlos syndrome and polycystic kidney disease patients [7]. Unruptured aneurysms are prevalent in 3% of the adult population and they are detected due to more frequent cranial imaging [8]. In a recent Delphi consensus, an UIA score was developed quantifying recently reported consensus data on factors associated with aneurysm rupture [9, 10]. This differs from the PHASES score, a model based on prospectively collected data from six cohort studies on risk of UIA rupture that provides absolute risks of rupture for the initial 5 years after aneurysm detection using six easily retrievable baseline characteristics (patient geographical location and age, aneurysm size and location, presence of arterial hypertension, and previous SAH from a different aneurysm).

Factors associated with an increased risk of aneurysm rupture can be categorized as patient related and aneurysm related.

Patient-related risk factors include being blacks and Hispanic, history of hypertension, smoking, alcohol abuse, use of sympathomimetic drugs, first degree relative or relatives with SAH. Aneurysm related risk factors include a size greater than 7 mm, irregularity or lobulation and vertebrobasilar aneurysmal location [10].

The exact etiology of aneurysm formation remains unclear. Hypertension and smoking-induced vascular changes are thought to have a major role. The most common histologic finding is a decrease in the thickness of tunica media, the middle muscular layer of the artery, causing structural defects. These defects, combined with hemodynamic factors, lead to aneurysmal formation at arterial branch points in the subarachnoid space. Aneurysms develop mostly at vascular bifurcations, because turbulent flow preferably develops at such sites. With an increase in the size of the aneurysm, wall compliance will decrease and wall tension increase that, in turn, render the aneurysm increasingly susceptible to rupture.

Location: The vast majority of aneurysms (80–90%) are located in the anterior (carotid) circulation, the anterior and posterior communicating, and the middle cerebral artery. The remaining 10–20% is located in the posterior (vertebrobasilar) circulation.

3 Pathophysiology

When an aneurysm ruptures, blood leaks into the subarachnoid space at the same pressure as intra-arterial pressure. This leads to a sudden increase in regional intracranial pressure (ICP) that might exceed systemic arterial pressure. Since cerebral perfusion pressure (CPP) is equal to Mean arterial pressure (MAP)-ICP, aneurysmal rupture may lead to at least a temporary cessation of blood supply to the brain. This causes sudden onset of severe headache and (transient or permanent) loss of consciousness [3]. The presence of blood and blood products in the subarachnoid space causes headache and meningism. Acute obstructive hydrocephalus develops due to the presence of intraventricular blood. Persistent hydrocephalus may develop even after intraventricular blood clears as a consequence of impaired reabsorption of cerebrospinal fluid (CSF).

The amount and location of blood and blood products seem to correlate with the incidence of cerebral vasospasm. Various studies have reported on inflammatory markers that might correlate with vasospasm but none of these markers have been shown to be causative. The expanding mass effect of the hemorrhage and the development of brain edema and hydrocephalus contribute to increase in intracranial pressure (ICP). Soon after SAH, ICP may approach systemic blood pressure. This phase is usually short-lasting (only a few minutes) and thought to be the limiting factor in further leakage of blood from the aneurysm. With recurrent episodes of bleeding, ICP may increase further because of a mass effect of clots, cerebral edema, or obstructive hydrocephalus. Intracerebral and intraventricular hematomas contribute to the increase in ICP in one-third of patients with SAH.

4 Grading of SAH for Management and Prognosis (Tables 1, 2, and 3)

5   Clinical Features and Diagnosis

Table 1
Hunt and Hess grading scale for SAH

Grade	Clinical description
I	Asymptomatic or minimal headache and slight nuchal rigidity
II	Moderate to severe headache, nuchal rigidity, no neurological deficit other than cranial nerve palsy
III	Drowsiness, confusion, or mild focal deficit
IV	Stupor, moderate-to-severe hemiparesis, and possibly early decerebrate rigidity and vegetative disturbances
V	Deep coma, decerebrate rigidity, and moribund appearance

Table 2
World Federation of Neurological Surgeons Grading Scale for aneurysmal SAH

Grade	GCS score	Motor deficit[a]
I	15	Absent
II	13 or 14	Absent
III	13 or 14	Present
IV	7–12	Present or absent
V	3–6	Present or absent

GCS Glasgow Coma Scale
[a]Excludes cranial neuropathies, but includes dysphasia

Table 3
Modified Fisher Scale versus Fisher scale and incidence of vasospasm. [15]

Grade	Criteria	Patients%	% with symptomatic vasospasm
Modified Fisher Scale			
1	Minimal/thin SAH, no IVH in both lateral ventricles	21.6	24
2	Minimal/thin SAH, *with* IVH in both lateral ventricles	10.8	33
3	Thick SAH,[a] no IVH in both lateral ventricles	33.9	33
4	Thick SAH,[a] *with* IVH in both lateral ventricles	33.7	40
Fisher scale			
1	No blood	8.1	21
2	Diffuse thin blood	10.9	25
3	Thick SAH present	67.7	37
4	ICH/IVH with either no or only diffuse SAH	13.3	31

[a]Completely filling ≥1 cistern or fissure

The presenting complaint of 80% of the patients presenting with SAH and can provide history is "the worst headache of my life," [11] This headache is characterized as being extremely sudden and immediately reaching maximal intensity (thunderclap headache). It may be associated with other red flag signs like nausea/vomiting, neck stiffness, and focal neurological deficits. Aneurysmal subarachnoid hemorrhage accounts for only 1% of all headaches evaluated in the emergency department. A high degree of clinical suspicion should exist in patients with acute severe headache [12].

A sentinel headache is present in 10–43% patients that might occur about 2–8 weeks prior to the SAH. This increases the odds of early rebleeding by tenfold [13]. Initial misdiagnosis of a sentinel headache increases the likelihood of mortality and morbidity by fourfold [14]. Aneurysmal rupture typically tends to occur during physical exertion or stress. Seizures may occur in the first 24 h particularly in patients with anterior communicating or middle cerebral artery involvement or when associated with intraparenchymal bleed.

5.1 Diagnosis

The sensitivity of a non-contrasted head CT remains nearly 100% for the first 72 h post bleed and declines sharply after 5–7 days. Thus, xanthochromia on lumbar puncture must be used to diagnose patients with delayed presentation or in CT negative patients, MRI with FLAIR, DWI, GRE, proton density sequences can be helpful in these cases as well. Once the SAH is diagnosed; severity grading scales should be used for risk stratification for vasospasm and prognostication.

6 Management: Pre-securing of Aneurysm

6.1 Stabilization: Airway Protection and Hemodynamics

The main goal prior to securing the aneurysm is preventing rebleeding and secondary neurological injury. The risk of ultra-early rebleeding within 24 h of initial SAH may be 15%, with recurrent hemorrhage fatality around 70%. Higher-admission Hunt and Hess grade and aneurysm size predict rebleeding.

6.1.1 Blood Pressure Control

The goal of blood pressure control in the pre-aneurysmal securing phase is balancing prevention of rebleeding and maintaining cerebral perfusion. Both the American Heart Association (AHA) and Neurocritical Care Society recommend that it is reasonable to maintain SBP <160 mm Hg. There is currently no clear evidence on what target blood pressure should be maintained. We usually maintain SBP between 100 and 140 mm Hg. Labetalol, nicardipine, or clevidipine drips or a combination can be used to achieve this target [16]. If these drips are not available, then target SBP can be achieved and maintained with pushes of intravenous labetalol alternated with hydralazine. This is initiated with either intermittent intravenous (IV) labetalol (10–15 mg every 15 min as needed [prn]) or hydralazine (10–20 mg every 20 min) if bradycardia is present. Sodium nitroprusside and nitroglycerin drips are generally avoided. These medications can worsen intracranial pressure due to venodilation. Carefully titrated analgesia may also be needed, balancing patient comfort and the intent to blunt the hyperadrenergic state with the need to continuously assess the neurologic status of the patient.

6.1.2 Antifibrinolytic Agents

Currently an early short course of antifibrinolytic therapy with epsilon aminocaproic acid (Amicar) or tranexamic acid prior to early aneurysm repair up to 72 h post ictus has been recommended by the Neurocritical Care Society [17]. This recommendation is based on a randomized controlled trial of Amicar ($n = 428$) patients and a couple of other case control studies.

Prior to 2002 when the treatment paradigm did not include early securing of aneurysm, antifibrinolytic agents would be continued for several days. Small cohort studies from the previous decade showed that despite a marked reduction in rebleeding, there was no benefit in clinical outcome as there was a significantly higher incidence of cerebral ischemia in the treated patients. Delayed (>48 h after the ictus) or prolonged (>3 days) antifibrinolytic therapy exposes patients to side effects of therapy when the risk of rebleeding is sharply reduced and should be avoided [17]. Relative contraindications include recent acute ischemic stroke, myocardial infarction, deep venous thrombosis (DVT) or pulmonary embolism. We usually obtain a baseline lower extremity duplex in all patients who receive antifibrinolytic therapy within the first 2 days of hospitalization.

6.1.3 Seizure Prophylaxis

Seizures have been reported in up to 20% of patients and are more common after SAH from MCA aneurysms, intraparenchymal hematoma, cortical infarcts high grade SAH and in patients with a history of hypertension,. Seizure-like movements have been reported in about 26% of patients; but it is unclear whether these movements are true seizures or posturing. Clinical seizures may occur in 1–7% according to one study and in 6–18% according to another retrospective review. Delayed seizures may occur in 3–7% of patients. If patients have a clinical seizure or present with a low grade SAH then continuous EEG monitoring should be used.

It is unclear what impact seizures have on outcomes in SAH patients. Given the theoretical risk of rebleeding with a seizure in the setting of an unsecured, recently ruptured aneurysm, our approach has been to administer on admission levetiracetam as anti-epileptic drug (AED) prophylaxis (1000 mg IV followed 1000 mg IV every 12 h). We stop AED prophylaxis 3–7 days after the aneurysm is secured unless the patient has had seizures. Routine ongoing use of anticonvulsant medication is not recommended after the aneurysm is secured. We monitor poor grade SAH patients who do not improve clinically and remain in a coma with periodic continuous EEG monitoring to diagnose non-convulsive status. In the International Subarachnoid Aneurysm Trial (ISAT) study, patients treated with coiling were found to have a lower incidence of seizures [18]. In two retrospective studies non-convulsive status epilepticus was found in 10–20% of patients and was a predictor of poor outcome [19, 20]. SAFARI score for predicting the risk of seizures based on 4 items (age $>$ =60, seizure occurrence before hospitalization, hydrocephalus requiring CSF diversion and ruptured aneurysm in anterior circulation) had an AUC of 0.77, 95% C.I. 0.56–0.73. It is a simple score that can be used clinically to individualize management of seizures in SAH. There have been no RCTs to determine the effectiveness of seizure prophylaxis in SAH.

6.1.4 Hydrocephalus

Acute ventricular enlargement with or without intraventricular blood occurs within 72 h of SAH in 20–30% of patients. If hydrocephalus is present and the patient is symptomatic (e.g., has a decreased level of consciousness), an external ventricular drain (EVD) should be placed. We keep the EVD open at 15–20 cm above the external auditory canal till the aneurysm is repaired. This is done to avoid reducing any tamponade effect around the thrombosed aneurysm. If neither a hematoma with a mass effect nor an obstructive element exists, cerebrospinal fluid drainage with serial lumbar puncture might be a good alternative for ventricular drainage [3]. Preoperative ventriculostomy is not associated with an increased risk of aneurysm rebleeding. Long-term outcomes in poor grade patients who improve after EVD placement are similar to patients with good grade hemorrhages. Microsurgical management like lamina terminalis fenestration diminishes the incidence of shunt-dependent hydrocephalus.

6.1.5 Intracranial Hypertension

Raised intracranial pressure may occur for a number of reasons, including hydrocephalus, re-rupture of aneurysm, and global cerebral edema. Some signs of raised intracranial pressure include somnolence, headache, nausea, vomiting, pupil changes such as dilation, and/or sluggish to no constriction to light stimulus. Monitoring may be performed by the placement of either an external ventricular drain or an intraparenchymal monitor. We treat sustained ICP $>$20 mm of Hg for $>$10 min that represent

Lundberg A waves with bolus hyperosmolar therapy. Usual ICP precautions are maintained throughout the course of ICU hospitalization including, head of the bed $> = 30$, prevention of hypercapnia, prevention of treatment of hyponatremia, management of CSF diversion, appropriate sedation and analgesia management (balancing the need to monitor the neurologic examination), induced normothermia. and higher risk of vasospasm [1]. Case-by-case decisions need to be made regarding decompressive surgeries in patients with intraparenchymal hemorrhage.

Summary of Pre-aneurysmal Securing Phase
- The blood pressure titrated to systolic blood pressure of 100–140 mm Hg with intravenous antihypertensive pushes or drips to prevent aneurysm re-rupture and maintain cerebral perfusion.
- Extraventricular drain should be inserted emergently for patients with symptomatic hydrocephalus.
- A short course ($\leq$24 h) of antifibrinolytic therapy that is discontinued after early aneurysm securing may be considered to reduce the risk of rebleeding.
- Anti-epileptic medication prophylaxis should be considered.

7 Management: Aneurysm Treatment

7.1 Coiling Versus Clipping

Early aneurysmal repair is recommended to reduce the risk of rebleeding and to facilitate treatment of cerebral vasospasm. Endovascular coiling was first described 20 years ago and involves detachable platinum coils of different thickness and length. These are introduced into the aneurysm via a microcatheter. The aneurysm is packed with coils to induce thrombosis which in effect eliminates its connection with the parent blood vessel.

In the ISAT trial, 2143 SAH patients for whom clipping and coiling were equally appropriate were randomized into endovascular or surgical aneurysm treatment. Recurrent SAH occurred after treatment at significantly higher annual rates in the coiling (2.9%) versus the clipping group (0.9%). However, at 1 year after treatment there were lower disability rates in coiled patients (15.6% coiling vs. 21.6% clipping), and there was no significant difference in mortality rates (8.1% coiling vs. 10.1% clipping). The 5-year mortality rate in the coiling group (11%) was significantly lower than that of the clipping group (14%) [18].

The decision on endovascular coiling of the aneurysm versus surgical clipping is complex and generally based on the individual patient and the characteristics of the aneurysm. Vertebrobasilar artery aneurysms and paraophthalmic aneurysms tend to be more easily accessed by endovascular coiling. Wide neck aneurysms, middle cerebral artery (MCA) aneurysms, and those associated with intraparenchymal hematomas tend to be more appropriate for surgical clipping. Advances in coil and stent technology may allow broader indications for an endovascular approach. When an aneurysm cannot be treated by either direct surgical clipping or coil embolization, the parent vessel can be occluded by either surgical clipping or coil embolization at the risk of ischemic sequelae. A careful evaluation during CTA or conventional angiogram of all cerebral vessels should be obtained as about 15–20% of patients will have multiple aneurysms. Even if other unsecured aneurysms are diagnosed, this would not impact the use of hypertensive treatment for cerebral vasospasm. The additional unruptured aneurysms can be addressed after the acute SAH phase.

A multidisciplinary team approach to the decision making involving the neurosurgeon, interventional neurologist or neuroradiologist, and neurointensivist is the best approach for reviewing all individual characteristics of patients and determining the best course for securing the aneurysm. If endovascular coiling is deemed appropriate, it should be performed at the time of the diagnostic angiogram, reducing the time to treatment and risk of rebleeding.

Summary Based on AHA Guidelines for Treatment of Ruptured Aneurysm [5]

- Ruptured aneurysms should be secured early by surgical clipping or endovascular coiling to reduce the risk of rebleeding.

- The decision to treat with surgical clipping or endovascular coiling should be made by teams with expertise in both techniques and should be based on characteristics of the individual patient and aneurysm.

- For patients with ruptured aneurysms for which treatment by surgical clipping and endovascular coiling are determined to be equally appropriate by a team of neurosurgeons and neurointerventionalists, endovascular coiling is the preferred technique.

8   Management Post-aneurysm Repair

8.1   Intracranial Hypertension

See previous section (6.1.5).

8.2   Pain Management

Headaches are best treated initially with acetaminophen (650 mg orally every 4–6 h prn) and, if needed, opioids such Dilaudid 0.2–0.4 mg IV every 3–4 h prn, or morphine sulfate 2–4 mg IV every 1 h prn or fentanyl may be used. Oversedation from opioids might make it difficult to distinguish signs of neurologic worsening from cerebral vasospasm or worsening hydrocephalus. Hence pain relief must be counterbalanced with the need for frequent neuro checks in these patients. We use low dose dexamethasone in patients with radicular pain to reduce inflammation although there is no evidence for this strategy. We also use Toradol 50 mg Q6hprn in patients with normal renal function.

8.3   Cerebral Vasospasm and Delayed Cerebral Ischemia (DCI)

Cerebral vasospasm is the most common neurologic complication after SAH. It typically occurs between days 4 and 14 after SAH and can be focal or diffuse, involving multiple cerebral vessels, and may occur suddenly or gradually. In some patients vasospasm might occur up to 21 days. Angiographic vasospasm has been reported in up to 30–70% of the patients with SAH and leads to stroke or death in 15–20%. Predictors of cerebral vasospasm include amount of blood on initial head CT, pulmonary complications, and cardiomyopathy. The modified Fisher scale found that the risk of vasospasm increased with increasing grade (Table 3). When not effectively treated, vasospasm can lead to delayed cerebral ischemia (DCI) and worsen clinical outcome.

Vasospasm can manifest in different ways, such as elevated blood flow velocity measured by transcranial doppler (TCD) ultrasound, conventional angiographic evidence of narrowed vessels, and decline in neurologic function on physical examination in the absence of other causes, or CT or MRI evidence of infarction with no other explanation. Symptomatic vasospasm is defined by either TCD ultrasound or angiographic evidence of vasospasm combined with new or worsening headache or alteration in the level of consciousness (confusion, delirium, or somnolence). Focal neurologic deficits, such as hemiparesis, aphasia, or other signs, may occur, indicating the involved vascular territory or territories. *Delayed cerebral ischemia* is defined as symptomatic vasospasm or CT or MRI evidence of infarction attributed to vasospasm. Prophylactic hypervolemia and prophylactic angioplasty are currently not recommended [5]. A recent meta-analysis reported no clinical benefit of statins in SAH. A phase 3 trial Simvastatin in Aneurysmal Subarachnoid hemorrhage (STASH) is currently in progress. Clazosentan to Overcome Neurological Ischemia and Infarction

Occurring after Subarachnoid hemorrhage (CONSCIOUS) group of studies using endothelin-1 receptor antagonist has failed to show any improvement in clinical outcomes. CONSCIOUS 3 being conducted in post coiling patients showed promising results in preliminary data but was terminated prior to completion based on the results of CONSCIOUS 2 trial.

8.3.1 Screening for Vasospasm

Some patients who have imaging evidence of vasospasm (TCD or angiography) do not exhibit clinical signs or imaging evidence of ischemia or infarction while some patients who clearly have clinical signs of focal brain ischemia do not have imaging evidence of vasospasm. This lack of overlap between the various manifestations of vasospasm makes interpreting clinical data and guiding treatment at the level of the individual patient challenging.

We screen for vasospasm with daily TCD ultrasound of the cerebral vessels. Mean velocities of TCD in the 120 s to 130 s cm/s for anterior (ACA) and MCA vessels suggest mild vasospasm, 140–170 s moderate vasospasm, and 180 and higher severe vasospasm when in conjunction with a Lindegaard ratio (MCA to extracranial internal carotid artery mean velocities) >3. Transcranial Doppler has limitations due to being operator dependent and technical factors such as inability to obtain temporal windows. Cerebral angiography can be used to confirm vasospasm and to potentially treat intra-arterially.

In patients whose baseline neurologic function is sufficient to examine cortical function on neurologic examination, we use the presence of changes in the examination to trigger hemodynamic and endovascular treatment for vasospasm. If such patients have elevated TCD velocities but no clinical changes, we increase the frequency of neurologic assessments, carefully maintain normal intravascular volume, and aim to keep the systolic blood pressure from dipping below 120 mm Hg.

High-grade patients with poor baseline neurologic examinations represent a challenge, as the clinical examination cannot be relied on to assess the adequacy of brain perfusion. In such patients, the addition of surveillance imaging using conventional angiography, CT or MR angiography and perfusion imaging and/or invasive physiologic monitoring of intracranial pressure, brain tissue oxygen tension (Licox; Integra™ Licox® Brain Oxygen Monitoring System; Integra Life Sciences Corporation, cerebral blood flow, or microdialysis during the highest risk period (post-bleed days 4–14) may be reasonable.

8.3.2 Nimodipine and Cerebral Vasospasm

Nimodipine (60 mg orally every 4 h for 21 days) is indicated to reduce poor outcome after aneurysmal SAH, although its effect may be modest and it has not clearly been demonstrated to prevent cerebral vasospasm. It may act more as a neuroprotectant. This recommendation for the use of nimodipine thus far has not been extended to other calcium channel antagonists. When blood pressure has been adversely affected by the nimodipine, dosage can be divided up to 30 mg orally every 2 h or even stopped completely with significant hypotension or when induced hypertensive treatment is introduced.

Hemodynamic augmentation, including hypervolemia, hypertension, and hemodilution (Triple-H therapy), to overcome the increased resistance to flow of spastic vessels had been the mainstay medical treatment for cerebral vasospasm. The effectiveness of this strategy has not been definitively established. Contemporary literature suggests euvolemia rather than hypervolemia should be maintained while hypovolemia has been associated with worse outcomes.

8.3.3 How Do We Induce Hemodynamic Augmentation?

Maintaining Euvolemia rather than induce hypervolemia is recommended for SAH patients. We use hemodynamic monitoring with as needed POCUS assessments to determine volume tolerance, 24 h fluid balance and in high grade patients we use PiCCO monitoring to target therapies to augment cerebral perfusion in patients with symptomatic vasospasm [21].

We use crystalloids normal saline (NS) or PlasmaLyte to replete any volume deficit that is present on admission with normal saline (NS), followed by the infusion of maintenance fluids (NS or PlasmaLyte at 1–1.5 ml/kg/h). We use a 24-h mildly positive to even fluid balance goal. Once the aneurysm is secured, we stop all antihypertensive medications and allow the blood pressure to rise to a systolic of 200 mm Hg before considering treatment. This removes the possibility of inadvertently exacerbating previously well-compensated vasospasm by lowering the blood pressure.

Once it is clear from clinical and/or imaging evidence that brain tissue perfusion is reduced due to vasospasm, it is necessary to act quickly to restore perfusion and avoid infarction.

- We immediately aim for slight hypervolemia (if patients are volume tolerant) with 1–2 l bolus of any isotonic fluid, usually NS or PlasmaLyte and induce hypertension with vasopressors (typically phenylephrine or norepinephrine) titrated to the resolution of the relevant clinical or imaging abnormality with a systolic blood pressure goal of 180–220 mm Hg.

- During hemodynamic augmentation, we monitor patients carefully for the previously mentioned complications including development of demand ischemia. In hypoalbuminemic patients we use 5% albumin as an adjunct to restore intravascular volume and to limit ongoing renal sodium and water loss.

- Sometimes cerebral perfusion pressure can be improved by reducing intracranial pressure is by a few mm Hg by lowering the EVD.

- Montreal Milrinone protocol: In selected patients we use the Montreal milrinone protocol to induce inotropy and improve cerebral perfusion [22].

Limitations of Hemodynamic Augmentation
- Existing evidence does not indicate that prophylactic Triple-H therapy, initiated before vasospasm is apparent, effectively prevents DCI or improves clinical outcome.

- Existing physiologic evidence suggests that hypertension, but not hypervolemia or hemodilution may cause desirable improvements in key physiologic parameters such as cerebral blood flow, brain tissue oxygenation, and cerebral delivery rate of oxygen.

- It is associated with a number of serious complications including exacerbation of cerebral edema, hemorrhagic transformation of cerebral infarctions, epidural hematoma formation, pulmonary edema, myocardial infarction, hyponatremia, coagulopathy related to the use of nonalbumin colloid solutions, and complications related to pulmonary artery catheters.

9   Endovascular Treatment for Cerebral Vasospasm

For the 25–50% of patients who do not respond to hemodynamic augmentation, urgent endovascular treatment is crucial [2]. The ideal timing for this intervention is not clear, but a number of studies suggest that the response is best when it occurs within 2 h of the onset of symptomatic vasospasm. Therefore, we should alert the neurointerventional team as soon as a patient develops symptomatic vasospasm. We may choose to do a CTH/CT angiography prior to endovascular treatment or in clearly symptomatic vasospasm proceed directly to the angiography suite.

Vasospasm can be treated endovascularly either via injection of vasodilators directly into the spastic vessel or vascular territory and balloon angioplasty.

- Intraarterial vasodilators including nicardipine, verapamil, and milrinone, are generally safer than angioplasty and are especially useful for improving vasospasm in smaller, distal vessels. Their duration of effect, however, can last for as little as 1–2 days, often necessitating multiple treatments.

- Angioplasty can safely be performed in only the largest cerebral vessels: internal carotid artery, M1 and M2 segments of the MCA, A1 segments of the ACA, P1 and P2 segments of the posterior cerebral artery, vertebral arteries, and the basilar artery.

- The major advantage of angioplasty over vasodilators is the improved durability of vessel widening which is particularly important in patients who develop vasospasm early. However, it is important to keep in mind that angioplasty is associated with a 5% risk of major complications including a 1% risk of vessel rupture.

Summary of Recommendations: Post-aneurysmal Securing

- An EVD should be urgently placed in patients with SAH with imaging evidence of hydrocephalus and altered consciousness.

- Intracranial pressure monitoring should be considered in those patients with SAH having somnolence, headache, nausea, vomiting, and pupil changes such as dilation and/or sluggish to no constriction to light stimulus.

- The choice of osmotic agent for the treatment of intracranial hypertension in patients with SAH should be based on the hemodynamic status of the individual patient. However, hypertonic saline may be more beneficial than mannitol as it does not lead to volume depletion.

- The prophylactic use of antiepileptic drugs such as phenytoin or levetiracetam should be considered in the pre-securing phase. Antiepileptic drugs should not be routinely used after the aneurysm is secured.

- The treatment of head and neck pain after SAH should be carefully treated, balancing patient comfort with avoidance of oversedation and inability to distinguish signs of neurologic worsening from cerebral vasospasm or other causes. Use of a tiered system of analgesics may help achieve this goal.

- The modified Fisher scale should be applied to the admission CT to determine the patient's risk of vasospasm.

- Frequent serial neurologic examinations and TCD studies should be used to detect the earliest evidence of vasospasm.

- Enteral nimodipine at a dose of 60 mg every 4 h should be used to prevent poor outcome in all patients with aneurysmal SAH.

- Euvolemia, as determined by fluid balance, urine output, and/or central venous pressure, should be strictly maintained with the use of isotonic fluids.

- Hemodynamic augmentation in the form of euvolemic hypertension should be used as the first-line treatment of symptomatic vasospasm.

- Urgent endovascular therapy is indicated for patients with symptomatic vasospasm who do not respond within 2–4 h of adequate hemodynamic augmentation.

10 Management of Medical Complications

10.1 Cardiac Complications

Patients with higher grade SAH are at an increased risk of developing neurogenic stunned myocardium. Various arrhythmias have been described in SAH patients. Bradycardia, relative tachycardia, and nonspecific ST- and T-wave abnormalities are strongly and independently associated with 3-month mortality after SAH. Baseline cardiac assessment with serial enzymes, electrocardiography, and echocardiography is recommended, especially in patients with evidence of myocardial dysfunction [17]. Clinical management may require more aggressive hemodynamic monitoring until cardiac function returns to normal [3, 17]. We generally use as needed POCUS assessments and PiCCO monitoring in all high-grade SAH patients. Patients may develop high troponin leaks and NSTEMI. Very rarely patients may develop true STEMI. We usually co-manage these patients with our cardiology colleagues and determine the threshold for activating the cardiac catheterization lab, and discuss the safety of using dual antiplatelets and heparin drip usually without a bolus with our neurosurgery colleagues.

10.2 Pulmonary Complications

Cardiogenic or non-cardiogenic pulmonary edema may develop in patients with high grade SAH. Pulmonary edema makes the management of volume status in patients at risk for vasospasm difficult. PiCCO monitoring, periodic POCUS, periodic CXR can be used for monitoring clinical changes. We use arterial lines in all SAH patients and maintain PaO_2 >60. We also use lung protective ventilation (8 cc/kg ideal body weight) with end-tidal CO_2 monitoring and daily arterial blood gases to prevent CO_2 retention.

10.3 Anemia and Transfusion

Post-operative anemia after SAH and greater systemic inflammatory response on admission potentially leading to cytokine-mediated inhibition of RBC production are the two main causes of anemia in SAH patients. Appropriate transfusion thresholds may vary depending on the presence or absence of clinical vasospasm. The use of PRBC transfusion to treat anemia might be reasonable in patients with SAH who are at risk of cerebral ischemia but whether transfusion is beneficial in mitigating ischemia cannot be determined from the available data [17]. The optimal hemoglobin goal has not been determined [5]. NCS guidelines recommend maintaining hemoglobin concentration above 8–10 g/dl. Transfusions are associated with medical complications like circulatory overload (TACO), transfusion related acute lung injury (TRALI) and it is important to weigh risk–benefit ratio prior to transfusing SAH patients.

10.4 Hypothalamic Dysfunction

Hypothalamic dysfunction remains underdiagnosed in all patients with acute brain injury.

Patients who do not respond to pressors should be assessed for hypothalamic dysfunction.

The optimal method of diagnosis remains unclear [5]. We assess hypothalamic pituitary axis function by checking random and AM cortisol, thyroid function panel as screening tests. The two randomized clinical trials using intravenous steroids in the treatment of DND showed no reduction in the incidence of symptomatic vasospasm but significant improvement almost in every aspect of neurological findings even a year after SAH.

From AHA guidelines for SAH [5].

- Administration of high dose corticosteroids is not recommended in acute SAH.

- Hormonal replacement with stress-dose corticosteroids for patients with vasospasm and unresponsiveness to induced hypertension may be considered.

- Long-term survivors of SAH frequently exhibit endocrine changes, predominantly growth and gonadal hormone deficiencies.

11   Fever Control

Refractory fever during the first 10 days after SAH is predicted by poor Hunt–Hess grade and presence of intraventricular hemorrhage. It is associated with increased mortality, increased length of stay, depressed consciousness, and more functional and cognitive disability among survivors. Early onset of fever usually predicts a non-infectious cause such as central fever. Our fever workup includes infectious workup with blood cultures, urine analysis and microscopy, CXR and non-infectious workup with lower extremity duplex for Deep venous thrombosis (DVT)/thrombophlebitis. For patients with EVD only if clinical suspicion is high for an infection we sample CSF.

From NCS and AHA guidelines on SAH [5, 17].

- Fever control is associated with reduced cerebral metabolic distress in patients with SAH, irrespective of intracranial pressure (ICP). Infectious causes of fever should always be sought and treated Aggressive control of fever to a target of normothermia by use of standard or advanced temperature modulating systems is reasonable in the acute phase of SAH.

- Treatment with an air-circulating cooling blanket is not beneficial and while the efficacy of most antipyretic agents (acetaminophen, ibuprofen) is low, they should be used as the first line of therapy.

- Arctic Sun and catheter-based cooling systems are superior to conventional cooling-blanket therapy for controlling fever.

- Patients should be monitored and treated for shivering. The Bedside Shivering Assessment Scale is a simple and reliable tool for evaluating the metabolic stress of shivering.

11.1 Hyperglycemia

Hyperglycemia at admission is associated with poor outcomes Predictors of high glycemic burden included age ≥ 54 years, Hunt and Hess grade III–V, poor Acute Physiology and Chronic Health Evaluation (APACHE)-2 physiological subscores, and a history of diabetes mellitus [15].

Nadir glucose <80 mg/dl is associated with cerebral infarction, vasospasm, and worse functional outcomes in multivariate models. A blood glucose target of 180 mg/dl or less resulted in lower mortality than did a target of 81–108 mg/dl. Current recommendation suggest that serum glucose should be maintained below 200 mg/dl and hypoglycemia (serum glucose <80 mg/dl) should be avoided [17].

11.2 Hyponatremia and Hypernatremia

Hyponatremia and hypernatremia after SAH are common. Hyponatremia is independently associated with poor outcomes. Clinically distinguishing between SIADH or Cerebral salt wasting (CSW) in these patients is difficult. We usually send a full workup including urine electrolytes, serum osmolality, urine osmolarity, serum uric acid to determine the diagnosis. While awaiting lab results we start these patients on a 2% drip at 50 cc/h. Irrespective of the clinical diagnosis of SIADH or CSW free water intake via intravenous and enteral routes should be limited We usually avoid using intermittent dosing of vasopressin receptor antagonists like intravenous conivaptan or oral tolvaptan given the risk of excess free water loss and hypovolemia. For patients with CSW inhibition of natriuresis with fludrocortisone can effectively reduce the sodium and water intake required for hypervolemia and prevent hyponatremia at the same time [5, 17]. Hypernatremia is also an independent predictor of poor outcomes [23]. Patients with Acom aneurysm may rarely develop frank diabetes insipidus.

11.3 Deep Venous Thromboprophylaxis

Deep venous thrombosis (DVT) is a relatively frequent complication after SAH. DVT prophylaxis strategy could be pharmacologic or a combination of mechanical and pharmacologic and should be employed for all SAH patients [17].

Summary of DVT prophylaxis from NCS guidelines.

- Routine venous Doppler ultrasonography is an efficient, noninvasive means of screening modality for DVT in both symptomatic and asymptomatic patients.

- Low molecular weight heparin or unfractionated heparin for prophylaxis should be withheld in patients with unprotected aneurysms and in those expected to undergo surgery.

- Use of unfractionated heparin for prophylaxis could be started 24 h after undergoing surgery.

- Unfractionated heparin and low molecular weighted heparin should be withheld 24 h before and after intracranial procedures.

12   Hospital Characteristics and Systems of Care

Low volume centers typically treat less than 18 cases/year while large volume centers treat more than 60 cases/year. Transferring patients to high volume centers improves outcomes. Mortality and long-term poor outcomes are substantially higher (by 10–20%) at small volume centers compared to high volume centers [5]. High volume centers should have appropriate specialty neurointensive care units, neurointensivists, vascular neurosurgeons and interventional neuroradiologists to provide the essential elements of multidisciplinary care [17]. We transfer all our SAH patients to our comprehensive stroke center which is also a high volume SAH center. We use dedicated protocols for pre-ICU or pre-ER stabilization to ensure safe and rapid transfer.

12.1 Ongoing Recovery, Functional, Behavioral, and Cognitive Outcome

Survivors of SAH experience deficits in memory, executive function, and language commonly. These cognitive impairments affect patients' activities of daily living, instrumental activities of daily living, return to work, and quality of life. Behavioral problems such as depression, anxiety, fatigue, and sleep disturbances compound cognitive impairments for a high proportion of patients. About 30–50% SAH survivors continue to suffer from long-term depression, PTSD, and anxiety.

Functional outcomes depend on initial grade of presentation, age, and secondary neurological injury. The FRESH score [24] is a simple and reliable score that includes initial Hunt Hess grade, APACHE-II physiologic scores on admission, age, and aneurysmal rebleed within 48 hours. Separate subscores help predict 1-year cognition (FRESH-cog) and quality of life (FRESH-quol) and were developed adjusting for education and premorbid disability.

In a study that evaluated the need for full neuropsychological batteries in SAH patients as compared to Telephone Interview for Cognitive Status (TICS), found that the TICS was adequate for identifying problems in various cognitive domains [25]. Factors associated with poor cognitive outcome include female gender, hydrocephalus, DCI, and older age [26]. It is important to assess all SAH survivors for cognitive and behavioral impairments at the time of follow-up.

12.2 Planning for Transition of Care to Rehabilitation or Long-Term Care Facilities

Tracheostomy: In Stroke-related Early Tracheostomy versus Prolonged Orotracheal Intubation in Neurocritical Care Trial (SETPOINT), a randomized pilot trial for patients with severe ischemic and hemorrhagic stroke were randomized to early tracheotomy tube (within days 1–3 from intubation) or to standard tracheostomy (between day 7 and 14 from intubation if extubation could not be achieved or was not feasible). This study reported similar ICU length of stay but a reduction in the use of sedatives and analgesics and a lower ICU and 6-month mortality in the early tracheostomy group [27]. Andriolo and colleagues [28] in a systematic review of tracheostomy in stroke patients defined early tracheostomy as 2–10 days after intubation, and late tracheostomy as >10 days after intubation. The pooled results showed that patients in the early tracheostomy had a lower risk of death and higher probability of discharge from the ICU. An important advantage of tracheostomy tube over endotracheal intubation is that it improves patient comfort, facilitates ongoing weaning and helps with early mobilization. An ongoing RCT; SETPOINT-2 might help answer the early versus late tracheostomy question satisfactorily for stroke patients [29]. As part of our Respiratory Recovery Pathway we advocate for early tracheostomy tube and daily pressure support trials at least three times per day. Only patients who are hemodynamically unstable or have refractory ICP crises do not undergo pressure support trials till they are stabilized medically.

Percutaneous endoscopic gastrostomy (PEG) tube: SAH patients with dysphagia or on mechanical ventilation receive enteral nutrition via nasogastric tubes (NGT) or orogastric tubes (OGT). Adverse effects of NGT include chronic sinusitis, gastroesophageal reflux, and aspiration pneumonia. PEG tube placement does not prevent aspiration pneumonia but increases patient comfort. The timing of PEG tube placement remains controversial. A systematic review of PEG tube placement in acute and subacute stroke reported that compared with NGT feeding, PEG tubes reduced treatment failure and gastrointestinal bleeding, and helped with more adequate feeding [30].

We currently encourage placement on tracheostomy tube and PEG tube either on the same day or within 24–48 h of each other. This strategy might help in a smoother transition of care to rehabilitation or long term care facilities when patients become medically stable.

Ventriculoperitoneal shunt (VP shunt): Factors associated with a need for VP shunt include female gender, older age, poor clinical grade on admission, and IVH. Additional factors include hyperglycemia at admission, findings on the admission CT scan (Fisher Grade 4, IV intraventricular hemorrhage, and bicaudate index ≥ 0.20), and development of nosocomial meningitis/ventriculitis [31]. It has been shown that compared with rapid EVD weaning gradual, multistep EVD weaning prolonged ICU or hospital stay and provided no advantage to patients in preventing the need for long-term shunt placement [32]. However, a recent multi-institutional survey based study found that most institutions favored a gradual weaning approach.

Incidence of chronic post-SAH hydrocephalus requiring shunt is 15–20%. There is no difference in the incidence of chronic post SAH hydrocephalus after clipping versus coiling. Hydrocephalus may develop days to weeks after subarachnoid hemorrhage in about one third of the patients and should be suspected in patients who have a good initial recovery followed by a plateau or decline in their condition [3].

Acknowledgements

We thank Sachin Agarwal.

Appendix: Subarachnoid Hemorrhage Steps in Management (Adapted from Handbook of Neurological Therapy) [33]

- Secure airway if needed.
- Maintain SBP 100–140 mm Hg till aneurysm is secured.
- Maintain head of bed at 30°.
- Increased ICP management: Boluses of mannitol 1 g/kg or hypertonics 23.4%.
- Head CT and CT-Angiography upon admission to Neurointensive care unit, if none prior.
- Consider reversal of antiplatelets/anticoagulants or any other underlying coagulopathy (e.g., liver or renal dysfunction).
- Consider Amicar if significant SAH (Hunt & Hess > = 3) or anticipate delay in treatment.
  - 4 g IV bolus over 60 min then 1 g/h, stop 4 h before angiography planned.
  - Contraindicated or caution in CAD, stroke, pulmonary embolism, thrombosis, or vasospasm.
- Keppra or fosphenytoin 20 mg/kg IV ×1 STAT (loading dose).
- Transfer to a high-volume SAH center.
- Admit to Neuro-ICU.
- Point of Care Ultrasound (POCUS) to establish a baseline for cardiac function and volume status.
- Place an arterial line.
- Contact Neurosurgery to arrange angiogram and possible EVD/ICP monitor if evidence of hydrocephalus or GCS </=8.

- Short acting, low dose sedation/analgesia prn (e.g., Tylenol, fentanyl, Precedex, propofol).

- Avoid NSAIDs, ASA, anti-platelets prior to surgery.

- Diet: usually NPO for angiogram, place Duotube if necessary for PO meds particularly from nimodipine.

- IV: NS at 1.0–1.5 cc/kg/h to maintain euvolemia.

- Goal SBP < 16.0 mm Hg, use nicardipine or labetalol drip prn for SBP <160.

- Nimodipine standing orders for all SAH patients for 21 days.
    - Nimodipine 60 mg PO/DT q4h for SBP >140.
    - Nimodipine 30 mg PO/DT q4h for SBP 120–140.

- Labs: CBC, BMP, Mg, Phos, Ion Ca++, LFTs, Coags, serial troponins, ß-hcg in young women, type and hold, UA, Urine toxicology.

- Pan Cultures (cultures blood, urine, sputum, stool, CSF as needed).

- Electrocardiogram.

- Chest X-ray.

- Transthoracic Echocardiogram, portable.

- If aneurysm protected, allow patient to autoregulate up to SBP 200 mm Hg 24 h post-clipping or coiling.

- No subcutaneous Heparin/Lovenox until 24 h post procedure (i.e., craniotomy).

- Maintain Venodyne at all times.

- Gastrointestinal prophylaxis: Pepcid 20 mg po daily or Proton pump inhibitor.

- Nicotine patch 14–21 mg daily (as needed).

- Zofran 4 mg IVSS q6h prn.

- Tylenol 650 PO/DT/PR q4h prn pain/fever.

- Aggressive bowel regimen: Senna 2 Tables PO/DT at bedtime and/or Colace 100 mg PO/DT TID.

- Lung Protective ventilation but with ETCO2 ordered for ventilated patients to prevent CO_2 retention.

- GOAL: normal PH, PCO_2 ($ETCO_2$)—35-45 if no ICP crises or 30–35 if ICP crises PaO_2: >60 mm Hg.

- Aggressive glucose, seizure, and fever control during vasospasm period.

- Daily transcranial Dopplers with assessment for Lindegaard ratio for 14 days post-SAH.

References

1. Taufique Z, May T, Meyers E et al (2016) Predictors of poor quality of life 1 year after subarachnoid hemorrhage. Neurosurgery 78:256–264. https://doi.org/10.1227/NEU.0000000000001042

2. Powell J, Kitchen N, Heslin J, Greenwood R (2002) Psychosocial outcomes at three and nine months after good neurological recovery from aneurysmal subarachnoid haemorrhage: predictors and prognosis. J Neurol Neurosurg Psychiatry 72:772–781

3. Lawton MT, Vates GE (2017) Subarachnoid hemorrhage. N Engl J Med 377:257–266. https://doi.org/10.1056/NEJMcp1605827

4. Ingall T, Asplund K, Mähönen M, Bonita R (2000) A multinational comparison of subarachnoid hemorrhage epidemiology in the WHO MONICA stroke study. Stroke 31:1054–1061

5. Connolly ES, Rabinstein AA, Carhuapoma JR et al (2012) Guidelines for the management of aneurysmal subarachnoid hemorrhage: a guideline for healthcare professionals from the american heart association/american stroke association. Stroke 43:1711–1737. https://doi.org/10.1161/STR.0b013e3182587839

6. Nieuwkamp D, Setz L, Algra A, Linn F (2009) Changes in case fatality of aneurysmal subarachnoid haemorrhage over time, according to age, sex, and region: a meta-analysis. Lancet 8(7):635–642

7. Broderick J, Brown R, Sauerbeck L, Hornung R (2009) Greater rupture risk for familial as compared to sporadic unruptured intracranial aneurysms. Stroke 40(6):1952–1957

8. Vlak M, Algra A, Brandenburg R, Rinkel G (2011) Prevalence of unruptured intracranial aneurysms, with emphasis on sex, age, comorbidity, country, and time period: a systematic review and meta-analysis. Lancet Neurol 10 (7):626–636

9. Etminan N, Beseoglu K, Barrow DL et al (2014) Multidisciplinary consensus on assessment of unruptured intracranial aneurysms. Stroke 45(5):1523–1530

10. Etminan N, Brown RD, Beseoglu K et al (2015) The unruptured intracranial aneurysm treatment score: a multidisciplinary consensus. Neurology 85:881–889. https://doi.org/10.1212/WNL.0000000000001891

11. Bassi P, Bandera R, Loiero M (1991) Warning signs in subarachnoid hemorrhage: a cooperative study. Acta Neurol (Napoli) 84 (4):277–281

12. Edlow JA, Caplan LR (2000) Avoiding pitfalls in the diagnosis of subarachnoid hemorrhage. N Engl J Med 342:29–36. https://doi.org/10.1056/NEJM200001063420106

13. Beck J, Raabe A, Szelenyi A et al (2006) Sentinel headache and the risk of rebleeding after aneurysmal subarachnoid hemorrhage. Stroke 37(11):2733–2737

14. Kowalski RG, Claassen J, Kreiter KT et al (2004) Initial misdiagnosis and outcome after subarachnoid hemorrhage. JAMA 291:866. https://doi.org/10.1001/jama.291.7.866

15. Frontera JA, Claassen J, Schmidt JM et al (2006) Prediction of symptomatic vasospasm after subarachnoid hemorrhage: the Modified Fisher Scale. Neurosurgery 59:21–27. https://doi.org/10.1227/01.NEU.0000218821.34014.1B

16. Ortega-Gutierrez S, Thomas J, Reccius A et al (2013) Effectiveness and safety of nicardipine and labetalol infusion for blood pressure management in patients with intracerebral and subarachnoid hemorrhage. Neurocrit Care 18:13–19. https://doi.org/10.1007/s12028-012-9782-1

17. Diringer M, Bleck T, Hemphill J, Menon D (2011) Critical care management of patients following aneurysmal subarachnoid hemorrhage: recommendations from the Neurocritical Care Society's Multidisciplinary. Neurocrit Care 15(2):211–240

18. Molyneux A, Kerr R, Yu L et al (2005) International subarachnoid aneurysm trial (ISAT) of neurosurgical clipping versus endovascular coiling in 2143 patients with ruptured intracranial aneurysms: a randomised comparison of effects on survival, dependency, seizures, rebleeding, subgroups, and aneurysm occlusion. Lancet 366(9488):809–817

19. Little AS, Kerrigan JF, McDougall CG et al (2007) Nonconvulsive status epilepticus in patients suffering spontaneous subarachnoid hemorrhage. J Neurosurg 106:805–811. https://doi.org/10.3171/jns.2007.106.5.805

20. Dennis L, Claassen J, Hirsch L, Emerson R (2002) Nonconvulsive status epilepticus after subarachnoid hemorrhage. Neurosurgery 51 (5):1136–1143. discussion 1144

21. Tagami T, Kuwamoto K, Watanabe A et al (2014) Optimal range of global end-diastolic volume for fluid management after aneurysmal subarachnoid hemorrhage. Crit Care Med 42:1348–1356. https://doi.org/10.1097/CCM.0000000000000163

22. Lasry O, Marcoux J (2014) The use of intravenous Milrinone to treat cerebral vasospasm following traumatic subarachnoid hemorrhage. Springerplus, New York, NY

23. Qureshi AI, Suri MFK, Sung GY et al (2002) Prognostic significance of hypernatremia and hyponatremia among patients with aneurysmal subarachnoid hemorrhage. Neurosurgery 50:749–756. https://doi.org/10.1097/00006123-200204000-00012

24. Witsch J, Frey H-P, Patel S et al (2016) Prognostication of long-term outcomes after subarachnoid hemorrhage: the FRESH-Score. Ann Neurol. https://doi.org/10.1002/ana.24675

25. S a M, Kreiter KT, Copeland D et al (2002) Global and domain-specific cognitive impairment and outcome after subarachnoid hemorrhage. Neurology 59:1750–1758. https://doi.org/10.1212/01.WNL.0000035748.91128.C2

26. Kreiter KT, Copeland D, Bernardini GL et al (2002) Predictors of cognitive dysfunction after subarachnoid hemorrhage. Stroke 33:200–208. https://doi.org/10.1161/hs0102.101080

27. Bosel J, Schiller P, Hook Y et al (2013) Stroke-related early tracheostomy versus prolonged orotracheal intubation in neurocritical care trial (SETPOINT): a randomized pilot trial. Stroke 44:21–28. https://doi.org/10.1161/STROKEAHA.112.669895

28. Andriolo BN, Andriolo RB, Saconato H et al (2015) Early versus late tracheostomy for critically ill patients. Cochrane Database Syst Rev. https://doi.org/10.1002/14651858.CD007271.pub3

29. Schönenberger S, Niesen W-D, Fuhrer H et al (2016) Early tracheostomy in ventilated stroke patients: Study protocol of the international multicentre randomized trial SETPOINT2 (stroke-related early tracheostomy vs. prolonged orotracheal intubation in neurocritical care trial 2). Int J Stroke 11:368–379. https://doi.org/10.1177/1747493015616638

30. Geeganage C, Beavan J, Ellender S, Bath PM (2012) Interventions for dysphagia and nutritional support in acute and subacute stroke. In: Bath PM (ed) Cochrane database system reviews. John Wiley & Sons, Ltd, Chichester, p CD000323

31. Rincon F, Gordon E, Starke RM et al (2010) Predictors of long-term shunt-dependent hydrocephalus after aneurysmal subarachnoid hemorrhage. J Neurosurg 113:774–780. https://doi.org/10.3171/2010.2.JNS09376

32. Klopfenstein JD, Kim LJ, Feiz-Erfan I et al (2004) Comparison of rapid and gradual weaning from external ventricular drainage in patients with aneurysmal subarachnoid hemorrhage: a prospective randomized trial. J Neurosurg 100:225–229. https://doi.org/10.3171/jns.2004.100.2.0225

33. Colosimo C, Gilhus N, Gil-Nagel A, Rapoport A (2015) Handbook of neurological therapy. Routledge, London

Chapter 8

Multimodality Monitoring of Aneurysmal Subarachnoid Hemorrhage

Michael E. Reznik and David J. Roh

Abstract

Initial treatment paradigms for patients suffering from aneurysmal subarachnoid hemorrhage focus on early surgical aneurysm treatment to prevent rebleeding. After aneurysm treatment, secondary treatment efforts focus on early detection and treatment of delayed cerebral ischemia to prevent secondary brain injury. Specialized neurocritical care centers can provide close neurological monitoring to assess for early signs of delayed cerebral ischemia to implement cerebral perfusion augmentation strategies. However, patients who do not have reliable neurological exams to follow are often those that are at higher risk of developing delayed cerebral ischemia and require surrogate measures of continuous monitoring to evaluate for early signs of secondary brain injury. Here we provide an overview of aneurysmal subarachnoid hemorrhage, delayed cerebral ischemia, as well as the role and use of multimodality monitoring in certain subarachnoid hemorrhage patients in efforts to detect and implement treatment strategies to prevent delayed cerebral ischemia and secondary brain injury.

Key words Subarachnoid hemorrhage, Delayed cerebral ischemia, Secondary brain injury, Multimodality monitoring, Cerebral perfusion

1 Introduction

The care of aneurysmal subarachnoid hemorrhage (SAH) patients has changed dramatically over the past few decades. This is due to both the surgical advances in aneurysm treatment in addition to the introduction of various invasive and noninvasive monitoring modalities. These multimodality monitoring (MMM) techniques have provided a greater understanding of cerebral physiology and a means of early detection of secondary brain injury, especially delayed cerebral ischemia (DCI). The physiologic information obtained from these monitoring tools allows for early opportunities to implement real-time precision-based management in order to prevent the accumulation of irreversible neurological injury. Many

Fawaz Al-Mufti and Krishna Amuluru (eds.), *Cerebrovascular Disorders*, Neuromethods, vol. 170,
https://doi.org/10.1007/978-1-0716-1530-0_8, © Springer Science+Business Media, LLC, part of Springer Nature 2021

of these tools have gained wide acceptance in clinical practice, though some remain restricted to specialized centers. This chapter aims to provide an overview of each individual modality, along with an approach to their use.

2   Pathophysiology and Clinical Presentation

Typical initial clinical presentations for SAH includes the classic "worst headache of life," along with some who describe a "sentinel" headache that precedes a more prominent symptom onset. However, other common symptoms include nausea, vomiting, meningismus, photophobia, focal neurological deficits including cranial neuropathies, and loss of consciousness (LOC). LOC at onset can occur for many reasons, including transient intracranial hypoperfusion from sudden elevation of intracranial pressure (ICP), global cerebral edema, seizure, or non-perfusing cardiac arrhythmia. LOC can often be the presentation of SAH patients regardless of severity. However, LOC is more common in patients with high-grade SAH along with patients that have higher blood burden, and is independently associated with worse outcomes [1].

Poor prognosis is determined by factors such as poor clinical grade, often designated by higher Hunt–Hess grades [2] (4–5, characterized by depressed consciousness), older age, higher volume of blood, and larger aneurysm size. Patients with low-grade SAH (typically designated as Hunt–Hess 1–3) often have uneventful hospital courses without any secondary complications. However, the rate of complications and in-hospital mortality for patients with high-grade SAH is significantly higher. Mortality remains as high as 71% for patients with the highest Hunt–Hess grade, though this has improved since grading systems were first introduced due to advances in modern medical and surgical management.

In addition to the known direct effects of the initial hemorrhage, aneurysm re-rupture, and DCI on poor outcomes after SAH [3], the presence of intraventricular hemorrhage (IVH), global cerebral edema, and other medical complications have become increasingly recognized as other significant contributing factors to poor outcome [4, 5]. Global cerebral edema is a common finding in high-grade SAH, occurring in up to 12% of all patients, and may be due to hypoxic–ischemic injury at ictus that may subsequently be exacerbated by rebound hyperemia from impaired cerebral autoregulation [4]. IVH itself can cause hydrocephalus from ventricular obstruction, but is also independently associated with the development of other neurological complications, most notably DCI [6]. All these factors and their complex interactions play a role in secondary brain injury after SAH.

3 Presentation and Pathophysiology of Delayed Cerebral Ischemia

DCI occurs in up to 20–40% of all SAH patients with the highest rates seen in high-grade SAH [6, 7]. DCI is perhaps the factor most responsible for additional morbidity for patients who survive their initial presentation and is currently defined as a neurological deterioration involving either new focal neurological impairment (hemiparesis, aphasia, neglect, etc.) or a change in level of consciousness (≥ 2 point decrease in Glasgow Coma Scale [GCS] score) lasting for at least 1 hour and not attributable to other causes [8]. Radiographic evidence of infarct is also considered by many to be evidence of DCI regardless of exam change, however this usually reflects fixed, irreversible injury. DCI typically occurs approximately 3–14 days after aneurysmal rupture, but can occur earlier [9], with a risk that declines rapidly after about 21 days.

The pathophysiology of DCI is largely thought to be triggered by blood degradation products coming into contact with the cerebral vasculature [10] resulting in vasospasm and perfusion deficits that lead to ischemia and secondary brain injury. Indeed, the degree of blood products present appears to have a correlation with the risk of developing subsequent DCI, as summarized by the modified Fisher scale, which takes into account SAH thickness and the presence of IVH [6, 7]. However, in reality the pathophysiology leading to DCI is likely multifactorial and incompletely understood. This is evidenced by the inconsistent association of large vessel, angiographic arterial narrowing (i.e., radiographic or angiographic vasospasm) to clinical symptoms consistent with DCI. Radiographic vasospasm is seen in a significant proportion of patients with SAH, but only 50% of patients with severe radiographic vasospasm go on to develop DCI [11].

Other factors independent of vasospasm that have been implicated in the development of DCI include microcirculatory dysregulation, microthrombosis, decreased levels of nitric oxide, increased levels of endothelin-1, generation of free radicals, oxidative stress, and inflammation secondary to blood–brain barrier breakdown [12]. More recently, cortical spreading depolarizations have also been linked to DCI, as they have been associated with metabolic and cellular dysfunction of underlying brain tissue, even in the absence of angiographic vasospasm [13].

4 Management Strategies of Multimodality Monitoring in Delayed Cerebral Ischemia

Given its high incidence and profound adverse effects on outcome, the risk of DCI—as well as other complications—necessitates high-level monitoring in patients with SAH in the acute period. The cornerstone of such monitoring continues to be neurological assessments, but unfortunately this is often not possible in the high-grade SAH patients who do not have a reliable exam to follow and have the highest risk of secondary brain injury from DCI. In these cases, the utilization of MMM tools can provide insight into sub-clinical physiological changes and progression towards secondary brain injury, allowing the opportunity to intervene before irreversible injury occurs (Fig. 1). The monitoring modalities available typically measure ICP/cerebral perfusion pressure (CPP), cerebral oxygenation (PbtO2/SjvO2), cerebral metabolism (microdialysis), cerebral blood flow (CBF and transcranial Doppler), and electrical activity (surface and depth EEG). The following section will discuss each modality individually and will specify the critical thresholds that have been established for each modality based on prior research and guideline statements (Table 1). However, the data from each modality should ideally be utilized in conjunction with the other modalities to gain a comprehensive picture, with trends rather than isolated values representing the standard method of data interpretation. The order in which the monitoring modalities are presented here reflect some clinicians' systematic approach to interpreting the data.

4.1   Intracranial Pressure and Cerebral Perfusion Pressure

ICP monitoring is the most widely utilized intracranial monitoring modality in SAH given the frequency in which ICP elevations are seen in SAH patients in addition to its association with worse outcome [14]. There are multiple devices capable of monitoring ICP, however the most frequently employed are extraventricular drains (EVD) and intraparenchymal monitors. Although they are relatively comparable in reliability and accuracy [15] and quoted with low complication rates, EVDs are the preferred monitoring modality as they can serve as an ICP monitor while also being able to therapeutically divert cerebrospinal fluid (CSF) [16]. The EVD's ability to divert CSF can improve cerebral blood flow [17] which is of use during periods of DCI, while it can also be utilized as a means to administer intrathecal medication to treat vasospasm [18] or lyse intraventricular clot [19]. Fiberoptic parenchymal ICP probes, although not as versatile as an EVD, provide advantages in their ease of placement (they do not require insertion within the ventricular system) and their ability to continuously provide ICP waveform monitoring, whereas EVDs only transduce pressure waveforms when the drainage system is closed. However, parenchymal probes cannot be calibrated once placed, meaning their values are subject to drift over time, and they are also susceptible to inaccuracies when compartmentalization of ICP across the dura mater occurs.

Although critical ICP thresholds of >20 mmHg are often treated when sustained, there is a lack of data supporting the use of ICP-targeted therapy in SAH patients. Rather, information provided by overall ICP trends and cerebral perfusion pressure

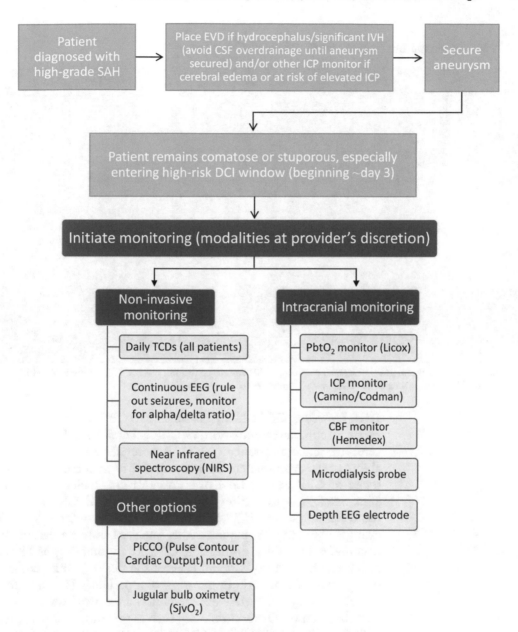

Fig. 1 Algorithm for early management of patients with high-grade subarachnoid hemorrhage (SAH). *EVD* external ventricular drain, *CSF* cerebrospinal fluid, *ICP* intracranial pressure, *DCI* delayed cerebral ischemia, *TCDs* transcranial dopplers, *EEG* electroencephalography, *PbtO2* brain tissue oxygen tension, *CBF* cerebral blood flow

(CPP) may provide a more comprehensive assessment than isolated ICP values alone. There is no direct monitor for CPP, but its values are derived from measurements obtained from ICP and arterial pressure monitors. CPP is defined as mean arterial pressure (MAP) minus ICP and serves a surrogate for cerebral blood flow (CBF). Clinicians often target a CPP of >60 mmHg by manipulating both MAP and ICP in order to optimize CBF and circumvent

Table 1
Multimodality monitoring devices and recommended target values

	Device	Target	Interpretation
Intracranial and cerebral perfusion pressure	EVD vs. Parenchymal	ICP <20–25 mmHg CPP >60 mmHg	Indirect surrogate for CBF
Oxygenation	(a) Parenchymal (PbtO2) (b) Jugular Bulb (SjvO2)	(a) PbtO2 > 20 mmHg (b) SjvO2 > 55%	(a) Regional brain tissue oxygenation (b) Global cerebral oxygen extraction
Cerebral blood flow	(a) Thermal diffusion probe (b) Transcranial Doppler	(a) CBF >20 cc/100 g/min (b) <200 cm/s anterior circ; <85 cm/s posterior circ; LR < 3	(a) Regional direct measurement of CBF (b) Vasospasm
Metabolism	Microdialysis	(a) LPR >25 or 40 (b) Glucose 1.5–2 mM (c) Glutamate <10μM (d) Glycerol <50μM	(a) Ischemia (b) Substrate availability (c) Excitotoxicity (d) Cellular injury

EVD extraventricular drain, *ICP* intracranial pressure, *CPP* cerebral perfusion pressure, *CBF* cerebral blood flow, *circ* circulation, *LR* Lindegaard ratio, *LPR* lactate-to-pyruvate ratio

DCI. However, CPP and its relationship to CBF is dependent upon intact cerebral autoregulation, as CBF is equal to CPP divided by cerebrovascular resistance (CVR). Autoregulation is the intrinsic ability of the arteriolar cerebral vasculature to maintain CBF over a range of CPP values. Loss of autoregulation, which is common after SAH, creates a direct linear correlation of CBF and CPP resulting in passive ICP changes that can be driven by MAP changes. Simply put, patients with impaired autoregulation may encounter difficulties optimizing CPP, as augmenting MAP may concurrently increase ICP and prevent the desired CPP goal from being achieved. Bedside monitoring of a patient's ICP waveform and ICP response to MAP augmentation are crucial ways to navigate these issues. Other modalities such as pressure reactivity index (PRx) and "optimal" CPP (CPPopt) are further quantitative methods of assessment, but these require specific expertise in data collection, management, and interpretation. Ultimately, the utilization of the trends (rather than isolated values) of both CPP and ICP and their relationship with other monitoring modalities trends are central to appropriate and comprehensive management strategies.

4.2 Brain Tissue Oxygenation

Adequate brain tissue oxygenation (PbtO2) maintains neuronal integrity, and suboptimal oxygenation values are used as markers to detect tissue at risk for secondary injury. Parenchymal regional PbtO2 catheters allow for continuous oxygen monitoring via a

Clark-type membrane electrode that measures oxygen content in adjacent white matter. Normal ranges using this technology are 20–35 mmHg, with current treatment guidelines suggesting a 20 mmHg threshold for treatment initiation [20]; cell death and infarction are associated with PbtO2 levels below 10 mm Hg. Poor PbtO2 may be due to either poor delivery of oxygen (either perfusion related or hypoxia) or increased consumption from hypermetabolic states (fevers, shivering, seizures). PbtO2 measurement can provide a useful tool to assist in evaluating whether the current CPP goals are truly providing adequate cerebral perfusion to the patient. If CPP is optimized and PbtO2 targets are not met, the practitioner may need to decide whether there still may be poor delivery, meaning the CPP needs to be further augmented, or if the low PbtO2 is being driven by increased consumption from a currently undetected hypermetabolic state such as seizures or infection, in which case the underlying cause should be identified.

Jugular bulb venous saturation (SjvO2), although not a direct parenchymal measurement of cerebral oxygenation like PbtO2, provides global cerebral utilization of oxygen. Utilizing a specialized catheter that is advanced cephalad through the internal jugular vein into the bulb, SjvO2 values can be obtained with normal values ranging between 55% and 75%. SjvO2 values below 55% signify increased oxygen extraction and are associated with tissue at risk for ischemia. Like ICP/CPP values, it is these tools' trends rather than isolated values that should drive clinician management decisions. Limitations of all regional probes are that their values are dependent on the location in which they are placed. Verification of suboptimal values by ensuring adequate placement (in white matter and not in infarcted tissue or ventricular system), in addition to providing oxygenation challenges by raising inspired fraction of oxygenation on the ventilator to assess if the probe is working, are both important components of troubleshooting prior to treatment decisions.

4.3 Cerebral Metabolism

Cerebral microdialysis allows for the analysis of intracerebral extracellular substrate from regional subcortical white matter to evaluate for signs of metabolic energy crises. This modality requires placement of a microdialysis catheter with a semipermeable membrane that allows molecules of a certain size (<20 kDa) to equilibrate down their concentration gradient via artificial CSF dialysate measured by a bedside analyzer. Substrates typically measure lactate, pyruvate, glucose, glutamate, and glycerol, and each molecule tests a different component important in energy metabolism. Lactate, pyruvate, and their ratio (LPR) are utilized as markers of anaerobic metabolism and energy failure, with LPR being more reliable than lactate alone. Normal lactate, pyruvate, and LPR thresholds are quoted as <4 mM, >120μM, and >25–40 respectively, with LPR thresholds being most consistently and reproducibly utilized for acute brain injury (rather than lactate or pyruvate alone). Glucose measures the amount of energy substrate available and delivered to the brain with normal levels quoted as 2 mM. Glutamate is an excitatory neurotransmitter associated with neuronal damage and is an early marker of vasospasm [21], with abnormal levels quoted to be $>10\mu M$. Finally, glycerol is a marker of cell membrane destruction and suggests that ischemia has progressed to cellular damage, with levels $>50\mu M$ being abnormal (Table 1).

Temporal relationships between microdialysis metabolite changes and DCI have shown that lactate and glutamate increase early in the course of DCI, followed by LPR and glycerol [22]. LPR and glutamate appear to have the highest sensitivity for detecting DCI and these changes can precede clinical DCI by approximately 12 h, providing a potential window for treatment prior to irreversible injury [23]. While utilization of microdialysis in isolation is limited, it represents a powerful tool to evaluate whether CPP goals are truly meeting a patient's cerebral energy needs when utilized in conjunction with the other components of MMM.

4.4 Cerebral Blood Flow

CBF in theory provides the most comprehensive information on energy delivery given its ability to account for autoregulation and perfusion. However, gold standard assessments are limited to neuroimaging modalities that cannot provide continuous monitoring. Transcranial Doppler (TCD) has become the most frequently utilized noninvasive means of approximating CBF by measuring blood flow velocity via high frequency energy. It is a staple of DCI monitoring in many centers and has the highest utility when examining the anterior circulation; it has a high degree of correlation with angiographic vasospasm when velocities reach >200 cm/s [24]. High TCD velocities are either due to decreased vessel diameter as seen in vasospasm or increased blood volume in cases of hyperemia, but these can be distinguished when using the Lindegaard Ratio (LR). This ratio is taken from the middle cerebral artery and external carotid artery velocities, with values >3 specifying that high velocities are due to vasospasm rather than hyperemia. However, similar to angiographic vasospasm, TCD elevations cannot be used in isolation to diagnose vasospasm, as not all TCD elevations lead to DCI and conversely, DCI can occur in patients with normal TCD velocities [25]. Typically, a patient's exam (or multimodality monitoring values if a patient is comatose) is used in conjunction with TCDs to evaluate whether the patient truly has evidence of DCI.

Invasive parenchymal regional thermal diffusion probes allow for continuous CBF monitoring. These probes measure heat dissipation in the subcortical white matter to extrapolate local CBF in units of cc/100 g/min. CBF values below 20 cc/100 g/min are associated with DCI and vasospasm [26]. However, just as in the other modalities, treating these values in isolation has not been well-established. In concept, continuous CBF would provide the most accurate reflection of a target of interest to prevent DCI, but there are limitations in its use given its regional nature, thus necessitating the use of CBF trends in conjunction with other modalities to drive treatment strategies.

4.5 Electroencephalography

Continuous surface electroencephalography (cEEG) is a cornerstone of multimodality monitoring given its noninvasive ability to provide global, temporal, and spatial real time changes of brain activity. Its primary use has been to detect nonconvulsive seizures in high-grade SAH, as this can occur in up to 12% of SAH patients [27]. However, cEEG has become increasingly utilized as a means of detecting cerebral ischemia via changes in background frequency and amplitude. Quantitative analysis of EEG (qEEG) via fast Fourier transform techniques, spectral analysis and alpha–delta ratio (ADR) have been shown to be associated with DCI [28]. Its use is recommended by guidelines for the detection of DCI in high grade SAH patients. However, some limitations for its widespread use include its requirement of center-specific expertise in interpretation, a relatively low signal-to-noise ratio, and the frequent presence of background artifact originating from both the patient and external stimuli elsewhere in the ICU.

Intracortical depth electrodes may provide a means to minimize these aforementioned limitations in order to detect seizures and cortical spreading depression not detected on scalp EEG [29]. Studies typically utilize the 8-contact Spencer depth electrode which ideally has 1 contact in the skull, 2–3 in the cortical gray matter, and the rest in the underlying white matter. This may provide a more sensitive means to detect vasospasm while providing a target for optimal CPP, although this avenue requires further research.

5   Multimodality Monitoring Strategies

The volume of data involved with MMM may be overwhelming, and therefore requires a streamlined approach for interpretation (Fig. 2). Although there are individual critical value thresholds proposed for each device (Table 1), focus should be paid to relationships between various modalities (for example, the effects of varying cerebral perfusion on metabolic status) and temporal trends representing an improving or worsening pattern. Appropriate interpretation of the data may then allow for management decisions that can prevent secondary brain injury (*see* example flow diagrams in Fig. 3). Since blood pressure management (including augmentation and tightening of parameters) is one of the foremost

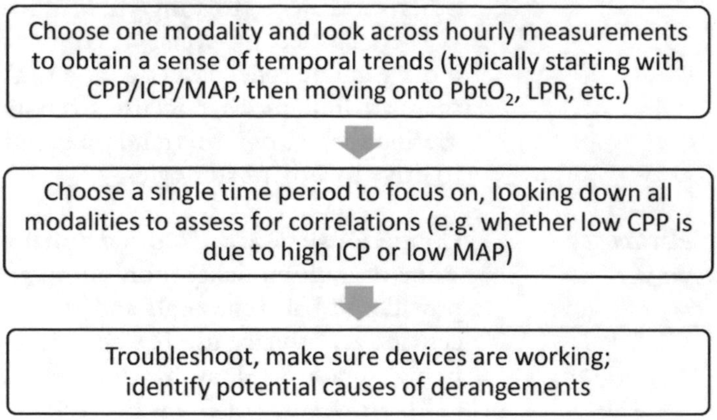

Choose one modality and look across hourly measurements to obtain a sense of temporal trends (typically starting with CPP/ICP/MAP, then moving onto PbtO$_2$, LPR, etc.)

Choose a single time period to focus on, looking down all modalities to assess for correlations (e.g. whether low CPP is due to high ICP or low MAP)

Troubleshoot, make sure devices are working; identify potential causes of derangements

Fig. 2 Approach to interpretation of intracranial monitoring data. *CPP* cerebral perfusion pressure, *ICP* intracranial pressure, *MAP* mean arterial pressure, *PbtO$_2$* brain oxygen tension, *LPR* lactate-pyruvate ratio

interventions available during this period, intracranial monitoring results can especially help navigate the fine line between cerebral hypoperfusion and hyperemia.

A major caveat to data interpretation, however, is that results are highly dependent on probe location. For that reason, choosing the proper placement beforehand is crucial; ideally, this means placing the probes in an area representing tissue at risk, either in key anterior circulation territories that are prone to vasospasm based on aneurysm location, or in penumbral areas around an infarct or hemorrhage that may be at risk of further cellular injury (but *not* directly inside the areas of dead tissue) [20]. Furthermore, because of probe "drift" over time, as well as a loss of calibration, these devices may eventually lose their effectiveness, rendering measurements uninterpretable. For these reasons, each abnormal result should be questioned as to whether it should be considered reliable.

In many cases these monitoring modalities can and should be supplemented by radiographic studies, particularly when there is concern for ongoing ischemia. Perfusion status can be confirmed with either CT- or MRI-based perfusion studies, while CT-, MR-, or conventional angiography can confirm the presence of arterial narrowing and other structural abnormalities; the latter also has the benefit of being able to provide treatment for vasospasm with intra-arterial vasodilator infusions or angioplasty. The information obtained from these tests—namely, finding a perfusion deficit in conjunction with radiographic vasospasm—can help further guide medical management through a critical period [30]. When all is said and done, such tests are also often used in conjunction with the neurological exam (when possible) to provide reassurance that a patient's vasospasm window has likely concluded, at which time monitoring strategies can be slowly phased out.

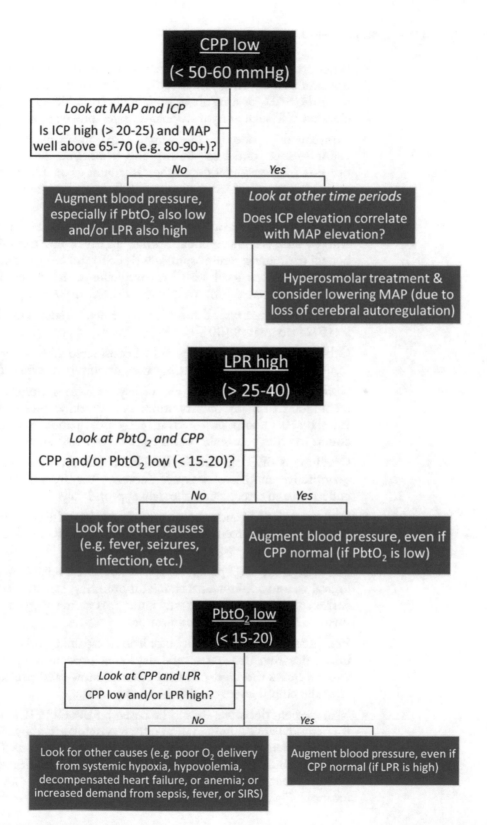

Fig. 3 Examples of management decisions based on intracranial monitoring data, after ensuring that devices are working correctly and are in a good location. *ICP* intracranial pressure, *MAP* mean arterial pressure, *CPP* cerebral perfusion pressure, *PbtO₂* brain tissue oxygen tension, *LPR* lactate-pyruvate ratio, *SIRS* systemic inflammatory response syndrome

- When an SAH patient has impaired consciousness on arrival, it is not immediately clear whether this is due to the initial injury from the SAH, or a result of potentially treatable etiologies of elevated ICP such as hydrocephalus or cerebral edema.

- Consequently, one of the initial steps in managing such patients—after initial respiratory and hemodynamic resuscitation—is the consideration of the placement of an EVD (especially if there is hydrocephalus or IVH) or a parenchymal ICP monitor.

- The information provided can then guide further medical and/or surgical management, while an EVD has the added benefit of draining cerebrospinal fluid (CSF) and intraventricular blood, which can itself treat hydrocephalus, assist in the management of elevated ICP, and allow for the optimization of CPP.

- At many centers, a typical cutoff for placement of such a monitor is a GCS less than 8 [30].

- Other monitoring modalities can be considered after the initial resuscitation period, first starting with noninvasive techniques.

- Serial TCD measurements are widely used as a noninvasive technique that many centers utilize as a screening tool to look for signs of vasospasm, starting early in a patient's hospital course in order to obtain baseline pre-vasospasm values.

- Continuous EEG can also predict DCI before it occurs, but the quantitative analysis techniques necessary require expertise in collecting and interpreting the data appropriately.

- Patients with SAH and other forms of acute brain injury who are comatose or stuporous should have prolonged EEG monitoring regardless, in order to rule out NCSE.

- For high-grade SAH patients who do not have a reliable neurological exam to follow, intracranial monitoring can be a useful surrogate that may provide vital clues portending the development of DCI and other complications.

- At centers with multimodality monitoring capability, this is often initiated within the first several days after presentation, as the patient enters the high-risk vasospasm window (and preferably after the culprit aneurysm has been secured).

- Management decisions should be based on the CPP, ICP. need for hemodynamic augmentation or hyperosmolar therapy. Additionally when able intracranial monitoring data such as LPR, Glucose PbtO2, etc. should be taken in consideration after ensuring that devices are working correctly and are in a good location (Fig. 3).

6   Conclusion

There remains a high burden of morbidity and mortality in patients with SAH despite advances in surgical and medical management, much of which is due to secondary neurological injury. Preventing the onset of DCI and other such neurological complications requires a systematic approach, and MMM represents an integrated monitoring strategy based on the acquisition of cerebral physiologic data. Studies attempting to validate its use on a broader scale are ongoing; in the meantime, however, its use is restricted to the most specialized, comprehensive centers.

References

1. Suwatcharangkoon S et al (2016) Loss of consciousness at onset of subarachnoid hemorrhage as an important marker of early brain injury. JAMA Neurol 73:28–35

2. Hunt WE, Hess RM (1968) Surgical risk as related to time of intervention in the repair of intracranial aneurysms. J Neurosurg 28:14–20

3. Kassell NF et al (1990) The International Cooperative Study on the timing of aneurysm surgery. Part 1: overall management results. J Neurosurg 73:18–36

4. Claassen J et al (2002) Global cerebral edema after subarachnoid hemorrhage: frequency, predictors, and impact on outcome. Stroke 33:1225–1232

5. Wartenberg KE et al (2006) Impact of medical complications on outcome after subarachnoid hemorrhage. Crit Care Med 34:617–623. quiz 624

6. Claassen J et al (2001) Effect of cisternal and ventricular blood on risk of delayed cerebral ischemia after subarachnoid hemorrhage: the Fisher Scale revisited. Stroke 32:2012–2020

7. Frontera JA et al (2006) Prediction of symptomatic vasospasm after subarachnoid hemorrhage: the modified fisher scale. Neurosurgery 59:21–27

8. Vergouwen MDI et al (2010) Definition of delayed cerebral ischemia after aneurysmal subarachnoid hemorrhage as an outcome event in clinical trials and observational studies: proposal of a multidisciplinary research group. Stroke 41:2391–2395

9. Al-Mufti F et al (2016) Ultra-early angiographic vasospasm associated with delayed cerebral ischemia and infarction following aneurysmal subarachnoid hemorrhage. J Neurosurg 126(5):1545–1551. https://doi.org/10.3171/2016.2.JNS151939

10. Macdonald RL, Weir BK (1991) A review of hemoglobin and the pathogenesis of cerebral vasospasm. Stroke 22:971–982

11. Crowley RW et al (2011) Angiographic vasospasm is strongly correlated with cerebral infarction after subarachnoid hemorrhage. Stroke 42:919–923

12. Budohoski KP et al (2014) The pathophysiology and treatment of delayed cerebral ischaemia following subarachnoid haemorrhage. J Neurol Neurosurg Psychiatry 85:1343–1353

13. Woitzik J et al (2012) Delayed cerebral ischemia and spreading depolarization in absence of angiographic vasospasm after subarachnoid hemorrhage. J Cereb Blood Flow Metab 32:203–212

14. Zoerle T et al (2015) Intracranial pressure after subarachnoid hemorrhage. Crit Care Med 43:168–176

15. Lescot T et al (2011) In vivo accuracy of two intraparenchymal intracranial pressure monitors. Intensive Care Med 37:875–879

16. Helbok R, Olson DM, Le Roux PD, Vespa P, Participants in the International Multidisciplinary Consensus Conference on Multimodality Monitoring (2014) Intracranial pressure and cerebral perfusion pressure monitoring in non-TBI patients: special considerations. Neurocrit Care 21(Suppl 2):S85–S94

17. Menon D, Weir B, Overton T (1981) Ventricular size and cerebral blood flow following subarachnoid hemorrhage. J Comput Assist Tomogr 5:328–333

18. Hänggi D et al (2015) NEWTON: nimodipine microparticles to enhance recovery while reducing toxicity after subarachnoid hemorrhage. Neurocrit Care 23:274–284

19. Hanley DF et al (2017) Thrombolytic removal of intraventricular haemorrhage in treatment of

severe stroke: results of the randomised, multi-centre, multiregion, placebo-controlled CLEAR III trial. Lancet Lond Engl 389:603–611

20. Le Roux P et al (2014) The International Multidisciplinary Consensus Conference on Multimodality Monitoring in Neurocritical Care: a list of recommendations and additional conclusions: a statement for healthcare professionals from the Neurocritical Care Society and the European Society of Intensive Care Medicine. Neurocrit Care 21(Suppl 2):S282–S296

21. Unterberg AW, Sakowitz OW, Sarrafzadeh AS, Benndorf G, Lanksch WR (2001) Role of bedside microdialysis in the diagnosis of cerebral vasospasm following aneurysmal subarachnoid hemorrhage. J Neurosurg 94:740–749

22. Helbok R et al (2011) Intracerebral monitoring of silent infarcts after subarachnoid hemorrhage. Neurocrit Care 14:162–167

23. Skjøth-Rasmussen J, Schulz M, Kristensen SR, Bjerre P (2004) Delayed neurological deficits detected by an ischemic pattern in the extracellular cerebral metabolites in patients with aneurysmal subarachnoid hemorrhage. J Neurosurg 100:8–15

24. Vora YY, Suarez-Almazor M, Steinke DE, Martin ML, Findlay JM (1999) Role of transcranial Doppler monitoring in the diagnosis of cerebral vasospasm after subarachnoid hemorrhage. Neurosurgery 44:1237–1247. discussion 1247–1248

25. Carrera E et al (2009) Transcranial Doppler for predicting delayed cerebral ischemia after subarachnoid hemorrhage. Neurosurgery 65:316–323.; discussion 323–324

26. Vajkoczy P, Horn P, Thome C, Munch E, Schmiedek P (2003) Regional cerebral blood flow monitoring in the diagnosis of delayed ischemia following aneurysmal subarachnoid hemorrhage. J Neurosurg 98:1227–1234

27. Claassen J et al (2006) Prognostic significance of continuous EEG monitoring in patients with poor-grade subarachnoid hemorrhage. Neurocrit Care 4:103–112

28. Vespa PM et al (1997) Early detection of vasospasm after acute subarachnoid hemorrhage using continuous EEG ICU monitoring. Electroencephalogr Clin Neurophysiol 103:607–615

29. Claassen J et al (2013) Nonconvulsive seizures after subarachnoid hemorrhage: Multimodal detection and outcomes. Ann Neurol 74:53–64

30. Connolly ES et al (2012) Guidelines for the management of aneurysmal subarachnoid hemorrhage: a guideline for healthcare professionals from the American Heart Association/american Stroke Association. *Stroke*. J Cereb Circ 43:1711–1737

Part III

Cerebral Aneurysms: Special Considerations

Chapter 9

Intracranial Blister Aneurysms

Abdulrahman Y. Alturki, Ajith Thomas, and Christopher S. Ogilvy

Abstract

Intracranial blister aneurysms are small (less than 3 mm), thin-walled, broad-based hemispherical expansions of the parent intracranial artery at a nonbranching site, that lack an identifiable neck. They represent <2% of all ruptured cerebral aneurysms. The clinical presentation, natural history, location, therapeutic challenges, and treatment strategies are considerably different from those of other intracranial aneurysms, and thus special attention is warranted toward this subgroup.

This chapter will review the pathophysiology, clinical presentations, and diagnostic evaluation of intracranial blister aneurysms. Finally, various management strategies for these aneurysms will be reviewed.

Key words Blister intracranial aneurysms, Blister-like aneurysms, Blood blister-like aneurysms, Clinical presentation, Clipping, Endovascular treatment

1 Introduction

Although a definitive diagnosis can only be made from histologic studies, blood blister-like aneurysms or blister aneurysms (BAs) are small (less than 3 mm), thin-walled, broad-based hemispherical expansions of the parent artery at a nonbranching site, that lack an identifiable neck. These aneurysms frequently arise at the proximal part of the basal arteries of the anterior circulation, commonly the supraclinoid internal carotid artery (ICA) [1]. However, the location of BAs is not restricted to the ICA alone; additional reported locations include the basilar and middle cerebral arteries [2, 3].

Although BAs comprise <2% of all ruptured intracranial aneurysms, they are an important cause of subarachnoid hemorrhage (SAH) as they are known to be associated with much higher rebleeding rates than congenital aneurysms [4]. BAs often escape initial angiographic detection and thus may be an important occult cause of angiographically-negative SAH [5–7].

Frequently, BAs are associated with hypertension, arteriosclerosis and ICA dissection. It has been reported that BAs occur

Fawaz Al-Mufti and Krishna Amuluru (eds.), *Cerebrovascular Disorders*, Neuromethods, vol. 170,
https://doi.org/10.1007/978-1-0716-1530-0_9, © Springer Science+Business Media, LLC, part of Springer Nature 2021

predominantly on the right side and in females. Due to the scarcity of such aneurysms, the natural history of BAs has not been well clarified [1, 8, 9].

In relation to the ICA, BAs have been referred to as follows [10].

- Dorsal.
- Distal-medial.
- Superior.
- Anterior wall.

Morphologically, BAs have been subcategorized as follows [11].

- Classic; part of the artery wall, without an identifiable sac.
- Berry-like; part of wall with a sac, in which the neck is not longer than the diameter of the artery.
- Longitudinal; in which the neck is longer than the diameter of the artery.
- Circumferential; encompassing a majority of the artery wall.

2   Pathophysiology

The pathogenesis of BAs remains unclear. Some reports have suggested that BAs are consequences of parent vessel dissection due to their localization at nonbranching, high shear stress segments of the ICA [1]. More common theory suggests that these aneurysms are associated with arteriosclerosis of the neighboring arterial wall [8, 12]. With hemodynamic stress, the border between the stiff arteriosclerotic and elastic normal arterial wall may be the portion that is most vulnerable to vessel injury and ultimate aneurysmal formation.

The tortuosity of the supraclinoid ICA may be of importance in explaining the common location of BAs, which is slightly distal to the greatest area of flow alteration [13]. However, this theory has been challenged by reported cases of BAs in different locations [4, 14].

Histologic analysis of BAs have suggested alternate microscopic findings compared to dissecting and/or saccular aneurysms. One report of BAs showed abrupt termination of the internal elastic lamina at the border between the normal and sclerotic carotid wall, as well as disappearance of the media. The authors also show that the dome of a BA is composed of fibrinous tissue and adventitia, with an absence of the usual collagenous layer of a saccular aneurysm. While inflammatory components are well known to be involved in saccular aneurysms, dissection and infiltration of the BA wall by inflammatory cells was absent [15].

Confirming some of aforementioned findings, results of an autopsy indicated that BAs are focal wall defects covered by a thin layer of fibrous tissue and adventitia, which lack the usual collagenous layer [16]. While other reports have postulated an infectious origin of BAs, these hypotheses have not confirmed histologically [6].

3 Clinical Presentations and Diagnostic Evaluation

The diagnosis and evaluation of BAs is crucial prior to intervention because of the difference in treatment strategies compared to other types of intracranial aneurysms.

Strict diagnostic evaluation is critical due to the fact that BAs typically present with SAH, and thus necessitate treatment.

Three-dimensional (3D) computed tomography angiography (CTA) may detect BAs; magnetic resonance (MR) studies of the aneurysm wall may also help to determine the presence of an associated dissection as the underlying etiology of the BA.

Due to the lack of a classic saccular-like bulge on noninvasive imaging, BAs are easily missed on MR and CTA. Rotational digital subtraction angiography (DSA) may be helpful in particular instances, mainly because of the small size and anterior location of BAs.

All vascular territories must be examined in which 3D imaging may be of further help detecting small arterial dilatations. In many cases, diagnosis will remain obscure due to neuroradiological features, which may not allow distinguishing BAs from normal variants. Thus in cases in which a readily apparent source of SAH is not identified, a high index of suspicion of BAs must be exercised. An atypical presentation of an aneurysm at a location other than an arterial branch site may be the only indicator of diagnosis.

Operators should be aware that SAH due to blister aneurysms may be easily mistaken for the more common perimesencephalic, nonaneurysmal SAH. Differentiation of the true etiology of SAH is imperative in these case as potentially devastating rebleeding will more commonly occur in the setting of BAs, while a more benign clinical presentation will ensue in the setting of perimesencephalic, nonaneurysmal SAH.

BAs reportedly exhibit rapid growth and changes in shape. Therefore, in patients with SAH and no evidence of perimesencephalic nonaneurysmal SAH and an initially negative DSA, a BA cannot be excluded until careful review of conventional, 3D CT, and follow-up angiography [1].

4 Management

Because of the very fragile walls and poorly defined necks of BAs, interventions (surgical or endovascular) are hazardous with a high rate of intraoperative or postoperative rupture. BAs require special consideration. Despite decades of advances in microsurgical and endovascular techniques, BAs remain formidable lesions without an established and reliable management strategy.

Once angiographic studies reveal a lesion that may be a BA, confirmation of collateral flow from the posterior or anterior and contralateral circulation is recommended. This strategy will help to determine whether the lesion can be trapped (via surgical or endovascular methods) if the BA cannot be treated by other modalities. If trapping is considered, a balloon occlusion test should be strongly considered prior to any intervention. (The authors understand that balloon test occlusion testing in SAH patients may be difficult and risky to preform due to the heparinization requirements, potentially uncooperative/comatose patients, vasospasm, etc.). In addition, the external carotid artery should be examined in case it is necessary to perform bypass surgery.

A conservative approach without invasive treatment of BAs results in growth and high rates of rebleeding, typically within the first week after the initial SAH [9].

The prognosis of patients presenting with SAH due to BAs is related to the clinical status on admission, evaluated by the Hunt and Hess clinical classification. However, because of the high incidence of intra-procedural or postoperative bleeding, the prognosis is markedly worse compared to those patients with saccular-type aneurysms.

4.1 Surgical Strategies

Although several surgical strategies are available to treat BAs, a general consensus does not exist on the safest treatment modality. The challenge for any surgeon is to determine the best surgical method (clipping, clipping onto wrapping, or trapping).

BAs have a high risk of intraoperative rupture due to the potential for large lacerations of the ICA during clipping. The intraoperative rupture rate is six times that of saccular aneurysms (41% vs. 7%) [17].

Blood loss and cerebral ischemia are major concerns and must be addressed during any preoperative planning. Given these risks, an emergency plan with access to the extracranial carotid artery for proximal control is imperative. The opening of the sylvian fissure for distal control should be considered before accessing the carotid and chiasmatic cisterns via craniotomy. The point of maximum dilatation of BA should be carefully studied on vascular imaging, and the relationship to the ICA must be determined:

Table 1
Recent reports of surgical management of blister aneurysms (≥10 patients)

Authors	Year	Number of patients	SAH %	Surgical method	Outcome
Owen et al. [18]	2016	17	88	12 clip, 1 clip/wrap, 3 bypass/ trap, 1 trap	10 good, 3 MD, 1 SD, 3 death
Bojanowski et al. [11]	2015	10	100	9 clip, 1 clip/wrap	10 good, no death
Kazumata et al. [22]	2014	20	100	6 clip, 14 bypass/trap	18 good, 2 SD
Kalani et al. [23]	2013	17	70	11 clip, 6 clip/wrap	14 good, 3 SD
Horiuchi et al. [24]	2011	26	29	23 clip, 2 clip/wrap, 1 bypass/ trap	24 good, 1 SD, 1 death
Lee et al. [25]	2009	18	100	16 clip/wrap, 1 trap, 1 suture repair	13 good, 1 MD, 2 SD, 2 death
Meling et al. [26]	2008	11	100	5 clip, 6 trap	5 good, 1 MD, 5 death
Sim et al. [10]	2006	19	100	6 clip, 3 clip/wrap, 1 trap	8 good, 1 MD, 1 death
Ogawa et al. [9]	2000	48	100	29 clip, 10 clip/wrap, 4 wrap, 2 bypass/trap, 3 trap	35 good, 5 SD, 8 death

SAH subarachnoid hemorrhage, *MD* moderate disability, *SD* severe disability

- Laterally oriented BAs tend to rupture during early dissection of the basal cisterns.

- Superiorly pointing lesions may make frontal retraction hazardous, sometimes necessitating subpial dissection to reach the medial portion of the neck safely and establish complete control of the ICA [18].

- Medially directed domes may be less prone to early intraoperative rupture, projecting away from the surgeon makes the limited view of the diseased segment and the neck a real challenge.

Direct clip application is attempted only after completely trapping the involved arterial segment with distal and proximal temporary clips.

Applying clips onto wrapping of various materials has been proposed as a viable method for the treatment of BAs [1, 8, 9]. However, reports on the surgical management of fusiform aneurysms challenge this notion. Autologous (muscle fascia, periosteum) and synthetic (cotton, muslin, Gore-Tex) materials used to boost fusiform aneurysms have been shown to be ineffective over time or tend to cause granulomatous reactions of the surrounding brain [19, 20]. Despite these arguments, this method of clipping may be feasible in cases with an advantageous morphology of a BA [11].

Other authors have proposed direct clipping or "tailored" clipping techniques according to BA morphology [11]. A clip's blade should be applied parallel to the parent artery and should catch the "normal" arterial wall beyond the lesion, just enough to avoid laceration of the base of the BA [8, 9]. Confirming the stability of clips is essential, done with permissive blood pressure elevation before closure. These principles apply to both direct clipping and clipping onto wrapping. If this method is not used, clip distortion may occur toward the end of the operation and rerupture can result.

Postoperative rupture of BAs is a major concern and can lead to death in more than 90% of cases [1, 8, 9]. Postoperative bleeding may occur as a result of clip torsion or slippage leading to aneurysm tear [10]. While surgical clipping is usually a very viable long-term solution, in one series, 8.3% of patients showed aneurysm regrowth after surgery [21].

Other potential surgical treatments may include the following.

- Wrapping the full circumference of the ICA using an encircling clip. Sundt clips can be used for wrapping because they allow encircling of the vessel with Dacron (Invista, Inc., Wichita, KS).

- ICA trapping with or without bypass. A low threshold for bypass is suggested [18], based on the dramatically increased rate of intraoperative aneurysm rupture and the difficulties of performing a bypass under stress or as a rescue procedure later.

- Direct suturing.

  Table 1 summarizes some of the surgical studies.

4.2 Endovascular Strategies

Given the challenges traditionally associated with surgical management of BAs, a recent interest in endovascular techniques has emerged and continues to evolve at a fast pace. Similar to surgical management, a consensus does not exist on the safest or most efficacious endovascular treatment method. Endovascular strategies include both deconstructive techniques (parent artery occlusion) as well as reconstructive techniques (coiling, stent-assisted coiling, balloon-assisted coiling, overlapped stent placement, flow diversion).

In a recent meta-analysis of endovascular treatments for BAs, deconstructive techniques had higher rates of complete occlusion on immediate post-treatment angiography compared to reconstructive techniques (77.3% vs. 33.0%,) but had a higher risk for perioperative stroke (29.1% vs. 5.0%) [27]. Deconstructive techniques should, at best, be viewed as a poor alternative and a last-resort treatment. Justifications for deconstructive treatment may include

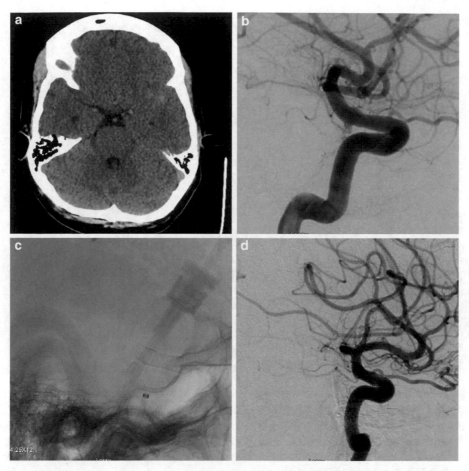

Fig. 1 (**a**) CT scan of the head in a patient with subarachnoid hemorrhage, with predominance of blood products in the left MCA cistern. (**b**) Digital subtraction angiogram (DSA) in frontal oblique projection showing blister aneurysm involving the dorsal communicating segment of the left internal carotid artery. (**c**) Unsubtracted image after flow diversion with pipeline embolization device across the neck of the aneurysm. (**d**) 8-month follow-up DSA showing vessel reconstruction with exclusion of blister aneurysm

repeated high-grade SAH, younger age of patients, concerns for vasospasm and secondary inadequate collateral circulation. The treating physician must take into consideration the increase in hemodynamic stress on additional cerebral vasculature, which may cause new aneurysms to form (up to 20% in patients after carotid sacrifice) [28].

Conventional endovascular approaches including primary coil embolization, stent-assisted and balloon-assisted coiling, although each of these methods have been associated with procedural morbidity and mortality, due to intraprocedural rupture and postoperative hemorrhage [29, 30]. Other authors have reported favorable results with layered-stent constructs with or without adjunctive coil use [31]. Despite the wide spectrum of conventional endovascular treatment strategies, a recent systematic review indicated that with

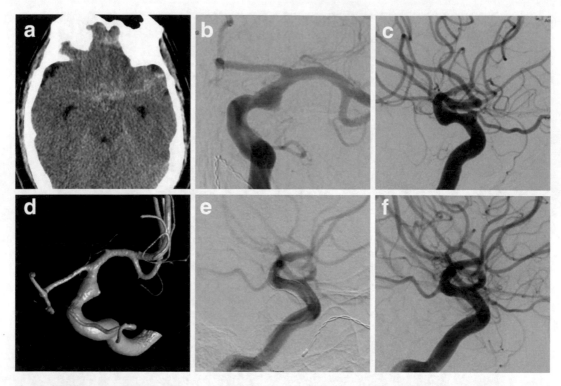

Fig. 2 (**a**) CT scan of the head in a patient with subarachnoid hemorrhage, with predominance of blood products in the left suprasellar cistern. Digital subtraction angiogram (DSA) in (**b**) frontal oblique and (**c**) lateral projections and (**d**) 3-D reconstruction showing blister aneurysm involving the dorsal supraclinoid segment of the left internal carotid artery. Post interventional angiogram in (**e**) early arterial and (**f**) late arterial phases after flow diversion with pipeline embolization device across the neck of the aneurysm showing vessel reconstruction with exclusion of blister aneurysm

endovascular methods, only 76.5% of patients achieved complete aneurysm occlusion, 15.9% of patients showed treatment-related complications, and 18.4% of patients needed additional intervention [21].

Recently, flow diverter (FD) stents have been developed to treat wide neck, complex, fusiform, and very small intracranial aneurysms, which have been traditionally difficult to treat by typical endovascular methods. FD stents have lower porosity and higher metal coverage ratios than conventional self-expandable intracranial stents (6–8% vs. 30–35% respectively). Because of these structural features, FD stents have much higher flow-diversion capacities than conventional self-expandable intracranial stents and a high capacity to reconstruct arterial blood flow in the parent artery, which diminishes flow and gradually induces thrombosis in the sac of the aneurysm. Histological studies following FD deployment revealed endothelialization of the device and reconstruction of the entire length of the parent vessel covered by the implant and across the aneurysmal neck. When compared to other reconstructive

Table 2
Recent reports on endovascular treatment of blister aneurysms (≥10 patients)

Authors	Year	Number of patients	SAH %	Endovascular method	Outcome
Linfante et al. [2]	2016	10	100	Pipeline Embolic Device	9 good, 1 death
Fang et al. [3]	2016	13	100	Willis covered stent	13 good
Aydin et al. [33]	2016	11	100	Flow diverter SILK stents	9 good, 1 MD, 1 death
Fang et al. [31]	2014	15	88	Stent/coil and overlapping stents	14 good, 1 MD
Yoon et al. [34]	2014	11	100	Pipeline Embolic Device	9 good, 1 MD, 1 death
Gonzalez et al. [35]	2014	12	100	Stent/coil (5), stent only (4), coiling and clipping (1), clip (1), conservative (1)	10 good, 1 MD, 1 death
Lim et al. [32]	2013	34	100	Overlapping stents with or without coiling (28), stent/coil (6)	25 good, 5 SD, 4 death
Meckel et al. [37]	2011	13	100	Stent/coil (11), parent artery occlusion (1), overlapping stents (1)	12 good, 1 death

SAH subarachnoid hemorrhage, MD moderate disability, SD severe disability

procedures in the treatment of BAs, FD stents had higher rates of mid to long-term complete occlusion (90.8% vs. 67.9%, $P = 0.03$) and a lower rate of retreatment (6.6% vs. 27.1%, $P = 0.0002$) [27]. Interestingly, in the same systematic review, there was no difference in perioperative morbidity, mortality, good clinical outcome rates or perioperative intracranial hemorrhage rates between the two groups [27]. Options for FD range from multiple overlapping conventional stents [32, 31], to different types of FD devices like the Willis covered stent (WCS) [3], SILK stent [33], and the Pipeline Embolic Device (PED) (Figs. 1 and 2) [2, 34].

When comparing PED, SILK, and WCS, the relative advantages and disadvantages of each device can be compared as follows [2, 3, 33, 34].

- PED has better flexibility, so navigation is easier.

- The WCS has a higher risk of side-branch occlusion, however PED also possesses similar risk due to high metal surface area coverage (approximately 30–35%).

- 15.4% of patients treated with WCS demonstrated a parent artery stenosis, while 9.1% of patients treated with the SILK device developed a minor stroke due to parent artery thrombosis. No patient developed parent artery stenosis due to PED use.
- All FD stents show a good BA occlusion rate at follow-up.
- All three FD devices reported satisfactory clinical outcomes, ≥90% mRS score 0–2.

Patients treated with FD devices for ruptured BAs are at high risk for hemorrhagic complications for several reasons.

- BAs lack all of the layers of the vessel wall. With disruption of the internal elastic lamina and media, the aneurysm is contained only by a thin wall composed of thrombosed blood products, fragmented adventitia, and fibrous tissue.

- Dual antiplatelet therapy is mandatory after FD stent deployment to prevent thrombotic complications. This has the potential to increase the risk for intracranial hemorrhage as patients may require placement or replacement of an external ventricular drain (EVD), or subsequent surgery for a ventriculoperitoneal shunt, gastrostomy, tracheostomy and/or decompressive craniectomy [3, 34, 35]..

- Vessel remodeling after FD stent deployment occurs in a delayed manner over the course of weeks or months, during which time the BA remains partially unsecured. During this interim period, the aneurysm may be at risk for rerupture [36].

- Medical treatment for vasospasm (i.e., hyperdynamic therapy) may place an additional hemodynamic stress on an unsecured aneurysm.

Despite these risks, some authors have postulated that the hemodynamic effects of FD devices may be enough to prevent a blister aneurysm from bleeding [33]. An increase in coverage by the implantation of overlapping stents may be an appealing solution but this technique is accompanied by additional risks, potentially resulting in occlusion or embolization of covered branch vessels. Monocular blindness and in-stent stenosis have been reported [33, 34]. A suggested approach to BAs will be discussed in the next section. Table 2 summarizes recent studies on the endovascular treatment of BAs.

4.3 Surgery Versus Endovascular Approach to Blister Aneurysms

Although no recommendations or randomized controlled trials exist to guide clinicians treating patients with BAs, management is largely dictated on an individual case basis. Many authors have attempted systematic reviews and meta-analyses on this rare disease without clear definition and thus existing guidelines do not exist on optimal management. Publications from earlier years are likely to be based on surgical management or coil embolization, while more recent studies will focus on new endovascular treatment methods such as flow diversion. Due to these heterogeneities in treatment strategy, there is a major concern about the methodologies and results [21, 27, 35, 38].

With these shortcomings in mind, some reviews have showed that endovascular treatment of BAs offers a lower morbidity and mortality compared with surgical approaches [35]. A favorable outcome (mRS 0–2) was reported in 67.4% and 78.9% of patients treated with surgery and with endovascular therapy, respectively ($P = 0.034$) [38]. Another review showed that neither surgical nor endovascular methods had an impact on clinical outcome, aneurysm regrowth, remote bleeding, or complication rate [21].

4.3.1 Suggested Management Approach to Blister Aneurysms

- Each aneurysm and patient must be considered individually when deciding upon a particular treatment strategy in a multi-disciplinary fashion, which should be tailored to the unique characteristics of each case and the comfort level of the treating physician with every treatment modality.

- Surgical or endovascular trapping should be viewed as a last resort.

- Due to potential instability of these lesions after treatment, follow up 3D-DSA is suggested 1 week after treatment, regardless of treatment method.

- As the majority of BAs are located in the supraclinoid internal carotid artery (>90%), FD stents represent a reasonable first option, especially in circumferential aneurysms. In the USA, PED was used with reasonable reported rates of success and complications [2, 34, 36]. SILK and WCS have been reported elsewhere [3, 33].

- In addition to the supraclinoid ICA, endovascular obliteration of BAs using PED may be attempted in other locations, especially in large arteries (vertibrobasilar complex and first segment of the anterior and middle cerebral arteries).

- Antiplatelet management is crucial; aspirin, ideally in combination with another short acting antiplatelet medication (biological half-life of a few hours). We prefer the use of tica-grelor as it has a binding site different from other ADP antagonists, leading to reversible blockage with platelet transfusion. This medication also avoid delays associated with platelet function testing.

- We recommend doing all necessary axillary procedures (central lines, EVD, etc.) before the initiation of the antiplatelet regiment; obtain verifying imaging after (x-rays and CT head).

- If it is necessary to do any invasive intracranial procedure (EVD removal, VP shunt insertion, etc.) hold aspirin and hold ticagrelor for 8–12 h. After performing the procedure, obtain a control CT scan then restart ticagrelor in 8–12 h.

- BAs in distal locations are preferentially managed by surgery, with a prepared bypass plan and early proximal control as primary objectives.

- In the rare occasions involving surgical management in patients with circle of Willis BAs (e.g., due to expanding hematoma), we suggest the following.
  - Expose the cervical ICA.

  - As a contingency plan, prepare the superficial temporal artery for potential bypass prior to the craniotomy if there is a lower threshold for bypass (determine the type of bypass depending on the BA location and collateral status).

  - All surgical equipment must be prepared prior to opening the dura. Such equipment includes wrapping material, microvascular suturing materials, bypass set and all types of clips (including the encircling clip graft [Sundt clip]).

  - Trap the BA segment prior to attempting any definitive obliteration maneuver.

  - Once the BA is obliterated, raise the blood pressure for 30-45 minutes and pay close attention to the parent artery–clip construct for evidence of hemorrhage.

  - Do not break the sterility of the operating room until the patient is extubated and examined.

  - Obtain a CT and CTA scan 24 h post-procedure.

5   Conclusion

Blister aneurysms are challenging lesions to treat whether by surgical or endovascular techniques. While multiple options exist, considerable uncertainty exists regarding the most efficacious management of blister aneurysms.

The choice of treatment method must be based on the initial clinical presentation and the radiologic features of the lesion. Considering the major limitations of recent literature regarding the treatment of BAs, endovascular treatment seems to have lower morbidity and mortality and provides a better outcome compared with surgical approaches. When an operator considers the reconstructive endovascular treatment, flow diversion appears to be a reasonable choice despite the need for antiplatelet treatment. Long-term follow-up, experience, and outcomes in larger studies are required to better define the role of different treatment modalities in the management of these difficult lesions.

References

1. Abe M, Tabuchi K, Yokoyama H, Uchino A (1998) Blood blisterlike aneurysms of the internal carotid artery. J Neurosurg 89 (3):419–424

2. Linfante I, Mayich M, Sonig A, Fujimoto J, Siddiqui A, Dabus G (2016) Flow diversion with Pipeline Embolic Device as treatment of subarachnoid hemorrhage secondary to blister aneurysms: dual-center experience and review of the literature. J Neurointerv Surg 2016: neurintsurg-2016-012287

3. Fang C, Tan H-Q, Han H-J, Feng H, Xu J-C, Yan S, Nie Z-Y, Jin L-J, Teng F (2016) Endovascular isolation of intracranial blood blister-like aneurysms with Willis covered stent. J Neurointerv Surg 2016:neurintsurg-2016-012662

4. Andaluz N, Zuccarello M (2008) Blister-like aneurysms of the anterior communicating artery: a retrospective review of diagnosis and treatment in five patients. Neurosurgery 62 (4):807–811

5. McLaughlin N, Laroche M, Bojanowski MW (2010) Surgical management of blood blister-like aneurysms of the internal carotid artery. World Neurosurg 74(4):483–493

6. Regelsberger J, Matschke J, Grzyska U, Ries T, Fiehler J, Köppen J, Westphal M (2011) Blister-like aneurysms—a diagnostic and therapeutic challenge. Neurosurg Rev 34 (4):409–416

7. Shigeta H, Kyoshima K, Nakagawa F, Kobayashi S (1992) Dorsal internal carotid artery aneurysms with special reference to angiographic presentation and surgical management. Acta Neurochir 119(1-4):42–48

8. Nakagawa F, Kobayashi S, Takemae T, Sugita K (1986) Aneurysms protruding from the dorsal wall of the internal carotid artery. J Neurosurg 65(3):303–308

9. Ogawa A, Suzuki M, Ogasawara K (2000) Aneurysms at nonbranching sites in the supraclinoid portion of the internal carotid artery: internal carotid artery trunk aneurysms. Neurosurgery 47(3):578–586

10. Sim SY, Shin YS, Cho KG, Kim SY, Kim SH, Ahn YH, Yoon SH, Cho KH (2006) Blood blister-like aneurysms at nonbranching sites of the internal carotid artery. J Neurosurg 105 (3):400–405

11. Bojanowski MW, Weil AG, McLaughlin N, Chaalala C, Magro E, Fournier J-Y (2015) Morphological aspects of blister aneurysms and nuances for surgical treatment. J Neurosurg 123(5):1156–1165

12. Ohara H, Sakamoto T, Suzuki J (1979) Sclerotic cerebral aneurysms. In: Suzuki J (ed) Cerebral aneurysms. Neuron Publishing, Co, Tokyo, pp 673–682

13. Shimizu H, Matsumoto Y, Tominaga T (2010) Non-saccular aneurysms of the supraclinoid internal carotid artery trunk causing subarachnoid hemorrhage: acute surgical treatments and review of literatures. Neurosurg Rev 33 (2):205–216

14. Andrews BT, Brant-Zawadzki M, Wilson CB (1986) Variant aneurysms of the fenestrated basilar artery. Neurosurgery 18(2):204–207

15. Kim J-H, Kwon T-H, Kim J-H, Park Y-K, Chung H-S (2006) Internal carotid artery dorsal wall aneurysm with configurational change: are they all false aneurysms? Surg Neurol 66(4):441–443

16. Ishikawa T, Nakamura N, Houkin K, Nomura M (1997) Pathological consideration of a "blister-like" aneurysm at the superior wall of the internal carotid artery: case report. Neurosurgery 40(2):403–406

17. Lawton MT, Du R (2005) Effect of the neurosurgeon's surgical experience on outcomes from intraoperative aneurysmal rupture. Neurosurgery 57(1):9–15

18. Owen CM, Montemurro N, Lawton MT (2016) Blister aneurysms of the internal carotid artery: microsurgical results and management strategy. Neurosurgery 80 (2):235–247

19. Kirollos R, Tyagi A, Marks P, Van Hille P (1997) Muslin induced granuloma following wrapping of intracranial aneurysms: the role of infection as an additional precipitating factor. Acta Neurochir 139(5):411–415

20. Andres RH, Guzman R, Weis J, Schroth G, Barth A (2007) Granuloma formation and occlusion of an unruptured aneurysm after wrapping. Acta Neurochir 149(9):953–958

21. Szmuda T, Sloniewski P, Waszak PM, Springer J, Szmuda M (2016) Towards a new treatment paradigm for ruptured blood blister-like aneurysms of the internal carotid artery? A rapid systematic review. J Neurointerv Surg 8 (5):488–494

22. Kazumata K, Nakayama N, Nakamura T, Kamiyama H, Terasaka S, Houkin K (2014) Changing treatment strategy from clipping to radial artery graft bypass and parent artery sacrifice in patients with ruptured blister-like internal carotid artery aneurysms. Neurosurgery 10:66–73

23. Kalani MYS, Zabramski JM, Kim LJ, Chowdhry SA, Mendes GA, Nakaji P, McDougall CG, Albuquerque FC, Spetzler RF (2013) Long-term follow-up of blister aneurysms of the internal carotid artery. Neurosurgery 73 (6):1026–1033

24. Horiuchi T, Kusano Y, Yako T, Murata T, Kakizawa Y, Hongo K (2011) Ruptured anterior paraclinoid aneurysms. Neurosurg Rev 34 (1):49–55

25. Lee J-W, Choi H-G, Jung J-Y, Huh S-K, Lee K-C (2009) Surgical strategies for ruptured blister-like aneurysms arising from the internal carotid artery: a clinical analysis of 18 consecutive patients. Acta Neurochir 151(2):125–130

26. Meling TR, Sorteberg A, Bakke SJ, Slettebø H, Hernesniemi J, Sorteberg W (2008) Blood blister-like aneurysms of the internal carotid artery trunk causing subarachnoid hemorrhage: treatment and outcome. J Neurosurg 108(4):662–671

27. Rouchaud A, Brinjikji W, Cloft H, Kallmes DF (2015) Endovascular treatment of ruptured blister-like aneurysms: a systematic review and meta-analysis with focus on deconstructive versus reconstructive and flow-diverter treatments. Am J Neuroradiol 36(12):2331–2339

28. Briganti F, Cirillo S, Caranci F, Esposito F, Maiuri F (2002) Development of "de novo" aneurysms following endovascular procedures. Neuroradiology 44(7):604–609

29. Nguyen TN, Raymond J, Guilbert F, Roy D, Bérubé MD, Mahmoud M, Weill A (2008) Association of endovascular therapy of very small ruptured aneurysms with higher rates of procedure-related rupture. J Neurosurg 108 (6):1088–1092

30. Park JH, In Sung P, Han DH, Kim SH, Oh CW, Kim J-E, Kim HJ, Han MH, Kwon O-K (2007) Endovascular treatment of blood blister-like aneurysms of the internal carotid artery. J Neurosurg 106(5):812–819

31. Fang Y-B, Li Q, Wu Y-N, Zhang Q, Yang P-F, Zhao W-Y, Huang Q-H, Hong B, Xu Y, Liu J-M (2014) Overlapping stents for blood blister-like aneurysms of the internal carotid artery. Clin Neurol Neurosurg 123:34–39

32. Lim YC, Kim BM, Suh SH, Jeon P, Kim SH, Ihn Y-K, Lee Y-J, Sim SY, Chung J, Kim DJ (2013) Reconstructive treatment of ruptured blood blister-like aneurysms with stent and coil. Neurosurgery 73(3):480–488

33. Aydin K, Arat A, Sencer S, Hakyemez B, Barburoglu M, Sencer A, İzgi N (2014) Treatment of ruptured blood blister-like aneurysms with flow diverter SILK stents. J Neurointerv Surg 2014:neurintsurg-2013-011090

34. Yoon JW, Siddiqui AH, Dumont TM, Levy EI, Hopkins LN, Lanzino G, Lopes DK, Moftakhar R, Billingsley JT, Welch BG (2014) Feasibility and safety of pipeline embolization device in patients with ruptured carotid blister aneurysms. Neurosurgery 75 (4):419–429

35. Gonzalez AM, Narata AP, Yilmaz H, Bijlenga P, Radovanovic I, Schaller K, Lovblad K-O, Pereira VM (2014) Blood blister-like aneurysms: single center experience and systematic literature review. Eur J Radiol 83 (1):197–205

36. Mazur MD, Taussky P, MacDonald JD, Park MS (2016) Rerupture of a blister aneurysm after treatment with a single flow-diverting stent. Neurosurgery 79(5):E634

37. Meckel S, Singh T, Undrén P, Ramgren B, Nilsson O, Phatouros C, McAuliffe W, Cronqvist M (2011) Endovascular treatment using predominantly stent-assisted coil embolization and antiplatelet and anticoagulation management of ruptured blood blister-like aneurysms. Am J Neuroradiol 32(4):764–771

38. Peschillo S, Cannizzaro D, Caporlingua A, Missori P (2016) A systematic review and meta-analysis of treatment and outcome of blister-like aneurysms. Am J Neuroradiol 37 (5):856–861

Chapter 10

Giant Intracranial Aneurysms

Abdulrahman Y. Alturki, Ajith Thomas, and Christopher S. Ogilvy

Abstract

Intracranial aneurysms are classified as giant when their largest diameter is equal to or greater than 25 mm; they represent approximately 5–7% of intracranial aneurysms. Giant intracranial aneurysms carry a poor natural history with a 68% mortality in 2 years and a 85% in 5 years for untreated cases. The clinical presentation, natural history, location, therapeutic challenges, and treatment strategy are considerably different from smaller intracranial aneurysms and thus warrant special attention.

This chapter will summarize the types of giant intracranial aneurysms, their pathophysiology, review clinical presentations and diagnostic evaluation. Different management strategies for giant intracranial aneurysms are also discussed.

Key words Giant intracranial aneurysms, Saccular, Fusiform, Serpentine, Clinical presentation, Clipping and endovascular treatment

1 Introduction

Giant intracranial aneurysms (GIA) possess, by definition, a minimum diameter of at least 25 mm along one axis. GIAs are a very heterogeneous group of aneurysms. A classification based solely on size as a criterion is considered inadequate by many authors, as this does not adequately consider the heterogeneous and varied subtypes that have distinct clinical, morphological, angiographic, and pathological features [1].

GIAs represent 5% of all intracranial aneurysms. GIAs are seldom encountered in multiple (7% of cases) instances, and they commonly become symptomatic between 40 and 70 years. A female predominance exists with a ratio of up to 3:1. Approximately 5% to 10% of GIAs present in the pediatric population [2].

Approximately two-thirds of GIAs are located in the anterior circulation and one-third in posterior circulation. GIAs predominantly involve the cavernous and ophthalmic segments of the internal carotid artery (ICA), followed by the middle cerebral artery (MCA). In the posterior circulation, there is a predominant

Fawaz Al-Mufti and Krishna Amuluru (eds.), *Cerebrovascular Disorders*, Neuromethods, vol. 170,
https://doi.org/10.1007/978-1-0716-1530-0_10, © Springer Science+Business Media, LLC, part of Springer Nature 2021

involvement of the BA bifurcation, P1 segments of posterior cerebral arteries (PCAs) and the superior cerebellar arteries (SCAs) [3].

GIAs have a poor natural history with a mortality rate of 68 and 85% at 2 and 5 years, respectively [4]. More than 50% of untreated GIAs rupture [3]. According to the International Study of Unruptured Intracranial Aneurysms (ISUIA), giant aneurysms demonstrate an 8% annual rupture risk in the anterior circulation and 10% annual rupture risk in the posterior circulation. In contrast, small aneurysms have a 0–3% annual rupture risk in the anterior circulation and 0.5–3.7% annual rupture risk in the posterior circulation [5]. In patients with SAH due to a ruptured giant aneurysm, the mortality rate is >50% [6].

Flemming et al. reported on vertebrobasilar fusiform aneurysms (i.e., dolichoectasia) and showed that the risk of cerebral infarction was 2.7%, 11.3%, and 15.9%, at 1, 5, and 10 years, respectively [7]. Median survival was 7.8 years and death was most commonly due to ischemia. Another report by the same group reported on a prospective study of hemorrhage risk with a mean follow-up of 4.4 years, and showed that the annual rupture rate was 0.9% overall and 2.3% in those with transitional or fusiform aneurysm subtypes. Aneurysm enlargement was a significant predictor of lesion rupture [8].

Giant aneurysms are classified into the following categories [2, 9].

1. Giant saccular aneurysms: those with a demonstrable neck with a sac-like dilatation arising from parent artery.

2. Giant fusiform aneurysms: characterized by circumferential dilatation of the parent artery. Sometimes referred to as dolichoectatic fusiform by some authors.

3. Giant serpentine aneurysms: partially thrombosed with residual serpiginous channel on angiography. These aneurysms have an irregular eccentric channel through intraluminal thrombus with a wavy sinusoidal course.

The clinical presentation, natural history, location, therapeutic challenges, and treatment strategy are considerably different from smaller intracranial aneurysms and warrant special attention.

2   Pathophysiology

Many theories exist regarding the pathophysiology and formation of GIAs, explained by morphology.

1. Saccular: These GIAs often occur at arterial bifurcations, probably the result of continuous hemodynamic stress. Cyclical periods of endothelial damage from turbulent flow, followed

by aberrant healing, contribute to aneurysmal growth. Some authors have postulated that giant aneurysms grow from repeated intramural hemorrhage within the aneurysm wall followed by thrombus formation and neovascularization. These aneurysms undergo continuous remodeling with weakening of wall in the zone of smooth muscle cell proliferation due to proteolysis. This mechanism can explain growth and expansion of completely thrombosed giant aneurysms either following treatment or spontaneously. Thus saccular GIAs adapt well to prolonged periods of increasing wall tension and local hemodynamic factors and do not rupture initially. They continue to undergo progressive remodeling, however this does not prevent progressive aneurysmal growth or subsequent rupture [2, 9, 10].

2. Fusiform: These aneurysms may result from atherosclerosis, congenital arteriopathies, or traumatic dissection. Fragmentation of the internal elastic lamina and thickening of the intima are seen on histopathology. Compared to focal defects of saccular aneurysms, fusiform GIAs demonstrate extensive defects in both the muscularis and internal elastic lamina. Many other pathological features have been described such as irregular thickness of tunica media, thickened fibrous tissue, hypertrophic and swollen connective tissue, and absence of intima [11].

   Arterial dissection is thought to be a factor in the formation of a significant percentage of this type of GIA, particularly those without elongation or tortuosity of the parent vessel; an acute dissection may be the inciting event in the pathogenesis of the chronic type. Fusiform aneurysms may progress due to their unique features followed by neoangiogenesis within a thickened intima, intramural thrombus formation, and repetitive intramural hemorrhage from newly formed vessels within the thrombus [12].

3. Serpentine: Scarce literature exists regarding on this rare entity. Findings include large globoid masses that contain an irregular serpentine channel coursing through a partially thrombosed aneurysm. The walls of the aneurysms have been found to be notably thicker than usual and are composed primarily of acellular, fibrous tissue without internal elastic lamina or endothelial lining. Theories of origin in the literature include: development from saccular aneurysms by a "contained expansion" process, by fusiform dilation of the parent vessel with thrombosis, by repeated dissection of the intrinsic vessel wall with intramural hemorrhages, by either congenital or hemodynamically induced degeneration of the vessel wall, or as a result of a degenerative connective tissue disease of the arterial wall [13, 14].

3   Clinical Presentations and Diagnostic Evaluation

In general, GIAs symptoms may arise from compression of neural structures, cerebral ischemia, or rupture. The symptoms related to mass effect depend on the aneurysm's location. In the anterior circulation, mass effect can be manifested as pain, visual field and acuity defects, epistaxis, confusion, and extraocular dysfunction. Dementia and mental disturbances, as well as hemiparesis and epilepsy, have also been described. In the posterior circulation, multiple cranial nerve dysfunctions may be present. If compression on the brainstem is significant, bulbar palsies and hemiparesis can also occur [15]. Specific presentations of individual subtypes of GIAs include the following.

Giant saccular aneurysms [15, 16].

- 30–50% of patients present with SAH.
- 50–70% of patients present with mass effect and/or brain edema caused by aneurysm thrombosis. (Acute thrombus formation within an aneurysm can lead to vasogenic edema in surrounding brain tissue similar to brain tumors and intracerebral hematomas).
- 7–8% of patients may present with thromboembolic symptoms from an intraluminal thrombus.

Giant fusiform and serpentine aneurysms [17, 18].

- Compared to saccular aneurysms, rupture is relatively uncommon as a presenting symptom, occurring in 18–40% of patients.
- As many as 50% of patients with giant fusiform aneurysms have signs of mass effect; 31% of patients have symptoms attributable specifically to brainstem or cerebellar compression.
- Ischemic symptoms occur in approximately 25% of patients, likely due to compromise of perforating vessels or embolization of intraluminal thrombus.

CT and CTA can identify SAH, dimensions of the lesion, presence of intraluminal thrombus, calcification of the wall, brain edema, as well as pertinent skull base anatomy (for surgical planning). On CTA, partially thrombosed GIAs show homogenous intense post-contrast enhancement of the central circulating channel, which can be more opacified than the acutely thrombosed peripheral portion of the aneurysm. Although the thrombus does not enhance, there is enhancement of the aneurysmal wall. This combination of densities is known to produce the "Target sign" [19].

High-quality six-vessel diagnostic cerebral angiography (DSA) with 3D reconstruction has long been considered the "gold standard." With the conformation of the aneurysm's location and anatomy, DSA also gives valuable information about adjacent branches, collateral circulation, status of extracranial circulation (for surgical planning) and distal cerebral perfusion. The treating physicians should understand that while DSA shows luminal filling, it may fail to reveal the aneurysm's true size if intraluminal thrombus is present.

Magnetic resonance imaging (MRI) is useful in delineating possible thrombosed portions of the aneurysm, which does not manifest on DSA. Furthermore, MRI is very useful in determining related injuries to the parenchyma adjacent to the GIA.

4   Management

If symptoms attributable to intraluminal thrombosis are present, medical management (consisting of blood pressure management, smoking cessation, and antiplatelet therapy) is appropriate for patients who are at high risk of complications with surgical or endovascular treatment. A conservative approach may be warranted in elderly patients or those with significant comorbidities or extremely complex lesions.

The formation of treatment strategies for unruptured GIAs is difficult as guidelines from large clinical trials are lacking. The giant intracranial aneurysm study group in a recent review found that the chances of good outcome after surgical or endovascular GIA treatment primarily depends on patient age and aneurysm location rather than on the type of treatment conducted [20]. This study reported that in unruptured GIAs, the proportion of good outcome was 79.7% (95% CI 71.5–87.8) after surgical treatment and 84.9% (79.1–90.7, $p = 0.54$) after endovascular treatment. The proportion of good outcomes was lower in high-quality studies and in studies presenting aggregate instead of individual patient data. The OR for good treatment outcome was 5.2 (95% CI 2.0–13.0) at the internal carotid artery compared to 0.1 (0.1–0.3, $p < 0.1$) in the posterior circulation.

Decompression of neural structures including the brain parenchyma and cranial nerves and/or other vascular structures may be essential for both GIAs causing mass effect and those that ruptured. The ultimate goal of treatment (regardless of the modality) is to exclude the aneurysm from the circulation with preservation of all other arteries not involved and distal to the aneurysm.

4.1 Surgical Strategies

Various investigators favor surgical treatment for giant intracranial aneurysms. The basic tenets of aneurysm surgery apply to GIAs: complete vascular control, wide aneurysm exposure, careful clip application, and selective use of alternative techniques when clipping is not possible [21].

Surgery for GIAs includes direct and indirect approaches:

Table 1
Common surgical approaches to giant intracranial aneurysms

Location of aneurysm	Surgical approach
Proximal and bifurcation of internal carotid artery	Pterional, orbitozygomatic
Proximal anterior cerebral artery	Pterional, orbitozygomatic
Distal anterior cerebral artery	Pterional, orbitozygomatic, interhemispheric
Middle cerebral artery	Pterional, orbitozygomatic
Vertebral artery and vertebrobasilar junction	Far lateral
Midbasilar artery	Petrosal, far lateral, orbitozygomatic
High basilar artery	Orbitozygomatic
Posterior inferior cerebellar artery	Far lateral, suboccipital
Anterior inferior cerebellar artery	Petrosal, far lateral, orbitozygomatic
Superior cerebellar artery	Orbitozygomatic

- A direct approach includes clipping of the neck of the aneurysm while preserving flow through the distal arteries and perforator vessels. This is possible in only 50–60% of cases [21].

- Indirect methods include trapping and bypass, Hunterian ligation, parent vessel sacrifice, and distal parent vessel occlusion with revascularization. Excision of the aneurysm with resuturing of the involved vessels is done to reconstitute the original flow pattern. Aneurysmorraphy is another method to treat GIAs which are not suitable for clipping or trapping, where the aneurysm is decompressed and clips are applied to reconstitute a vessel wall as close to physiologic shape as possible.

Ligation of the internal carotid artery without prior collateral flow studies results in a 25% morbidity and a 12% mortality, whereas when ligation is performed following collateral flow studies, morbidity is 4.7% and mortality is 0%. Therefore, when a patient shows intolerance for balloon test occlusion (BTO), a high-flow bypass is required prior to occlusion of the internal carotid in order to maintain cerebral perfusion [22].

The optimal surgical approach to GIAs should offer a wide degree of exposure with visualization of the aneurysm's origin, outflow, perforators, adjacent vessels, and neural structures. Location of the aneurysm dictates the most appropriate surgical approach. Table 1 summarizes common surgical approaches to GIAs [21, 23]. Other considerations include the use of intraoperative electrophysiologic monitoring, intraoperative angiography, and specialized anesthesia techniques.

Table 2
Recent Surgical series for all types of GIAs

Study/year	N	Most common location (%)	Mortality (%)	Morbidity (%)	Outcomes	Occlusion	Note/Technique
Chen et al. [24], 2011	28	C-2 segment (32%)	0%	NA	GOS 4-5:85.7% GOS 2-3:14.3%	Complete obliteration 100%	Includes large plus giant
Eliava et al. [25], 2010	83	All paraclinoid	3.6%	Overall morbidity NA, vision-related 25%, motor-related 18.8%, CN-related 10%	GOS 4-5: 83.1% GOS 3: 13.3% GOS 1: 3.6%	NA	Includes large plus giant
Doormaal et al. [26], 2010	32	ICA (50%)	0%	15.6%	Favorable 88% Unfavorable 12%	94% bypass remained patent	Clipping under laser-protected bypass
Cantore et al. [27], 2008	99	Supraclinoid (52.5%)	8%	22.2%	Favorable 85.8%	NA	Clipping + other surgical EC/IC techniques
Hauck et al. [28], 2008	62	Proximal ICA (48%)	15%	Neurological 44% Systemic 30%	GOS 4-5: 68% GOS 2-3: 17% GOS 1: 15%	Complete 90% Near complete 5% incomplete 5%	Clipping only
Sharma et al. [29], 2008	177	Cavernous ICA (32%)	9%	Hydrocephalus 5.6% Meningitis 5% Postoperative infarcts 6.7%	GOS 4-5: 86.5% GOS 2-3: 4.5% GOS 1: 9%	Complete obliteration 89.8%	Clipping, EC/IC, ICA occlusion and ligation, trapping + ligation, wrapping, aneurysmorraphy

(continued)

Table 2
(continued)

Study/ year	N	Most common location (%)	Mortality (%)	Morbidity (%)	Outcomes	Occlusion	Note/Technique
Kato et al. [30], 2003	139	MCA 43%	8%	5.75%	GOS 5: 67.2%, GOS 4: 15.1%, GOS 3: 7.9%, GOS 2: 1.4%, GOS 1: 8%	NA	Trapping-clipping-evacuation, aneurysmectomy, ECIC-coiling, conservative, VA ligation, ECIC + ligation, clipping + ECIC
Piepgras et al. [31], 1998	105	Anterior circulation (74.2%), posterior circulation (25.7%)	8.6%	Undetermined	Favorable 72% Unfavorable 28%	NA	Clipping; repair with vessel reconstruction, trapping, proximal ligation, other surgical EC/IC techniques
Drake et al. [32], 1997	120	Basilar trunk (26.6%)	13.3%	26.6%	Excellent 60% good 15.8% Poor 10.8% Died 13.3%	NA	Only giant fusiform. Explored, clipped, wrapped, reconstruction of artery, proximal occlusion, trapped, excised

Factors that can complicate surgery include wide neck, afferent and efferent vessels encompassed by the aneurysm, the presence of intrasaccular thrombus, and calcifications in the aneurysm neck and the source artery. Paraclinoid aneurysms and those located in the posterior circulation represent a greater challenge [16, 21]. Table 2 summarizes several recent surgical studies.

4.2 Endovascular Strategies

Endovascular therapy has emerged as a potential alternative to surgery for treatment of giant aneurysms. However, problems regarding endovascular management include incomplete obliteration, aneurysm recanalization, procedure-related complications, distal location, and cost.

Endovascular modalities such as BTO, Wada testing, operative suction decompression, operative balloon occlusion, and intraoperative angiography may be adjuncts to definitive surgical therapies. Endovascular approaches may also provide definitive treatment, such as endovascular parent vessel deconstruction, coiling with or without stent/balloon assistance, Onyx (ev3, Irvine, California) embolization, and, recently, flow diversion.

Parent vessel deconstruction is an established approach for the treatment of GIAs. BTO is a necessary procedure during the workup for carotid lesions. Once a patient has successfully passed a BTO, the procedural risk and economics of the procedure become more manageable. The treating physician must take into consideration the increase in hemodynamic stress on other cerebral arteries, which can cause new aneurysms to form (up to 20% in patients after carotid sacrifice) [33].

Coil embolization is one of the earliest used endovascular techniques to treat GIAs, although this management technique may be challenging given the long procedural time, sub-optimal packing density and the exceedingly high recurrence rate. Moreover, in GIAs with wide necks, there can be difficulty in maintaining coil position within the aneurysm and stent-assistance is often required to prevent coil prolapse. One study reported incomplete aneurysm occlusion in 69% of large and giant aneurysms ($\geq$20 mm) treated with coil embolization at 6 month follow-up and retreatment was required in 41.4% [34]. On the other hand, another study reported a complete occlusion rate of 71% for GIAs treated with coil embolization [35]. Chalouhi et al. demonstrated high rates of recurrence (52%) and retreatment (47%) for aneurysms >25 mm, and showed rates of morbidity and mortality of 17% and 8%, respectively [36].

Some operators recommend parent vessel deconstruction over primary coil embolization as the first endovascular option. Gruber et al. reported a hemorrhage rate of 6.4% after coiling, whereas no hemorrhages occurred after parent vessel sacrifice [35]. Another study found an annual rebleeding rate of 1.9% for large and giant aneurysms [36].

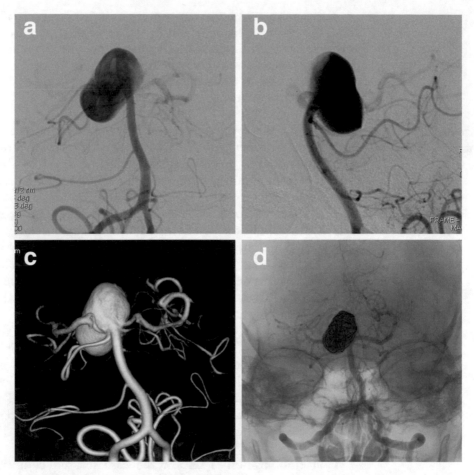

Fig. 1 (**a**) AP and (**b**) lateral views with (**c**) 3D reconstruction showing giant 27 mm basilar terminus aneurysm. (**d**) Post treatment DSA after flow diversion extending from left PCA into basilar artery, with concurrent coil embolization

Flow diverting devices are important and promising tools in the treatment of intracranial aneurysms (Fig. 1). In one review, a complete occlusion rate of 76% was reported. The same report showed ischemic stroke rate of 6%, with higher rates in posterior circulation aneurysms and large/giant aneurysms [37]. In a retrospective multicenter study that studied the Pipeline embolization device (PED) in the treatment of 50 unruptured large and giant aneurysms (>20 mm), complete occlusion was achieved in 61.5% of cases; specifically in 70% of saccular aneurysms and in 58.6% of fusiform aneurysms. There was no significant correlation between the morphology of the aneurysm or the number of PEDs placed on the rate of complete aneurysm occlusion. Retreatment was performed in 16% of cases. Symptomatic thromboembolic complications occurred in 12% of cases, while symptomatic hemorrhagic complications occurred in 8% [38].

Table 3
Some endovascular series (≥10 patients)

Series, year	Number of aneurysms	Number of GIA	Fusiform morphology %	SAH %	Method Used	Immediate complete occlusion %	Recurrence and/or Retreatment	Morbidity (%)	Mortality (%)	Notes
Derrey et al. [44], 2015	79	79	12.7%	32.9%	Coiling, SAC, PVO, 36.7% surgical	51.7%	NM	48.1%	16.5%	Multiple aneurysms in 21.5%
Huang et al. [45], 2015	11	11	54.5%	18.2%	Coiling, SAC, PVO	72%	12.5%, NM retreatment	54%	0	All MCA aneurysms
Chalouhi et al. [36], 2013	334	32	4.1%	NM	Coiling, SAC, PVO, BAC	≥95%	52% and 47.6%	9.4%	0.3%	–
Gao et al. [46], 2012	106	31	NM	61.3%	Coiling, SAC, BAC	48.1%	46.4%, NM retreatment	7.5%	2.8%	–
Hauck et al. [47], 2009	15	15	NM	0%	Coiling, SAC	27% (100% occlusion)	80% retreatment	0	0	–

(continued)

Table 3
(continued)

Series, year	Number of aneurysms	Number of GIA	Fusiform morphology %	SAH %	Method Used	Immediate complete occlusion %	Recurrence and/or Retreatment	Morbidity (%)	Mortality (%)	Notes
Jahromi et al. [48], 2008	39	39	NM	26%	Coiling, SAC, PVO, BAC	36% (100% occluded), 64% (≥95% occluded)	NM	26%	29%	–
Standhardt et al. [49], 2008	202	19	NM	0%	Coils	10.5% complete, 57.9% neck remnant	NM	3.5% overall, NM for GIA	0.5% overall, NM for GIA	–
Li et al. [50], 2007	20	20	30%	5%	Coiling, PVO	55%	15% retreatment	NM	5%	3 had intracranial covered stent

SAC stent-assisted coiling, *PVO* parent vessel occlusion, *BAC* balloon-assisted coiling, *NM* not mentioned

Table 4
Some endovascular series using flow diverters (≥10 patients)

Series, year	Number of aneurysms	Number of GIA	Fusiform morphology %	SAH %	Device Used	Complete occlusion[a]	Recurrence and/or Retreatment	Morbidity (%)	Mortality (%)	Notes
Adeeb et al. [38], 2016	50	NM (all ≥20 mm)	74%	0	PED	61.5%	16% retreatment	20%	6.6%	22% added coils
Hanel et al. [51], 2015	207	21	23.8%	0	PED	73%	NM	23.8%	9.5%	–
Zhou et al. [40], 2014	28	8	NM	0	Tubridge	72%	NM	0	0	64% added coils
Berge et al. [52], 2012	77	18	32%	13%	Silk	84.5%	0	7.8%	3%	–
Briganti et al. [53], 2012	295	NM (>15 mm (? GIA) in 138)	11.8% (? GIA)	0	PED and silk	85%	NM	3.7%	5.9%	None of the data were specific for GIA
Lubicz et al. [54], 2010	34	4	50%	0	Silk	NA	25%	25%	0%	–
Lylyk et al. [55], 2009	63	8	12% (? GIA)	13.2%	PED	100%	0	0	0%	–

PED Pipeline Embolization Device, *NM* not mentioned
[a]At last follow-up

Similarly, other flow diversion devices like Silk have shown similar angiographic cure rates (~70%) but with higher overall complication rates (42.9%). The results were particularly unfavorable for giant fusiform aneurysms of the posterior circulation as high rates of complications (57%) were related to brainstem ischemia secondary to perforator occlusion [39]. Comparable complete occlusion rates have also been reported for the Tubridge stent [40].

Economically, the cost of initial treatment of large and giant aneurysms with PED is favorable compared to traditional embolization techniques. The potential cost benefit depends on aneurysm volume, coil type, and number of PEDs used. Thus, PED therapy is considered a cost-saving intervention in aneurysms >0.9 cm^3 or when a single device is used [41].

The majority of studies comparing different endovascular techniques in aneurysm treatment include GIAs in the overall assessment. In unruptured, ≥20 mm saccular aneurysms, the complete occlusion rate for PED vs. coil embolization was 87.5% vs. 23.5% with complication rates of 0% vs. 7.7%, respectively [42]. When comparing safety and efficacy between flow diversion and stent-assisted coiling in large and giant unruptured aneurysms, the rate of complete occlusion was significantly higher in the flow diversion cohort compared to the stent-assisted coiling cohort at a 6 month follow-up. The flow diversion cohort achieved greater improvement and a lower rate of recurrence. The rate of periprocedural complications and outcome was similar [43]. Table 3 summarizes several recent endovascular series, and Table 4 summarizes several recent endovascular series using flow diverters.

4.3 Surgery Versus Endovascular Approach to GIAs

Readers must be aware that neither recommendations nor randomized controlled trials exist to guide clinicians in the treatment of patients with GIAs. Nor do studies exist which compare these treatment modalities head to head. To further complicate matters, the associated literature possesses a large amount of heterogeneity in techniques and inconsistent reporting of results, morbidities and mortality, regardless of the treatment modality.

Flow-diverting stents are a promising advance in the treatment of GIAs, with good short- and medium-term results. However, surgical treatment of these lesions remains a reasonable therapeutic modality in selected centers, due to the durability, costs, low rates of recanalization, and good clinical outcomes.

One study (despite the significant limitations of such type of studies) compared the direct cost of surgical and endovascular treatment of unruptured GIAs and found no difference in the costs of hospital stay between the 2 groups. Imaging costs were significantly higher in the surgical group, as were the costs of the intervention room and personnel involved in the intervention. Implants used per patient were more expensive in the endovascular group. The total direct treatment costs were higher in the endovascular group. Treatment costs were associated with the type of treatment and GIA location but not with patient age, sex, or GIA size [56].

4.3.1 Suggested Management Approach to GIAs

Each aneurysm and patient must be considered individually in deciding upon a particular treatment strategy in a multidisciplinary fashion, which should be tailored to the unique characteristics of each case and the comfort level of the treating physician with every treatment modality. For each lesion, the aneurysm specific and patient specific factors must be carefully weighed in the decision to treat and the modality to use.

Typically, flow diversion is a reasonable first option in patients that are elderly, with a fusiform morphology and para-clinoidal or posterior circulation location. For middle cerebral and anterior cerebral artery lesions, bypass with flow diversion or vessel occlusion may be the best solution. GIAs represent some of the highest risk lesions when facing treatment and a team approach should be utilized considering all surgical and endovascular techniques available.

5   Conclusion

GIAs are formidable lesions with a dismal natural history. The high morbidity and mortality risks mandate treatment in these patients whenever possible. A multidisciplinary approach for their treatment is advised. Exclusion of an aneurysm from the cerebral circulation and secondary elimination of mass effect are the main goals of treatment.

Surgically, GIAs require vascular control, temporary clipping, and complex techniques to reconstruct the aneurysm neck. Preservation of branch and perforating arteries may be difficult with GIAs, and alternative techniques may be indicated if direct clipping cannot preserve a normal vascular anatomy. Occlusions of the parent artery and trapping and bypass procedures are important surgical contingency strategies. Endovascular procedures must also be considered, both as adjuncts to traditional surgical techniques and as stand-alone procedures. Whatever intervention is undertaken, careful consideration of aneurysm size, morphology, location, and surrounding vasculature is necessary to choose the optimal treatment strategy. The treating physician must critically evaluate the risks associated with any invasive procedure and its efficacy over the patient's remaining functional lifetime.

References

1. Krings T, Alvarez H, Reinacher P, Ozanne A, Baccin C, Gandolfo C, Zhao W-Y, Reinges M, Lasjaunias P (2007) Growth and rupture mechanism of partially thrombosed aneurysms. Interv Neuroradiol 13(2):117–126

2. Parkinson RJ, Eddleman CS, Batjer HH, Bendok BR (2006) Giant intracranial aneurysms: endovascular challenges. Neurosurgery 59(5):S3–S103

3. Choi IS, David C (2003) Giant intracranial aneurysms: development, clinical presentation and treatment. Eur J Radiol 46(3):178–194

4. Drake C (1979) Giant intracranial aneurysms: experience with surgical treatment in 174 patients. Clin Neurosurg 26:12

5. Wiebers DO, Investigators ISoUIA (2003) Unruptured intracranial aneurysms: natural history, clinical outcome, and risks of surgical and endovascular treatment. Lancet 362 (9378):103–110

6. Barrow D, Alleyne C (1994) Natural history of giant intracranial aneurysms and indications for intervention. Clin Neurosurg 42:214–244

7. Flemming KD, Wiebers DO, Brown RD Jr, Link MJ, Huston J III, McClelland RL, Christianson TJ (2005) The natural history of radiographically defined vertebrobasilar nonsaccular intracranial aneurysms. Cerebrovasc Dis 20 (4):270–279

8. Flemming KD, Wiebers DO, Brown RD Jr, Link MJ, Nakatomi H, Huston J III, McClelland R, Christianson TJ (2004) Prospective risk of hemorrhage in patients with vertebrobasilar nonsaccular intracranial aneurysm. J Neurosurg 101(1):82–87

9. Lonjon M, Pennes F, Sedat J, Bataille B (2015) Epidemiology, genetic, natural history and clinical presentation of giant cerebral aneurysms. Neurochirurgie 61(6):361–365

10. Krings T, Mandell DM, Kiehl T-R, Geibprasert S, Tymianski M, Alvarez H, Hans F-J (2011) Intracranial aneurysms: from vessel wall pathology to therapeutic approach. Nat Rev Neurol 7(10):547–559

11. Little JR, St Louis P, Weinstein M, Dohn DF (1981) Giant fusiform aneurysm of the cerebral arteries. Stroke 12(2):183–188

12. Nakatomi H, Segawa H, Kurata A, Shiokawa Y, Nagata K, Kamiyama H, Ueki K, Kirino T (2000) Clinicopathological study of intracranial fusiform and dolichoectatic aneurysms insight on the mechanism of growth. Stroke 31(4):896–900

13. Christiano LD, Gupta G, Prestigiacomo CJ, Gandhi CD (2009) Giant serpentine aneurysms. Neurosurg Focus 26(5):E5

14. Segal HD, McLaurin RL (1977) Giant serpentine aneurysm: report of two cases. J Neurosurg 46(1):115–120

15. Pia H, Zierski J (1982) Giant cerebral aneurysms. Neurosurg Rev 5(4):117–148

16. Lawton M, Spetzler R (1998) Surgical strategies for giant intracranial aneurysms. Neurosurg Clin N Am 9(4):725–742

17. Anson JA, Lawton MT, Spetzler RF (1996) Characteristics and surgical treatment of dolichoectatic and fusiform aneurysms. J Neurosurg 84(2):185–193

18. Resta M, Gentile M, Di Cuonzo F, Vinjau E, Brindicci D, Carella A (1984) Clinical-angiographic correlations in 132 patients with megadolichovertebrobasilar anomaly. Neuroradiology 26(3):213–216

19. Pinto RS, Kricheff II, Butler AR, Murali R (1979) Correlation of computed tomographic, angiographic, and neuropathological changes in Giant cerebral aneurysms 1. Radiology 132 (1):85–92

20. Dengler J, Maldaner N, Gläsker S, Endres M, Wagner M, Malzahn U, Heuschmann PU, Vajkoczy P, Group GIAS (2016) Outcome of surgical or endovascular treatment of Giant intracranial aneurysms, with emphasis on age, aneurysm location, and Unruptured Aneuryms-a systematic review and meta-analysis. Cerebrovasc Dis 41(3-4):187–198

21. Hanel RA, Spetzler RF (2008) Surgical treatment of complex intracranial aneurysms. Neurosurgery 62(6):SHC1289–SHC1299

22. de Sousa AA, de Sousa Filho JL, Dellaretti Filho MA (2015) Treatment of Giant intracranial aneurysms: a review based on experience from 286 cases. Arq Bras Neurocir 34 (04):295–303

23. Lemole GM, Henn J, Spetzler RF, Riina HA (2000) Surgical management of giant aneurysms. Oper Tech Neurosurg 3(4):239–254

24. Chen S, Kato Y, Subramanian B, Kumar A, Watabe T, Imizu S, Oda J, Oguri D, Sano H (2011) Retrograde suction decompression assisted clipping of large and giant cerebral aneurysms: our experience. Minim Invasive Neurosurg 54(01):1–4

25. Eliava SSFYM, Yakovlev SB, Shekhtman OD, Kheireddin AS, Sazonov IA, Sazonova OB, Okishev DN (2010) Results of microsurgical treatment of large and giant ICA aneurysms using the retrograde suction decompression

(RSD) technique: series of 92 patients. World Neurosurg 73:683–687

26. van Doormaal TPC, van der Zwan A, Verweij BH, Regli L, Tulleken CAF (2010) Giant aneurysm clipping under protection of an excimer laser–assisted non-occlusive anastomosis bypass. Neurosurgery 66(3):439–447. https://doi.org/10.1227/01.neu. 0000364998.95710.73

27. Cantore G, Santoro A, Guidetti G, Delfinis CP, Colonnese C, Passacantilli E (2008) Surgical treatment of giant intracranial aneurysms: current viewpoint. Neurosurgery 63(4):279–290

28. Hauck EF, Wohlfeld B, Welch BG, White JA, Samson D (2008) Clipping of very large or giant unruptured intracranial aneurysms in the anterior circulation: an outcome study: clinical article. J Neurosurg 109 (6):1012–1018

29. Sharma BS, Gupta A, Ahmad FU, Suri A, Mehta VS (2008) Surgical management of giant intracranial aneurysms. Clin Neurol Neurosurg 110(7):674–681

30. Kato Y, Sano H, Imizu S, Yoneda M, Viral M, Nagata J, Kanno T (2003) Surgical strategies for treatment of giant or large intracranial aneurysms: our experience with 139 cases. Minim Invasive Neurosurg 46(06):339–343

31. Piepgras DG, Khurana VG, Whisnant JP (1998) Ruptured giant intracranial aneurysms. Part II. A retrospective analysis of timing and outcome of surgical treatment. J Neurosurg 88 (3):430–435

32. Drake CG, Peerless SJ (1997) Giant fusiform intracranial aneurysms: review of 120 patients treated surgically from 1965 to 1992. J Neurosurg 87(2):141–162

33. Briganti F, Cirillo S, Caranci F, Esposito F, Maiuri F (2002) Development of" de novo " aneurysms following endovascular procedures. Neuroradiology 44(7):604–609

34. Sluzewski M, Menovsky T, Van Rooij WJ, Wijnalda D (2003) Coiling of very large or giant cerebral aneurysms: long-term clinical and serial angiographic results. Am J Neuroradiol 24(2):257–262

35. Gruber A, Killer M, Bavinzski G, Richling B (1999) Clinical and angiographic results of endosaccular coiling treatment of giant and very large intracranial aneurysms: a 7-year, single-center experience. Neurosurgery 45 (4):793

36. Chalouhi N, Tjoumakaris S, Gonzalez L, Dumont A, Starke R, Hasan D, Wu C, Singhal S, Moukarzel L, Rosenwasser R (2014) Coiling of large and giant aneurysms:

complications and long-term results of 334 cases. Am J Neuroradiol 35(3):546–552

37. Brinjikji W, Murad MH, Lanzino G, Cloft HJ, Kallmes DF (2013) Endovascular treatment of intracranial aneurysms with flow diverters a meta-analysis. Stroke 44(2):442–447

38. Nimer Adeeb CG, Shallwani H, Shakir HJ, Foreman P, Moore J, Dmytriw A, Alturki A, Siddiqui A, Levy E, Snyder K, Harrigan M, Ogilvy C, Thomas A (2016) Pipeline Embolization Device in treatment of unruptured large and giant aneurysms. World Neurosurg 105:232–237

39. Strauss I, Maimon S (2016) Silk flow diverter in the treatment of complex intracranial aneurysms: a single-center experience with 60 patients. Acta Neurochir 158(2):247–254

40. Zhou Y, Yang P, Fang Y, Xu Y, Hong B, Zhao W, Li Q, Zhao R, Huang Q, Liu J (2014) A novel flow-diverting device (Tubridge) for the treatment of 28 large or giant intracranial aneurysms: a single-center experience. AJNR Am J Neuroradiol 35 (12):2326

41. El-Chalouhi N, Jabbour PM, Tjoumakaris SI, Starke RM, Dumont AS, Liu H, Rosenwasser R, El Moursi S, Gonzalez LF (2014) Treatment of large and giant intracranial aneurysms: cost comparison of flow diversion and traditional embolization strategies. World Neurosurg 82(5):696–701

42. Chalouhi N, Tjoumakaris S, Starke RM, Gonzalez LF, Randazzo C, Hasan D, McMahon JF, Singhal S, Moukarzel LA, Dumont AS (2013) Comparison of flow diversion and coiling in large unruptured intracranial saccular aneurysms. Stroke 44(8):2150–2154

43. Zhang Y, Zhou Y, Yang P, Liu J, Xu Y, Hong B, Zhao W, Chen Q, Huang Q-H (2015) Comparison of the flow diverter and stent-assisted coiling in large and giant aneurysms: safety and efficacy based on a propensity score-matched analysis. Eur Radiol 26:2369–2377

44. Derrey S, Penchet G, Thines L, Lonjon M, David P, Bataille B, Emery E, Lubrano V, Laguarrigue J, Bresson D (2015) French collaborative group series on giant intracranial aneurysms: current management. Neurochirurgie 61(6):371–377

45. Huang L, Cao W, Ge L, Lu G, Wan J, Zhang L, Gu W, Zhang X, Geng D (2015) Endovascular management of giant middle cerebral artery aneurysms. Int J Clin Exp Med 8(5):7517

46. Gao X, Liang G, Li Z, Wei X, Cao P (2012) A single Centre experience and follow-up of patients with endovascular coiling of large and

giant intracranial aneurysms with parent artery preservation. J Clin Neurosci 19(3):364–369

47. Hauck EF, Welch BG, White JA, Replogle RE, Purdy PD, Pride LG, Samson D (2009) Stent/coil treatment of very large and giant unruptured ophthalmic and cavernous aneurysms. Surg Neurol 71(1):19–24

48. Jahromi BS, Mocco J, Bang JA, Gologorsky Y, Siddiqui AH, Horowitz MB, Hopkins LN, Levy EI (2008) Clinical and angiographic outcome after endovascular management of giant intracranial aneurysms. Neurosurgery 63 (4):662–675

49. Standhardt H, Boecher-Schwarz H, Gruber A, Benesch T, Knosp E, Bavinzski G (2008) Endovascular treatment of Unruptured intracranial aneurysms with Guglielmi detachable coils short-and long-term results of a single-Centre series. Stroke 39(3):899–904

50. Li M-H, Li Y-D, Fang C, Gu B-X, Cheng Y-S, Wang Y-L, Gao B-L, Zhao J-G, Wang J, Li M (2007) Endovascular treatment of giant or very large intracranial aneurysms with different modalities: an analysis of 20 cases. Neuroradiology 49(10):819–828

51. Hanel R, Bonafe A, Fischer S, Diaz O, Kallmes D, Barnwell S, Woo H (2015) O-020 treatment of giant intracranial aneurysms with pipeline: aspire (aneurysm study of pipeline in an observational registry) results. J NeuroInterventional Surg 7(Suppl 1):A11

52. Berge J, Biondi A, Machi P, Brunel H, Pierot L, Gabrillargues J, Kadziolka K, Barreau X, Dousset V, Bonafé A (2012) Flow-diverter silk stent for the treatment of intracranial aneurysms: 1-year follow-up in a multicenter study. Am J Neuroradiol 33(6):1150–1155

53. Briganti F, Napoli M, Tortora F, Solari D, Bergui M, Boccardi E, Cagliari E, Castellan L, Causin F, Ciceri E (2012) Italian multicenter experience with flow-diverter devices for intracranial unruptured aneurysm treatment with periprocedural complications—a retrospective data analysis. Neuroradiology 54 (10):1145–1152

54. Lubicz B, Collignon L, Raphaeli G, Pruvo J-P, Bruneau M, De Witte O, Leclerc X (2010) Flow-diverter stent for the endovascular treatment of intracranial aneurysms a prospective study in 29 patients with 34 aneurysms. Stroke 41(10):2247–2253

55. Lylyk P, Miranda C, Ceratto R, Ferrario A, Scrivano E, Luna HR, Berez AL, Tran Q, Nelson PK, Fiorella D (2009) Curative endovascular reconstruction of cerebral aneurysms with the pipeline embolization device: the Buenos Aires experience. Neurosurgery 64 (4):632–643

56. Familiari P, Maldaner N, Kursumovic A, Rath SA, Vajkoczy P, Raco A, Dengler J (2015) Cost comparison of surgical and endovascular treatment of unruptured giant intracranial aneurysms. Neurosurgery 77(5):733–743

Chapter 11

Atypical Aneurysms: Mycotic Aneurysms, Dissecting Aneurysms, and Pseudoaneurysms

Ram Gowda, Timothy R. Miller, and Nicholas A. Morris

Abstract

Atypical aneurysms, including intracranial mycotic aneurysms, dissecting aneurysms, and pseudoaneurysms, present distinct challenges for diagnosis and management. We review the pathogenesis, diagnosis, and management of these atypical aneurysms. While newer vessel wall imaging techniques promise greater sensitivity, catheter angiography is often necessary for diagnosis. Ruptured atypical aneurysms can cause significant morbidity; unfortunately risk of rupture can be difficult to predict. Endovascular and surgical intervention can preserve the underlying vessel or require parent vessel occlusion. Treatment is individualized to the patient's clinical situation and anatomy.

Key words Mycotic aneurysm, Dissecting aneurysm, Pseudoaneurysm, Subarachnoid hemorrhage

---

1 Mycotic Aneurysms

1.1 Introduction

Mycotic aneurysms (MAs) can be intracavitary (thorax and abdomen) or peripheral (extremities, neck, cranium). William Osler originated the term *mycotic endarteritis* for an aortic aneurysm associated with endocarditis, though the entity had been previously described by Rudolf Virchow and others [1, 2]. "Mycotic" referred to the aneurysm's resemblance to a fungal growth, not a fungal infection; thus the more accurate term *infectious aneurysm* has come into use to denote aneurysms of infectious etiology, from either bacteremia or contiguous spread. MAs account for about 1% of aortic aneurysms and about 1–5% of intracranial aneurysms [2, 3]. Although rare, MAs often go undetected until autopsy, and their incidence is likely underestimated.

Intracranial mycotic aneurysms (ICMAs) are the most common type of MA and are usually associated with infectious endocarditis (IE), although other etiologies have become more common with more widespread use of antibiotics. ICMAs are seen in 2–10% of cases of IE [1, 3]. Contiguous spread of infection from the meninges, dural venous sinuses, or paranasal sinuses is a significant

Fawaz Al-Mufti and Krishna Amuluru (eds.), *Cerebrovascular Disorders*, Neuromethods, vol. 170,
https://doi.org/10.1007/978-1-0716-1530-0_11, © Springer Science+Business Media, LLC, part of Springer Nature 2021

cause of ICMAs [4]. Iatrogenic or traumatic vessel infection are rare causes.

In the modern era, *Staphylococcus aureus* and *Streptococcus viridans* species are the most common causes of IE and thus MAs. However, MAs have been associated with a variety of pathogens causing IE, including other Staphylococcus and Streptococcus species, Enterococcus, HACEK bacteria, and other gram-negative bacteria. ICMA formation from contiguous spread of an intracranial infection can arise from anaerobic bacteria, mycobacteria, and fungi, in addition to the more common bacterial pathogens [1]. There are also reports of ICMA formation caused by varicella zoster virus [5] and HIV [3].

Some studies have reported a slight male predominance among ICMA patients [6] while another large pooled case series showed equal numbers of males and females [3].

1.2 Pathophysiology and Clinical Presentation

Whether from contiguous or hematogenous spread, arterial wall infection results in muscular and elastic lamina destruction and aneurysmal ballooning of the affected vessel. Contiguous spread of infection from the meninges, dural venous sinuses, and paranasal sinuses tends to affect the adjacent proximal vessels, while septic emboli usually lodge at cortical vessel branching points, most often in the distal branches of the middle cerebral artery [3]. Bacteria from septic microemboli gain entrance to the adventitia via the vasa vasorum; the arterial wall infection and consequent inflammatory response spreads to the muscular tunica media and internal elastic membrane [2]. Both infiltrating neutrophils and bacteria secrete enzymes that break down elastin and collagen, but the inflammatory response may be more destructive [7]. In large arteries such as the aorta, occlusion of the vasa vasorum by septic microemboli contributes to arterial wall ischemia and destruction, but this is less important in small arteries such as the distal cerebral vessels, which are less dependent on the vasa vasorum. Nevertheless, microemboli permit bacterial access to the artery wall via the vasa vasorum, which likely explains the rarity of MA formation in bacteremia alone [2].

Usually the chief presenting symptoms of ICMAs are those of the underlying cause, whether intracranial infection or IE. As with noninfectious aneurysms, ICMAs often remain asymptomatic unless rupture occurs. There are no specific signs or symptoms of ICMA; fever, delirium, headache, or seizures do not distinguish ICMA from other neurological complications of IE such as stroke or cerebral abscess [3, 8]. Even focal neurological deficits, seen in 23% of IE patients with ICMA in one case series, did not distinguish these patients from IE patients without ICMA [8]. Nevertheless, the usual red flags of intracranial aneurysmal rupture—thunderclap headache, meningeal signs, cranial nerve and other focal deficits, or

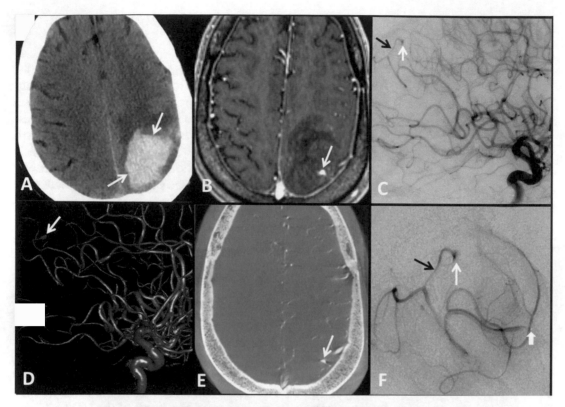

Fig. 1 A 74-year-old male presented with altered mental status and bacteremia. Initial non-contrast head CT (**a**) demonstrated an intraparenchymal hematoma in the left parietal lobe (arrows **a**). A follow up brain MRI with contrast (**b**) demonstrated a small, enhancing outpouching (arrow **b**) centered in the hematoma, which was suspicious for aneurysm. A lateral view from a left internal carotid artery angiogram (**c**) confirmed the lesion (white arrow **c**) arising from a distal cortical branch of the left middle cerebral artery (black arrow **c**). 3D rotational angiography (**d**) and flat panel computed tomography (**e**) performed from the left internal carotid artery better delineated the lesion in the high left parietal lobe (white arrows **d**, **e**). Given the distal cortical location of the lesion, as well as the patient's history of bacteremia, these findings were thought be most consistent with a mycotic aneurysm. An attempt was made to treat the lesion by liquid embolization. A flow directed microcatheter was navigated into the distal posterior division of the left middle cerebral artery. A subsequent microcatheter injection (**f**) demonstrated the small mycotic aneurysm (white arrow **f**) and its distal parent cortical branch (black arrow **f**). However, several uninvolved distal left middle cerebral artery cortical branches were also visualized on this microcatheter injection (block arrow **f** indicating microcatheter tip). Consequently, it was decided to not proceed with endovascular embolization and the patient subsequently underwent successful microsurgical repair of the lesion (not shown)

sudden deterioration of mental status—should prompt workup for ICMA in patients with predisposing disease.

Unlike with noninfectious saccular (berry) aneurysms, intraparenchymal hemorrhage is more common than subarachnoid hemorrhage (SAH) with ruptured ICMAs, given the latter are more often located in distal vessels instead of the circle of Willis [3]. An example is shown in Fig. 1. Ischemic stroke can also result from septic emboli in the parent vessel. Symptomatic vasospasm after SAH can also cause strokes and has been observed in both IE and meningitis [4].

1.3 Management Strategies

1.3.1 Diagnostic Approach

Imaging

- As there are no specific presenting signs and symptoms for ICMA, clinical suspicion should prompt cerebrovascular imaging. Current American Heart Association (AHA) and European Society of Cardiology (ESC) guidelines recommend cerebrovascular imaging to detect ICMA in all patients with IE with neurological symptoms [1, 9]. Patients with contiguously spreading intracranial infections should also receive cerebrovascular imaging. The available evidence suggests the following.
  - Computed tomography angiography (CTA).
  - Magnetic resonance angiography (MRA).
  - Digital subtraction angiography (DSA).
    It remains the gold standard for small aneurysms (<5 mm) for which CTA and MRA are less sensitive.

    However, there are conflicting data on the extent of DSA's superiority [1, 10].

- Data specific to ICMAs are scarce: a small series reported sensitivities of 43% and 33% for CTA and MRA, respectively [11], and another small series reported a sensitivity of 46% for CTA or MRA [12]. Repeat imaging, whether DSA or noninvasive, may be more sensitive purely because of growth of the ICMA. Thus, negative CTA or MRA should prompt repeat imaging and/or DSA in patients with high clinical suspicion for ICMA.

- As intraparenchymal hemorrhage, SAH, or ischemic stroke can suggest the presence of an ICMA, magnetic resonance imaging (MRI) of the brain is also helpful at finding abnormalities missed on computed tomography (CT).

Radiographic Features

Some radiographic aspects of ICMAs distinguish them from berry aneurysms.

ICMAs

- ICMAs in IE are typically located in small distal vessels; most often branches of the middle cerebral artery, with a minority in posterior cerebral artery branches and rarely in anterior cerebral artery branches.

- The location of ICMAs arising from contiguous spread of intracranial infection are more often proximal, reflecting the location of the infected meninges, dural venous sinuses, or paranasal sinuses, and can also be found in the vertebrobasilar system [3].

- ICMAs in IE are most often saccular but without a defined neck, less often fusiform, and are frequently multiple.

- ICMAs from other causes are more often fusiform than saccular and are usually single [3].

Berry Aneurysms

- Berry aneurysms are typically found in the circle of Willis.

- Berry aneurysms typically have a well-defined neck and are often single.

1.3.2 Treatment Approach

The natural history of ICMAs is unpredictable as there is no reliable determination of rupture risk. Ruptured ICMAs are consistently associated with worse outcomes, with a mortality rate of 12–90% [1]; the rates are necessarily wide-ranging given the variability of patient factors that likely control outcome more than the ICMA rupture itself. Unlike berry aneurysms, size of ICMAs is not reliably associated with risk of rupture: rapidly growing ICMAs may rupture at small sizes while slow growing ICMAs may be able to attain larger sizes [13]. Fungal infection, multiple ICMAs, vertebrobasilar system location, and meningitis are all associated with worse outcomes [1].

Given their rarity and variability, it is not surprising that there are no randomized controlled data to guide treatment of ICMAs. Treatment should target ICMAs at high risk of rupture, but as previously noted, there is no accurate method of identifying these lesions. Current treatment methods are as follows.

- Antimicrobial therapy alone or in conjunction with endovascular or surgical treatment. Antimicrobial therapy should be guided by pathogen identification and antimicrobial sensitivities when available.

  – In a pooled analysis of several case series, about 30% of ICMAs resolved on antimicrobial therapy alone, while 20% increased in size [3]. Serial imaging to monitor ICMA response to treatment is reasonable. ICMAs that shrink or remain stable during antimicrobial therapy are overall less likely to rupture, though the possibility of rupture remains [3].

- Along with ruptured ICMAs, unruptured ICMAs that appear unstable despite antimicrobial therapy should prompt consideration of endovascular or surgical treatment. Endovascular treatment has become increasingly common in light of technical advances.

  – Its chief advantages over surgery are decreased procedural risk, the ability to treat multiple and/or surgically inaccessible ICMAs, and shortened delay to cardiac surgery [14].

  – Endovascular treatment can involve direct embolization of the ICMA or indirect closure via parent artery occlusion. Direct embolization uses coiling, flow diversion, and/or liquid embolic agents (e.g., Onyx, N-butyl cyanoacrylate). As ICMAs usually lack a well-defined neck, coiling them can be technically challenging [6]. Distally located aneurysms may not be directly accessible. There is also a risk of aneurysm rupture during coil deployment, as ICMAs are friable.

Parent artery occlusion uses liquid embolic agents and/or coils to eliminate proximal blood flow to the ICMA. For ICMAs in inaccessible locations or with shapes not conducive to coiling, parent artery occlusion may be the only endovascular treatment option. However, its use is limited to noneloquent vascular territory, as the parent artery must be sacrificed. Collateral circulation should be assessed prior to vessel occlusion.

Both coiling and liquid embolic agent methods have been used successfully in published case reports [14]. Previous concerns regarding the placement of stents and coils into infected vessels have not been borne out, as there is no evidence that the use of endovascular devices prolongs infection [3].

- Surgical treatment of ICMAs comprises multiple techniques, including clipping, resection, and/or parent artery ligation. Surgery remains crucial for ruptured ICMAs requiring hematoma evacuation and unruptured ICMAs with significant mass effect.
    - Surgical treatment offers a particular advantage over endovascular treatment for ICMAs in eloquent tissue, as flow preservation through anastomosis or bypass can be done.

    - The technical challenges are highly variable; ICMAs in IE may be quite distal and relatively accessible while vertebrobasilar system ICMAs from meningitis or other local infection may not be operable. Unlike berry aneurysms, saccular ICMAs can be risky to clip as their friable walls tend to bleed [15].

    - The chief disadvantages of surgical treatment are its higher procedural risk and delay of cardiac surgery. Anticoagulating patients during cardiac surgery after neurosurgical treatment of ICMAs risks new intracranial bleeding.

    Based on case series, postponing cardiac surgery at least 2 weeks reduces this risk, with better safety after 4 weeks, though significant neurologic morbidity can still occur [16]. In accordance with this limited data, the AHA recommends postponing cardiac surgery for a minimum of 2 weeks but preferably 3–4 weeks after neurosurgical treatment of ICMAs; the ESC recommends waiting at least 1 month [1, 9].

    The AHA also recommends considering bioprosthetic cardiac valves instead of mechanical valves in patients with ICMAs to avoid the need for lifelong anticoagulation [1].

It is difficult to determine the effect of endovascular or surgical treatment of ICMAs on outcomes, as patients have a serious underlying infection among other medical comorbidities that drives their prognosis. In a study from the Nationwide Inpatient Sample, conservatively managed patients with ICMAs and IE had a 27% mortality rate, compared with 15% in patients who underwent endovascular or surgical treatment [6]. However, lower mortality rates may reflect selection bias among candidates for intervention.

2 Dissecting Aneurysms

2.1 Introduction

Arterial dissection results from spontaneous or traumatic separation of the arterial wall layers, permitting blood to enter the newly created false lumen from an intimal tear or disrupted vasa vasorum. A dissecting aneurysm typically forms from dissection beneath the adventitia, which allows the intramural hematoma to stretch the arterial wall outward. Subintimal dissection, on the other hand, results in luminal stenosis or occlusion as the false lumen bows inward. As the dissecting aneurysm does not involve all of the arterial wall layers, it is not a true aneurysm.

Intracranial dissecting aneurysms (ICDAs) are uncommon, only accounting for roughly 1–2% of treated aneurysms, but are diagnosed at higher rates postmortem in cases of fatal subarachnoid hemorrhage (SAH) [17]. Intracranial artery dissection is less common than cervical artery dissection but appears more likely to cause dissecting aneurysm formation; the posterior circulation accounts for the majority of cases, but this may represent a higher likelihood of symptomatic dissection [18, 19]. While cervical carotid artery dissection was 1.8 times more likely to cause aneurysmal dilatation than cervical vertebral artery dissection in the largest published series [20], it is unclear if this same relationship holds in the intracranial circulation. In one series, vascular risk factors such as hypertension and tobacco use were not associated with aneurysm formation among patients with intracranial dissection [19]. Inherited connective tissue diseases such as Ehlers–Danlos syndrome, Marfan syndrome, and Loeys–Dietz syndrome have been implicated in case reports of intracranial dissection and ICDAs. Fibromuscular dysplasia is associated with both dissection and aneurysms but not specifically ICDAs [17]. While blunt trauma is a well-described cause of cervical artery dissection and pseudoaneurysm formation, it is not as well-associated with dissection of the intracranial vessels, though several case reports of traumatic ICDAs have been published [21]. Iatrogenic vessel wall injury can also cause dissecting aneurysm formation. Minor trauma from sudden head and neck movements is often later identified as a precipitant in cases of spontaneous cervical artery dissection, but this has not been well-associated with intracranial artery dissection.

2.2 Pathophysiology and Clinical Presentation

Intramural hematoma formation and disruption of the internal elastic lamina and media are pathologic hallmarks of ICDAs [18]. The thinner adventitia of the intracranial vessels compared to that of the cervical vessels may predispose them to subadventitial dissection, which in turn is more likely to cause aneurysmal dilatation [19]. It is unclear whether the initial intramural hematoma arises from bleeding in the adventitial vasa vasorum.

Subintimal dissection with a single entry point from the true lumen can create a rapidly expanding intramural hematoma that in turn causes focal aneurysmal dilatation. This mechanism of formation may be more unstable and predispose to rupture compared with subintimal dissection with entrance and exit points between the true to false lumens [22].

Over time, nonaneurysmal intracranial dissection can transform into chronic aneurysms (Fig. 2). In one histopathologic analysis, granulation tissue replaced the intramural hematoma 2 weeks after dissection, followed by intimal hyperplasia around the false vessel lumen. One month afterward, true lumen reformation was observed within the intima [18]. As the vessel recanalized, the irregular intima and disrupted elastic lamina and media contributed to the development of vessel ectasia. This chronic fusiform aneurysm—a true aneurysm unlike the aneurysm formed from acute dissection—appeared up to 7 years after the initial dissection [18]. ICDAs can also transform into true fusiform aneurysms over time via a similar process, as the outward aneurysmal dilation created by the intramural hematoma is incorporated into the vessel lumen [23].

There are rare reports of infections causing ICDAs, particularly varicella zoster virus, which causes vasculopathy through invasion of the arterial wall and disruption of the internal elastic lamina [5]. The pathology and management of these ICDAs overlaps with those of ICMAs, covered in the previous section.

There are no symptoms and/or signs specific to intracranial dissection over cervicocephalic dissection; similarly, ICDAs lack specific presenting features. Headache and neck pain are the most common symptoms; patients can also present with cranial nerve palsies or other focal neurologic deficits. Even when presenting with SAH, patients with intracranial dissection do not usually report thunderclap headache [17].

Most ICDAs are found after SAH is diagnosed. ICDAs are more often found in cases of intracranial dissection with SAH than cases without SAH [22]. A large case series of dissecting intracranial anterior circulation aneurysms found 65% were associated with SAH, while the remaining 35% were associated with ischemic stroke [24]. A high proportion carried a poor clinical grade on admission: 38% of those presenting with SAH were classified as Hunt and Hess grade IV or V [24].

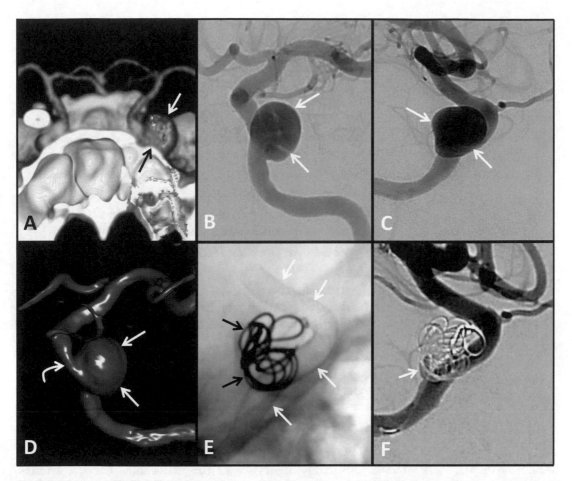

Fig. 2 A 29-year-old male with a remote history of blunt head trauma following a motorcycle crash underwent CTA (a) for an unrelated acute stab wound to the neck and was found to have a large 16 mm saccular aneurysm arising from the cavernous left internal carotid artery (arrows a). AP and lateral views from a left internal carotid artery angiogram (b, c), as well as 3D rotational angiography (d), better delineated the lesion (white arrows b, c, d). 3D rotational angiography from the left internal carotid artery also demonstrated an area of dysplasia and narrowing involving the parent vessel just distal to the aneurysm (curved arrow d). Given the patient's history of remote head trauma, as well as the associated dysplasia of the parent left internal carotid artery, the aneurysm was thought to most likely be traumatic in etiology. The lesion was successfully treated by flow diversion using the Pipeline Embolization Device (Covidien, Irvine, California) as well as adjunctive coiling. Non-subtracted and subtracted lateral views from a left internal carotid artery angiogram performed following treatment (e, f) demonstrate both the coils (black arrows e) as well as the flow diverter device (white arrows e). There was only minimal residual filling of the posterior aspect of the lesion (white arrow f), which thrombosed on follow-up angiography performed 6 weeks later (not shown)

2.3 Management Strategies

While SAH can be seen on CT and MRI, definitive diagnosis of ICDAs requires neurovascular imaging.

2.3.1 Diagnostic Approach

Imaging

- Obliteration and irregularity of the vessel lumen, creating a "string sign," may be seen on CTA or MRA, along with an intimal flap or false lumen.
  - While MRA and CTA appear roughly equivalent in sensitivity for cervical artery dissections, CTA was more sensitive at detecting dissecting aneurysms in a small retrospective study [25].
- DSA remains the gold standard for diagnosis of ICDAs.
  - Though, in a case series of anterior circulation ICDAs, MRA findings corresponded with DSA findings in all patients [24].
- Data specifically comparing MRA, CTA, and DSA in ICDAs are lacking.
- MRI, especially fat-suppressed T1 sequences to detect intramural hematoma, can be very helpful; MRI also visualizes the external vessel diameter, which is not captured by intraluminal imaging [17].
  - Recent advances in vessel wall imaging utilizing high resolution MRI have improved detection of even subtle radiological signs of dissection, including intimal tears, intramural hematomas, arterial wall enhancement, and aneurysmal dilatation [26].

2.3.2 Treatment Approach

Unlike dissecting aneurysms of the extracranial cervical arteries, which generally have a benign course [23], ICDAs carry a high risk of bleeding. Anterior and posterior circulation ICDAs presenting with SAH had a 44% and 71% risk of rebleeding before intervention, respectively, based on two case series [24, 27]. Data are very limited on ICDAs presenting with ischemic stroke; SAH has been reported in up to 22% of these patients [17].

No randomized controlled studies exist to guide management of ICDAs; small case series and expert opinion have guided practice.

- Because of the high risk of bleeding, ICDAs presenting with SAH usually necessitate endovascular or surgical treatment. Endovascular and surgical treatments exclude blood flow to the ICDA, either with sacrifice of the parent artery (deconstruction) or preservation of parent artery patency (reconstruction).
  - There has been no systematic comparison between endovascular and surgical treatment, though the former has become more common with recent technical advances.

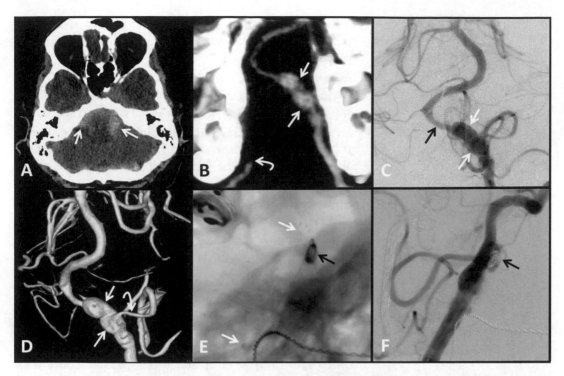

Fig. 3 A 55-year-old male recently treated for a type A aortic dissection following a motor vehicle collision presented with worst headache of life. An initial non-contrast head CT (**a**) demonstrated diffuse subarachnoid hemorrhage, most pronounced in the posterior fossa surrounding the brainstem (white arrows **a**). Subsequent CTA (**b**) demonstrated a saccular and fusiform dissecting aneurysm involving the intracranial V4 segment of the left vertebral artery (white arrows **b**), as well as dissection of the intracranial V4 segment of the right vertebral resulting in occlusion of the vessel (curved arrow **b**). Due to the patient's aortic dissection, catheter angiography was performed via a left brachial approach. An AP view from a left vertebral artery angiogram (**c**) demonstrates the saccular and fusiform aneurysm (white arrows **c**), as well as narrowing of the vertebral artery just distally (black arrow **c**). 3D rotational angiography (**d**) performed from the left vertebral artery better delineates the lesion (white arrows **d**), as well as demonstrates the left posterior inferior cerebellar artery arising from the aneurysm (curved arrow **d**). Given the occlusion of the right vertebral artery, as well as the left posterior inferior cerebellar artery originating from the aneurysm, the decision was made to attempt to preserve the left vertebral artery by performing stent assisted coil embolization of the lesion. Two Enterprise® (Cordis Neurovascular, Miami, Florida, USA) endovascular stents were placed across the lesion, followed by coil embolization of an anteriorly oriented saccular component. Non-subtracted and subtracted lateral views from a left vertebral artery angiogram performed following embolization (**e, f**) demonstrate one set of stent tines (white arrows **e**), as well as coils in the anterior saccular component (black arrows **e, f**)

- Deconstructive Techniques.

  Proximal occlusion of the parent artery by surgical clipping or endovascular coil placement, sometimes with use of liquid embolic agents.

  Trapping of the dissection and ICDA by proximal and distal clipping or intralesional coiling.

- As retrograde blood flow into ICDAs risks rebleeding, trapping is preferred over sole proximal occlusion of the parent vessel [17].

- As sacrifice of the parent artery risks infarction of the supplied vascular territory, assessment of collateral blood supply should be done prior to vessel deconstruction.

- However, given the high morbidity and mortality of ICDAs with SAH, the risk of infarction may still be acceptable compared to the benefit of intervention, even in eloquent tissue.

  - Reconstructive Techniques.

  Traditional aneurysm closure by surgical clipping or endovascular coiling.

  Endovascular stenting.

  - Unlike with berry aneurysms, traditional aneurysm closure is often technically infeasible for ICDAs, as the typically fusiform aneurysmal dilatation lacks a well-defined neck. Flow diversion stenting with or without coiling of the ICDA has been performed successfully (*see* Fig. 3), though the need for antiplatelet therapy raises concern for rebleeding [28].

  Wrapping of the aneurysmal dilatation.

  - Wrapping of the ICDA can reduce the risk of rerupture but does not ensure vessel patency.

  Surgical ICDA resection and reanastomosis.

  - ICDA resection and reanastomosis is rarely feasible but has been performed for middle cerebral artery branch dissections [17].

  Extracranial–intracranial bypass.

  - Extracranial–intracranial bypass, usually in conjunction with ICDA trapping or parent artery occlusion, is sometimes an option if stenting is infeasible and loss of the affected vascular territory is unacceptable.

- For ICDAs presenting without SAH, conservative management is an increasingly used option unless there is significant mass effect or repeated ischemic event occur [17].
  - Although data specific to ICDAs are lacking, antithrombotic treatment is recommended as patients with intracranial dissection without SAH have a high rate of recurrent ischemic stroke but a low rate of hemorrhage [18].

Despite treatment, patients with ICDAs and SAH face high rates of morbidity and mortality. In one series, among 32 patients with anterior circulation ICDAs and SAH, 23 of whom underwent endovascular or surgical treatment, 31% were dead and 33% were severely disabled or vegetative at 3 months [24]; the 31% mortality rate was matched in another series of 42 patients with vertebrobasilar ICDAs and SAH [27]. Mortality was driven by frequent rerupture, consistent with high rates of repeat hemorrhage in intracranial dissection [18].

By contrast, patients with ICDAs without SAH have generally favorable outcomes based on the limited data available. In the above case series of anterior circulation ICDAs, out of 17 patients without SAH, 76% had good recovery, 12% had moderate disability, 12% had severe disability, and none remained in a vegetative state or died at 3-month follow-up [24]. These rates approach those of cervicocephalic dissection.

3 Pseudoaneurysms

3.1 Introduction

Complete traumatic disruption of the vessel wall, including the adventitia, can cause the formation of a false lumen contained by hematoma or surrounding tissue. This is termed a pseudoaneurysm and is distinguished from a true aneurysm, in which the vessel wall layers are distended but remain intact.

Intracranial pseudoaneurysms (IPAs) are rare, accounting for less than 1% of cerebral aneurysms [29]. The etiology is usually traumatic or iatrogenic injury. As previously noted, traumatic blunt cerebrovascular injury most often affects the cervical vessels, but traumatic IPAs do occur frequently in the setting of penetrating brain injury [21]. Endovascular procedures such as mechanical thrombectomy or neurosurgical procedures such as transsphenoidal hypophysectomy can also rarely damage vessels and cause IPA formation. There are also reports of IPA formation as a complication of gamma knife surgery [30].

Patients with IPAs tend to be young and male, as expected for traumatic etiology. However, over half of patients with posterior circulation IPAs in a pooled case series were under age 16, suggesting children and adolescents may have higher shear forces acting on their vessels [21].

3.2 Pathophysiology and Clinical Presentation

The initial presentation of traumatic IPAs usually reflects the underlying head injury. SAH from an IPA will present similarly to SAH from berry aneurysm rupture: headache, meningeal signs, cranial nerve palsies and other focal neurological deficits, as well as decreased level of consciousness. A sufficiently large IPA may cause focal deficits from mass effect on surrounding structures. Extension of a ruptured internal carotid artery IPA to the cavernous segment can cause massive epistaxis and unilateral blindness [29]. Frequently the SAH occurs after several days or longer, reflecting a period of IPA formation and growth; rebleeding rates are high [21]. Iatrogenic IPAs can similarly present in a delayed fashion.

3.3 Management Strategies

3.3.1 Diagnostic Approach

Imaging

- Diffuse SAH, particularly in a cisternal pattern distinguished from focal traumatic SAH, may be the first sign of IPA on CT or MRI.

- While CTA and/or MRA are the usual initial cerebrovascular studies, DSA is often required to characterize IPAs, as contrast filling may be subtle and visualization through the arterial and venous phases is helpful [21].

- IPAs have a highly variable and irregular appearance and may be difficult to distinguish from ICDAs.

3.3.2 Treatment Approach

Treatment of IPAs is similar to that of ICDAs.

- As they usually present with SAH and often rebleed, endovascular or surgical treatment is preferred.

- Reconstructive techniques such as aneurysm clipping or coiling are even more difficult for IPAs, as there is rarely a lesional neck or dilatation to occlude, though flow diverter stents have been used successfully [29].
  - Sacrifice of the parent vessel is often required (*see* Fig. 4).

Patient outcome depends highly on the degree of predisposing injury, though SAH from IPAs can be severe. In the above pooled case series of 26 traumatic posterior circulation IPAs, 27% died and 19% could not live independently, in line with previously reported mortality rates of 30–50% for traumatic IPAs [21]. Secondary injury from treatment is not likely as important, as the typically younger IPA patients better tolerate parent vessel occlusion due to superior collateral circulation [29]. Early detection and treatment before rebleeding is thus vital.

3.4 Conclusion

Atypical aneurysms, including ICMAs, ICDAs, and IPAs, present distinct challenges in diagnosis and management. Treatment is individualized based on clinical context, location, morphology, vascular supply, and collateral flow. Hence, they escape a simplified algorithmic approach. We recommend multidisciplinary evaluation in caring for patients with these relatively rare aneurysms.

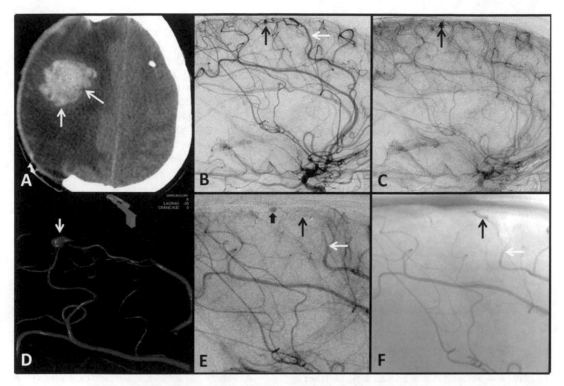

Fig. 4 A 37-year-old male in a motorcycle accident developed blunt head trauma and cervical artery dissections (not shown) complicated by right middle cerebral artery infarction requiring decompressive craniectomy. The patient subsequently developed altered mental status several weeks after the initial injury and a non-contrast head CT (**a**) demonstrated a new right frontal lobe intraparenchymal hematoma (white arrows **a**). Follow up early and delayed lateral views from a right internal carotid artery angiogram (**b, c**) identified an approximately 4 mm pseudoaneurysm with stagnant flow (black arrows **b, c**) arising from a distal cortical branch of the right callosomarginal artery (white arrow **b**). The lesion was demonstrated to better effect on 3D rotation angiography (**d**, white arrow). The pseudoaneurysm was successfully treated by sacrifice of the distal right callosomarginal artery using a liquid embolic agent (Onyx Liquid embolic system, Covidien, Irvine, California). Subtracted and non-subtracted lateral views from a right internal carotid artery angiogram performed immediately following embolization (**e, f**) demonstrated the Onyx cast (black arrows **e , f**), callosomarginal stump (white arrows **e, f**) and contrast stasis in the aneurysm sac (block arrow **e**). The aneurysm subsequently thrombosed

References

1. Wilson WR, Bower TC, Creager MA, Amin-Hanjani S, O'Gara PT, Lockhart PB et al (2016) Vascular graft infections, mycotic aneurysms, and endovascular infections: a scientific statement from the American Heart Association. Circulation 134:e412–e460

2. Deipolyi AR, Rho J, Khademhosseini A, Oklu R (2016) Diagnosis and management of mycotic aneurysms. Clin Imaging 40:256–262

3. Ducruet AF, Hickman ZL, Zacharia BE, Narula R, Grobelny BT, Gorski J et al (2010) Intracranial infectious aneurysms: a comprehensive review. Neurosurg Rev 33:37–46

4. Ramos-Estebanez C, Yavagal D (2014) Meningitis complicated by mycotic aneurysms. Oxf Med Case Reports 2014:40–42

5. Nagel MA, Gilden D (2014) Update on varicella zoster virus vasculopathy. Curr Infect Dis Rep 16:407

6. Singla A, Fargen K, Blackburn S, Neal D, Martin TD, Hess PJ et al (2016) National treatment practices in the management of infectious

intracranial aneurysms and infective endocarditis. J Neurointerv Surg 8:741–746

7. Buckmaster MJ, Curci JA, Murray PR, Liao S, Allen BT, Sicard GA et al (1999) Source of elastin-degrading enzymes in mycotic aortic aneurysms: bacteria or host inflammatory response? Cardiovasc Surg 7:16–26

8. Salgado AV, Furlan AJ, Keys TF (1987) Mycotic aneurysm, subarachnoid hemorrhage, and indications for cerebral angiography in infective endocarditis. Stroke 18:1057–1060

9. Habib G, Lancellotti P, Antunes MJ, Bongiorni MG, Casalta JP, Del Zotti F et al (2015) ESC guidelines for the management of infective endocarditis: the task force for the management of infective endocarditis of the European Society of Cardiology (esc). Endorsed by: European Association for Cardio-thoracic Surgery (EACTS), the European Association of Nuclear Medicine (EANM). Eur Heart J 36:3075–3128

10. White PM, Teasdale EM, Wardlaw JM, Easton V (2001) Intracranial aneurysms: CT angiography and MR angiography for detection prospective blinded comparison in a large patient cohort. Radiology 219:739–749

11. Hui FK, Bain M, Obuchowski NA, Gordon S, Spiotta AM, Moskowitz S et al (2015) Mycotic aneurysm detection rates with cerebral angiography in patients with infective endocarditis. J Neurointerv Surg 7:449–452

12. Walkoff L, Brinjikji W, Rouchaud A, Caroff J, Kallmes DF (2016) Comparing magnetic resonance angiography (MRA) and computed tomography angiography (CTA) with conventional angiography in the detection of distal territory cerebral mycotic and oncotic aneurysms. Interv Neuroradiol 22:524–528

13. Kannoth S, Thomas SV (2009) Intracranial microbial aneurysm (infectious aneurysm): current options for diagnosis and management. Neurocrit Care 11:120–129

14. Zanaty M, Chalouhi N, Starke RM, Tjoumakaris S, Gonzalez LF, Hasan D et al (2013) Endovascular treatment of cerebral mycotic aneurysm: a review of the literature and single center experience. Biomed Res Int 2013:151643

15. Phuong LK, Link M, Wijdicks E (2002) Management of intracranial infectious aneurysms: a series of 16 cases. Neurosurgery 51:1145–1151; discussion 1151–1142

16. Eishi K, Kawazoe K, Kuriyama Y, Kitoh Y, Kawashima Y, Omae T (1995) Surgical management of infective endocarditis associated with cerebral complications. Multi-center

retrospective study in Japan. J Thorac Cardiovasc Surg 110:1745–1755

17. Debette S, Compter A, Labeyrie MA, Uyttenboogaart M, Metso TM, Majersik JJ et al (2015) Epidemiology, pathophysiology, diagnosis, and management of intracranial artery dissection. Lancet Neurol 14:640–654

18. Ono H, Nakatomi H, Tsutsumi K, Inoue T, Teraoka A, Yoshimoto Y et al (2013) Symptomatic recurrence of intracranial arterial dissections: follow-up study of 143 consecutive cases and pathological investigation. Stroke 44:126–131

19. Badve MS, Henderson RD, O'Sullivan JD, Wong AA, Mitchell K, Coulthard A et al (2014) Vertebrobasilar dissections: case series comparing patients with and without dissecting aneurysms. J Clin Neurosci 21:2028–2030

20. Debette S, Grond-Ginsbach C, Bodenant M, Kloss M, Engelter S, Metso T et al (2011) Differential features of carotid and vertebral artery dissections: the cadisp study. Neurology 77:1174–1181

21. deSouza RM, Shah M, Koumellis P, Foroughi M (2016) Subarachnoid haemorrhage secondary to traumatic intracranial aneurysm of the posterior cerebral circulation: case series and literature review. Acta Neurochir 158:1731–1740

22. Mizutani T, Kojima H, Asamoto S, Miki Y (2001) Pathological mechanism and three-dimensional structure of cerebral dissecting aneurysms. J Neurosurg 94:712–717

23. Guillon B, Brunereau L, Biousse V, Djouhri H, Lévy C, Bousser MG (1999) Long-term follow-up of aneurysms developed during extracranial internal carotid artery dissection. Neurology 53:117–122

24. Ohkuma H, Suzuki S, Ogane K, Study Group of the Association of Cerebrovascular Disease in Tohoku Jp (2002) Dissecting aneurysms of intracranial carotid circulation. Stroke 33:941–947

25. Vertinsky AT, Schwartz NE, Fischbein NJ, Rosenberg J, Albers GW, Zaharchuk G (2008) Comparison of multidetector CT angiography and mr imaging of cervical artery dissection. AJNR Am J Neuroradiol 29:1753–1760

26. Jung SC, Kim HS, Choi CG, Kim SJ, Lee DH, Suh DC et al (2016) Quantitative analysis using high-resolution 3T MRI in acute intracranial artery dissection. J Neuroimaging 26:612–617

27. Mizutani T, Aruga T, Kirino T, Miki Y, Saito I, Tsuchida T (1995) Recurrent subarachnoid hemorrhage from untreated ruptured

vertebrobasilar dissecting aneurysms. Neurosurgery 36:905–911; discussion 912-903

28. Zhao KJ, Fang YB, Huang QH, Xu Y, Hong B, Li Q et al (2013) Reconstructive treatment of ruptured intracranial spontaneous vertebral artery dissection aneurysms: long-term results and predictors of unfavorable outcomes. PLoS One 8:e67169

29. Moon TH, Kim SH, Lee JW, Huh SK (2015) Clinical analysis of traumatic cerebral pseudoaneurysms. Korean J Neurotrauma 11:124–130

30. Sunderland G, Hassan F, Bhatnagar P, Mitchell P, Jayakrishnan V, Forster D et al (2014) Development of anterior inferior cerebellar artery pseudoaneurysm after gamma knife surgery for vestibular schwannoma. A case report and review of the literature. Br J Neurosurg 28:536–538

Part IV

Non-aneurysmal Cerebrovascular Malformations

Chapter 12

Brain Arteriovenous Malformations: Surgical, Endovascular, and Radiosurgical Techniques

Michael Crimmins, Daniel Ikeda, and Jeremy Karlin

Abstract

Brain AVMs (arteriovenous malformations) are congenital lesions with a low annual risk but high lifetime risk of morbidity and mortality which is due to rupture. This chapter provides an overview of the epidemiology, pathophysiology, natural history, clinical presentation, diagnosis, and management of AVMs. It also provides a review of the literature with presentation of the risks and benefits of the different management options available for AVM such as medical management, endovascular embolization, microsurgical resection, and stereotactic radiosurgery. Furthermore, the clinical presentation of patients with AVM and how it relates to the preferred management option are discussed in detail.

Keywords Cerebral arteriovenous malformation, A Randomized Trial of Unruptured Brain Arteriovenous Malformations (ARUBA), Spetzler-Martin grade, Microsurgery, Embolization, Stereotactic radiosurgery (SRS), Onyx, nBCA, Nidus

1 Introduction

Brain AVMs are congenital lesions with a low annual risk but high lifetime risk of morbidity and mortality which is due to rupture causing intraparenchymal and/or subarachnoid/ventricular hemorrhage, seizures, or progressive deficits related to arteriovenous shunting and perilesional hypoperfusion. Not all patients with brain AVMs should be treated. When an AVM is discovered, either incidentally or due to rupture, seizures, progressive neurological decline, or other symptoms, a discussion of risks and benefits of treatment versus conservative management needs to be had. Many patients will have a firm desire to treat, or not to treat, prior to the discussion on risks/benefits of treatment versus conservation management. While a trained and experienced neurovascular surgeon should have confidence in removing most AVMs, a discussion of individualized risks of the patient's lesion based on the prospective natural history studies will allow the patient to decide based on the information given [1]. The patient will typically choose the

Fawaz Al-Mufti and Krishna Amuluru (eds.), *Cerebrovascular Disorders*, Neuromethods, vol. 170,
https://doi.org/10.1007/978-1-0716-1530-0_12, © Springer Science+Business Media, LLC, part of Springer Nature 2021

treatment path preferred by the expert physician. This responsibility should not be taken lightly, and a multidisciplinary discussion with other surgeons and non-surgical physicians with appropriate expertise (vascular neurologists, neuroradiologists, radiosurgeons, interventional neuroradiologists) should be had. The composition of this team is dependent on the institution, but there should be a treatment decision-by-committee approach to these patients.

2 Pathophysiology and Clinical Presentation

Fundamentally, AVMs are congenital vascular lesions that arise from a disruption of normal vascular morphogenesis during fetal development. Arteriovenous malformations are deficient in an intervening capillary bed, but the precise mechanism of pathogenesis remains unknown. Some authors suggest that AVMs represent a persistence of the congenital vascular plexus with failure of the necessary venous/arterial remodeling, while others have suggested that AVMs are a dynamic product of a proliferative capillaropathy [2]. Arteriovenous malformations are further differentiated anatomically based on the involvement of brain parenchyma. A compact nidus is an organized malformed capillary bed that displaces normal brain parenchyma. A diffuse nidus is loosely organized with sparse, abnormal AV channels such that normal brain parenchyma persists [2].

The velocity of blood flow is considerably higher through AVMs than through normal brain parenchyma. As a result of the abnormal hemodynamic condition, feeding arteries and draining veins become progressively dilated and tortuous. High flow may result in saccular aneurysm formation, progressive stenosis, and eventual occlusion of feeding arteries. Draining veins can be single or multiple and deep or cortical and are also exposed to increased intraluminal stress. Direct shunting of blood at arterial pressure causes dilatation, thickening, and tortuosity in the involved veins [2].

The clinical presentation of AVMs can vary according to size and location and include signs of intracranial hemorrhage (focal deficits, nausea, vomiting, etc.), seizures, and headaches. Left untreated, AVMs possess an annual rupture rate of 3–5%, with higher rates of bleeding in AVMs that have previously ruptured or contain intranidal aneurysms. Intracranial hemorrhage is a common cause of symptomatic presentation and often is the presenting feature in the second through fourth decades of life. Seizures at presentation have been reported to occur in about 30% of patients. Headaches have been reported in 14% of patients [1, 3].

3 Management Strategies

3.1 Diagnostic Approach

The gold standard for brain AVM treatment is surgery when appropriate. The surgical treatment of AVMs has evolved significantly with the development of advanced microsurgical techniques, preoperative endovascular embolization, and standardization of postoperative neurocritical care. Despite the innovations in surgical technology and advances in medical care, patient outcomes after surgical treatment are predicated largely on patient selection. Numerous classification systems exist which are designed to assist neurosurgeons and neurointerventionalists in evaluating patients with AVMs for the most appropriate treatment.

- The seminal 1986 publication from Spetzler and Martin outlined an enduring approach to categorizing AVMS upon three variables: size of the AVM, eloquence of adjacent brain tissue, and deep or superficial venous drainage (Table 1) [4].

- Predictive of postoperative morbidity, the authors found only 1 patient of 44 patients treated with Spetzler-Martin Grade (SMG) I and II AVMs had a new postoperative deficit, whereas 31% of patients operated on with SMG V AVMs had a new neurological injury after surgery.

- Lawton further defined SMG III AVMs into lesions with different surgical risks [5]. The author posited that small AVMs in eloquent regions with deep venous drainage had operative risks similar to lower-grade lesions; however, larger AVMs (3–6 cm)

Table 1
Spetzler-Martin grading with the supplementary variable system

Spetzler-Martin grade	Points assigned	Supplementary grade
Size of AVM		Age at presentation
Small (<3 cm)	1	<20 years
Medium (3–6 cm)	2	20–40 years
Large (>6 cm)	3	>40 years
Eloquence of adjacent brain[a]		Prior hemorrhage
Noneloquent	0	Yes
Eloquent	1	No
Pattern of venous drainage		Angiographically compact
Superficial only	0	Yes
Any deep	1	No

[a]Sensorimotor, language, and visual cortex; thalamus and hypothalamus; internal capsule; brain stem; deep cerebellar nuclei and cerebellar peduncles

in eloquent regions with superficial drainage had risks similar to higher-grade lesions.

- In addition to the SMG, patient factors including age, hemorrhagic presentation, and angiographic diffuseness were recommended to be incorporated as additional potential operative risks and evaluated with a supplementary grade scale [6].

The recent study, A Randomized trial of Unruptured Brain Arteriovenous malformations (ARUBA), aimed to find the potential benefit of interventions aimed at treating AVMs as opposed to the natural history with medical management [3]. The study was halted early because of the superiority of the medical management group. The primary endpoint of new symptomatic stroke or death was met in 10.1% of patients treated in the medical group and 30.7% of the interventional therapy group.

- ARUBA is the largest prospective study to evaluate the treatment outcomes for AVMs. However the study possessed some design limitations including lack of uniformity of treatment (only 16% undergoing surgery), short follow-up, and a large portion of patients (24%) treated outside the randomization process.

- Based on published clinical series, AVMs 3 cm and smaller have an angiographic obliteration rate of 94–100% when treated surgically and a major surgical morbidity rate of 7.4% [7, 8].

- With the debatable methods and design of ARUBA, an ongoing trial, Treatment Of Brain AVMS (TOBAS), is currently enrolling patients and is hoping to show a benefit in preventive interventions for AVMs [9].

3.2 Management Strategies: Treatment Approach

3.2.1 Embolization

Pre-Embolization Considerations

- All lines should be flushing (groin, guide catheter, distal access catheter (DAC), microcatheter). It is not everyone's practice to attach the microcatheter to a flush line during catheterization.

- Heparinization to 2–3 times the initial ACT is done prior to intracranial catheterizations in all cases. (This is the author's recommendation, even for ruptured AVMs due to the high risk of clot formation on catheters or within partially embolized (casted) veins.)

Anesthesia Considerations (All AVMs)

Special Monitoring Considerations

- Arterial beat-to-beat blood pressure monitoring prior to intubation. End-tidal CO_2 capnography during induction.

- Maintain normotension and normocapnea throughout the procedure.
- Maintain normotension and normocapnea during induction to avoid rapid alterations in cerebral hemodynamics. Patients with AVMs have altered CBF and cerebral autoregulation that puts them at risk of watershed infarction or hemorrhage without proper assessment and maintenance of these factors.
- Maintain 0–1 twitches with frequent monitoring for paralysis during the case to avoid inadvertent movement of the catheter during intracranial catheterization or groin/neck dissection complications if the patient were to move.

Field of View Considerations

- Do not place the twitch monitor on the head as this can obstruct the fluoroscopic views and will require movement later.
- Place the monitors, tubes, and lines in a way so that the lateral plane can be moved freely in and out without catching on a line. ET tubes and temp probes should be below the tragus of the ear to avoid obstruction of the intracranial view.
- Some patients may need to be scanned immediately afterward. It is some surgeons' protocol to keep patients sedated and intubated after the procedure if there is a significant alteration in cerebral blood flow. The patient may need to go directly to the operating room for resection if alterations in blood flow due to embolization are considered life-threatening.

Pre-Surgical Embolization Procedure

Preparation

- What is the goal of the embolization? Nidal reduction, deep feeder embolization, shunt flow reduction, or management of high-risk component of the AVM?
- How many stages are planned for embolization? Large AVMs or AVMs with multiple feeders will typically require multiple embolizations prior to going to surgery.
- Is there a vascular neurosurgeon available if the patient requires urgent surgical treatment following embolization?
- What is the best vascular access for this patient? Is all required equipment available?
- Plan to spend a significant amount of time during the first procedure evaluating the anatomy of the AVM. A full 6-vessel diagnostic cerebral angiogram should be done before embolization.

- Plan to selectively catheterize and inject multiple intracranial vessels prior to the first embolization, to better understand the relational anatomy and to determine en passage vessels, feeder vessels, nidal flow, and venous drainage.

- Preoperative imaging with CT or MRI should be available at the time of the embolization to correlate anatomy on noninvasive imaging with procedural images. CT or MR angiography are the preferred imaging techniques as they will likely be necessary to cross-reference vessels during an angiogram to reduce the risk of embolization of normal brain territory (most important when working in or around eloquent territories).

- All lines and flushes should have 5000 international units (I.U.) of heparin in 1 liter of NS.

- All flush lines should be pressurized (300 mmHg) and checked (by the operator and the nurse/technician) for air bubbles prior to attaching to the patient to reduce the risk of clot formation and air embolus.

- Plan to recheck ACT and rebolus heparin every hour during intracranial catheterization.

Procedure

- Arterial puncture with placement of sheath sized for procedure (5F or 6F is typically adequate for embolizations).

- Check baseline ACT.

- Extracranial catheterization is performed with catheter-over-wire technique. Commonly used extracranial wires are the 0.035 inch angled Glidewire (Terumo Interventional Systems) or Bentson wire (Cook Group Inc.).

- Perform a 6-vessel angiogram utilizing a diagnostic catheter.

- If access is straightforward (most cases), a 5-French guide catheter is typically large enough to provide adequate support for embolization. A 6-French guide catheter may be preferred as better roadmapping and angiography can be performed, even with a DAC in the lumen of the guide catheter.

- Place the guide catheter into the origin of the extracranial feeding artery (typically ICA or vertebral artery). Check that the catheter is non-occlusive.

- Heparinize (50 I.U./kg is a common starting dose) the patient, and recheck an ACT in 7–10 min to ensure ACT is 2–3 times the initial. Some operators have a goal of 250–300 on ACT. Pediatric and ruptured AVM patients may require more heparin due to alterations in metabolism and distribution.

- A DAC can assist with needed support for distal embolizations. A frontal AVM with minimal tortuosity will probably not require a DAC. A parietal AVM originating from a posterior division

MCA branch is more likely to have multiple turns and loops that will be difficult to traverse without a DAC. DAC placement within the cavernous or supraclinoid ICA is common. More distal placement is feasible but likely has a higher risk of damaging perforator vessels at the tip. Many operators will steam or heat shape the tip into a hockey stick to allow for better distal tracking. DACs are typically advanced over a microcatheter and microwire to avoid the higher risk of dissection associated with tracking over a guidewire. Smaller guidewires are an alternative if additional support is needed to advance the DAC. Utilizing the shortest DAC necessary to meet your target DAC tip location will reduce the risk of running out of microcatheter length in distal embolizations.

- Selective angiography of feeder vessels should be performed with meticulous detail to understand relevant anatomy and embolization planning. A larger microcatheter (0.017 or 0.021 inch I.D.) provides better injections and mapping than what is typically used for embolization purposes (0.013 inch I.D.).

- Decisions on prioritizing goals of vessel embolization are highly debated, and there is no standardized approach (*see* Table 2 for vessel embolization triage principles).

Table 2
Vessel embolization prioritization

Features to embolize (order of importance for preoperative embolization)	Author's theory (no definitive literature exists on this topic)
Intranidal aneurysm	Changes in flow pattern can increase shear stress but redirecting flow into the remaining vessels. If intranidal aneurysms are left for later embolizations, they likely rupture at greater rates than other feeders
Deep feeders	The vessels furthest from the surface are the hardest to access during surgery. Early embolization of these feeders should be done as complications may abort subsequent embolizations
AV shunt	Reducing flow in fast-flow fistulas likely reduces risk of surgical perioperative hemorrhage
Peripheral feeders	Embolic casting of peripheral feeders can assist in intra-operative surgical identification of lesion margins
Superficial feeders	These are considered to be low risk and should not be embolized unless flow reduction is necessary
Flow-related aneurysms	These aneurysms typically do not need to be treated prior to AVM resection because they frequently resolve after resection

- After selective catheterization of the feeding artery with the embolization microcatheter, selective angiography to understand flow and the position of the draining veins is key. Effort needs to be made to have the microcatheter tip distal and below (gravity wise) the last en passage vessel to avoid reflux/occlusion into this vessel. A few cm is desirable as reflux may prevent successful liquid embolization. Typically tip location is just peripheral to the AVM nidus to allow for surgical localization. A standardized approach will allow for better coordination between the interventional and surgical team.

- Gravity positions are typically more effective at nidal penetration than anti-gravity positions.

- Each embolisate will be discussed separately below.

Onyx

- The most typically used embolisate used currently is Onyx (ethylene vinyl alcohol dissolved in DMSO with suspended tantalum powder, Medtronic). This comes in Onyx 18 (6% EVOH) and 34 (8% EVOH) (Fig. 1).

- Onyx is a copolymer that is cohesive but non-adhesive. It has a long injection time as it does not polymerize as quickly as n-BCA glue.

- Microcatheters must be flushed with DMSO (typically 0.5–1.0 mL over 1–3 min). DMSO injections faster than 0.3 mL/min can cause toxicity.

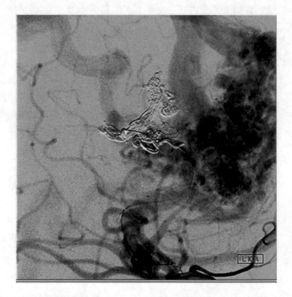

Fig. 1 SEQ figure * ARABIC 1-stage 1 AVM embolization with *Onyx* 18. There was significant nidal penetration without occlusion of either of the large draining veins

- Onyx 18 has better nidal penetration than Onyx 34 but has higher rates of retrograde flow. Onyx 18 is better for smaller feeders and when deeper penetration is desirable. Onyx 34 is better at forming the reflux plug and for large shunting arteries.

- Commonly taught methods are tapping (mini-pushes on the syringe for a few minutes to slowly build the plug), push and wait (building the plug at once), and push until reflux (building the plug and embolizing distally at the same time until reflux occurs).

- Reflux >1 cm should be avoided (unless the detachable-tip Apollo microcatheter is used) to avoid microcatheter retention.

- Injection every 1–2 min should occur to avoid clogging the microcatheter.

- It is frequently necessary to do multiple injections with frequent stops when Onyx refluxes, flows into en passage feeders, or fills the vein. The greatest risk of liquid embolization is occluding the main draining vein, causing a pressure gradient across the AVM bed (and possibly normal brain parenchyma), which greatly increases the risk of rupture. Vein occlusion is a neurosurgical emergency, requiring immediate surgical resection of the AVM.

- Injections usually take 10–25 min and can be upward of an hour. Patience is necessary as reflux can continue to occur on multiple sequential injections and at some point will often suddenly move into a desirable direction.

Key Onyx Principles

- Utilize DMSO-compatible microcatheters and syringes only.

- Onyx must be shaken for a minimum of 5 min and preferably for 30 min as the tantalum precipitates, potentially making the embolisate less radio-opaque.

- Care on syringe-catheter connections needs to be made to ensure that a DMSO meniscus has been formed to prevent catheter-syringe adhesion.

- Negative roadmaps with subtraction angiography prior to each injection can aid in determining where Onyx is going during the current injection.

- Periodic angiography during embolization from the guide catheter or DAC is advisable to determine when the vessel is occluded and to find latent en passage feeders which may not have been visible prior to embolization.

- Apollo microcatheters allow for greater reflux with minimal loss of trackability compared to other 0.013 microcatheters. The

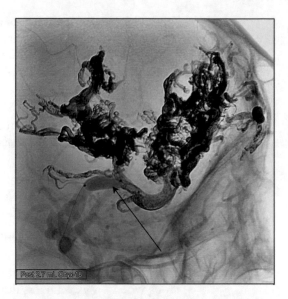

Fig. 2 The last stage of embolization for the same patient as in Fig. 1. Unsubtracted images in a similar plane as Fig. 1 demonstrate a Scepter-C balloon (arrow) in the main pedicle of the AVM which enabled good control and embolization of this pedicle without vein occlusion

detachable tip can be retained while removing the catheter with good effect.

- DMSO-compatible balloons are now available which are beneficial for large feeders and shunts (Fig. 2). This allows for continuous infusion and maximal penetration with minimal reflux. This is not useful for small vessels or tortuous feeders as it is large and not highly trackable. Vessel rupture can also occur with balloon overinflation.

n-BCA Glue

- n-BCA used to be the mainstay of AVM embolization and is used in many places both inside and outside of the United States.

- n-BCA (n-butyl cyanoacrylate) is a rapidly polymerizing substance when it comes in contact with ionic material (blood). n-BCA is typically used for vessel takedowns or single-vessel AVMs as it is fast and effective.

- n-BCA must be mixed with tantalum as it is not radio-opaque and is difficult to visualize on surgical fields when tantalum is not used.

- Ethiodized oil is used to thin the glue and must be mixed with tantalum prior to adding the glue. A ratio of 70% glue to Ethiodol is used for fast-flow (<1 s) vessels. More dilute mixture (50% or 40%) is used for slower-flow vessels to allow for greater penetration.

- 5% dextrose solution (3 mL) is thoroughly infused into the microcatheter immediately prior to attaching the glue syringe.

- A negative roadmap should be made prior to attaching the glue syringe. This can be done during dextrose infusion to ensure the catheter is cleared of ionic material.

- A dextrose meniscus should be formed on the catheter hub to avoid early polymerization of the glue in the hub.

- Infusion of glue should be done over 3–10 s. Techniques used include column building (for vessel takedown), wedge technique (tip embedded in the nidus for nidal penetration), and embolic technique (tapping very small drops to embed distally in the nidus).

Key N-BCA Principles

- Avoid polycarbonate syringes and microcatheters as they can melt.

- Careful avoidance of saline, contrast, and blood contamination must be had. Operators should utilize a dedicated prep space, dry towels, new gloves to avoid premature polymerization.

- Catheter retention risk is high as embolization time is measured in brief seconds.

- Ethiodol increases the polymerization time but also increases the viscosity of glue. Glacial acetic acid (GAA) can be utilized at low volumes (20μL in 1 mL) which increases polymerization time but does not increase viscosity.

3.2.2   Surgical Technique

Once the decision has been made to operate on a patient with brain AVM, the surgeon must consider many factors including, but not limited to, premorbid patient conditions, anatomic specifics of the AVM, whether the lesion has had preoperative embolization, and the capability of the support staff and the anesthesia team. While not exhaustive, this portion of the chapter serves to highlight the general procedural principles of the microsurgical treatment of patients with brain AVMs.

Preparation

- Anesthesia: Surgery for brain AVMs is performed with the patient under general anesthesia. A skilled neuro-anesthesiologist is as necessary as a skilled surgeon. Patients are placed under general anesthesia with arterial catheter monitoring of blood pressure and large-bore intravenous access. Brain relaxation is often performed with mild hyperventilation, diuretics, and/or cerebrospinal fluid drainage. Mannitol 0.25–0.75 g/kg is dosed prior to the craniotomy. Normotension to slight hypotension is usually preferred with a mean arterial pressure goal around 80 mmHg.

- Neurophysiological monitoring: Corticography with stimulation has been used to delineate the eloquent regions of the brain. Somatosensory evoked potentials (SSEPs) are often used to help identify or prevent injury to sensory cortex or deeper nuclei. SSEP is additionally helpful in identifying the central sulcus between motor and sensory cortexes with the "phase reversal" potential.

Procedure

- Positioning: The patient's head should be fixed in a three-point head clamp above the level of the heart. The axis of surgery should be perpendicular to the lesion, so it is optimal to have the surface of the AVM parallel to the surface of the floor.

- Craniotomy and exposure: Frameless stereotaxy can guide the appropriate size and area prepared for the skin incision and craniotomy. The craniotomy should be generous and encompass the AVM as well as a safe margin around normal cortex to ensure adequate exposure. If the draining vein is superficial, care must be taken to not injure it with bony elevation as some may traverse the dura. Brain relaxation helps with opening of the dura. The dura is opened sharply, and, once again, care must be taken not to injure any draining veins. Once the AVM is exposed, intraoperative ultrasonography may be used to ensure the craniotomy has encompassed the limits of the AVM. Intraoperative fluorescent indocyanine green may be used to assess flow toward the nidus of the AVM.

- AVM dissection and removal: Once the AVM is adequately exposed, dissection of the malformation is performed under surgical magnification, opening the thickened arachnoid at the periphery of the lesion sharply. Often surrounding cortical sulci need to be opened to identify the limits of the AVM. If preoperative embolization was performed with liquid embolic agent, the agent is often identified and can assist in recognizing the brain/ AVM interface. The most dominant draining vein must be kept intact until the surgeon is absolutely certain that all feeding arteries have been cauterized and divided (Fig. 3). For noneloquent cortex, the AVM can be resected by finding a gliotic plane surrounding the AVM without entering the margin of the nidus. *En passage* arteries, or vessels feeding the normal cortex transiting across the nidus, must be preserved. Thorough coagulation of the arteries and veins at the margin of resection is necessary prior to division. For AVMs in eloquent regions, care is taken to divide the nidus sharply from the cortex without injury to the surrounding brain. This requires even more fastidious attention to coagulation as a feeding artery that retracts within the brain can be devastating. Once the AVM has

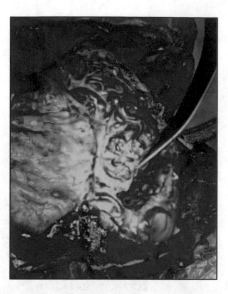

Fig. 3 AVM resection. Wide craniotomy with careful cauterization and division around the nidus, leaving the veins intact until the nidus is devascularized

circumferentially been dissected, the vein is observed for a decrease or absence of flow. Finally, it may be cauterized and divided.

- Intraoperative angiography: Once the AVM has been removed and adequate hemostasis has been obtained, intraoperative angiography is performed to ensure no remnant of AVM remains. A metallic AVM clip may be left intentionally to help serve as guide in case there is a portion of the AVM remaining.

- Closure: Attention to postoperative hemostasis is necessary to prevent postoperative hematoma formation. Blood pressure is allowed to rise to preoperative baseline during hemostasis. Oxidized cellulose is used to assist in obtaining hemostasis within the resection cavity. A watertight closure is performed of the dura, and the bone flap is reaffixed with low-profile titanium plates and screws. The galea is reapproximated, and staples are used in the superficial layer. A tightly fitting head wrap is applied to assist with postoperative hemostasis at the superficial layer.

Postoperative Care and Complications

Universally, patients are admitted to the postoperative intensive care unit. Intra-arterial and large-bore intravenous access is maintained in the immediate postoperative period. Strict fluid balance is monitored and maintained to ensure euvolemia. In some cases, especially large AVMs or when postoperative hemostasis is in question, it may be prudent to keep the patient intubated and sedated for the first 24 h after surgery. In addition, aminocaproic acid may be used for a short period postoperatively if there is a concern for bleeding. Patients are maintained on antiepileptic medication for

7 days after surgery if they have no prior history of seizure. The postoperative complication profile of microsurgery for AVM is broad, but many can be mitigated by anticipating potential pitfalls.

- Intraoperative hemorrhage: The most dreaded complication is difficult to control or uncontrollable intraoperative hemorrhage. This often represents inadvertent injury to the nidus or rupture of the AVM from injury or occlusion to the draining vein. At this time, the surgeon and team need to work quickly to remove the AVM as quickly as possible to control the bleeding while safely maintaining hemostasis.

- Postoperative hemorrhage: This likely represents a failure of hemostasis. In a rare case, this could represent residual AVM. Intraoperative angiography should decrease the risk of this. This has been reported in as many as 34% of cases; however, more modern studies show that this should be a rare complication [10].

- Cerebral ischemia: New cerebral ischemia after surgery likely represents injury to normal vessels such as *en passage* arteries during surgery or potential retrograde thrombosis of parent arteries. Though a rare complication, large, high-flow AVM may cause thrombosis in large feeding vessels as flow is abruptly decreased with AVM removal.

- Normal perfusion pressure breakthrough: Normal perfusion pressure breakthrough is a debated complication after removing large, high-flow AVMs in which patients may suffer postoperative cerebral edema or hemorrhage in adjacent cerebral cortex. It is postulated that long-standing abnormalities in blood from the AVM may cause impairment to autoregulation in adjacent cerebral cortex after the AVM is removed [4, 11]. One theory to decrease the risk of this phenomenon is to decrease flow to the AVM in a stepwise fashion with staged endovascular embolization.

- Seizures: Postoperative epileptic seizures can complicate AVM resection and increase the risk of postoperative hemorrhage. The authors continue patients with a history of epilepsy on their preoperative antiepileptic medication. If the patient has not had seizures, levetiracetam is dosed for 7 days postoperatively.

3.2.3   Stereotactic Radiosurgery

Stereotactic radiosurgery (SRS) is a widely accepted, noninvasive treatment approach for AVM. In experienced hands, SRS is potentially effective for all surgical grades with minimal morbidity. Over the past two decades, we have seen refinements in neuroimaging and treatment delivery allowing for clinically relevant improvements in the therapeutic ratio of SRS. The goal of SRS is to completely obliterate the nidus and ultimately eliminate any future

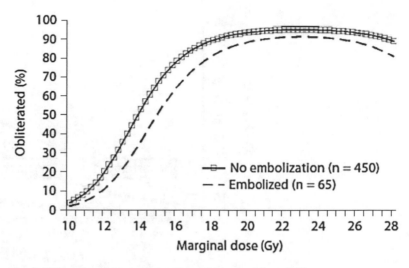

Fig. 4 Obliteration rates as a function of marginal dose [12]

risk of hemorrhage. The following sections will attempt to provide an introduction of the general principles related to treatment delivery and posttreatment management.

Dose Considerations

The efficacy of SRS is directly related to the dose prescription. There is a 50% predicted rate of obliteration at 14 Gy, and there appears to be no advantage to treating beyond 20 Gy (Fig. 4) [12]. For reasons unknown, prior attempts at embolization appear to confer a relative resistance to SRS. As the treatment volume increases, the risk of complications increases due to the amount of brain tissue exposed to high doses of radiation. The best available evidence would suggest that the amount of brain tissue, AVM included, exposed to $\geq$12 Gy is a good surrogate for predicting the risk of symptomatic radiation injury. Another important consideration is the location of the AVM. The brain stem and basal ganglia appear to be most sensitive to radiation exposure, followed by the cerebellum. The cerebrum appears to be the least sensitive, without much regard for which lobe is specifically affected (Fig. 5). A common dose used in routine practice is 17.5 Gy, and at this dose, successful obliteration should be achieved in >70% of patients. Some institutions prefer to tailor the dose to the target volume, an example being 21, 18, and 15 Gy for lesional volumes $\leq$7, 7–14, and >14 cm^3.

Treating large lesions with single-fraction SRS results in a proportionally higher risk of complications due to the increasing amount of brain tissue exposed to high doses of radiation. Two common solutions have been volume-staged and dose-staged (i.e., hypofractionated) SRS. Volume-staged SRS will usually divide the lesion into two to three abutting compartments and treat each individually with several months in between sessions. Reports of

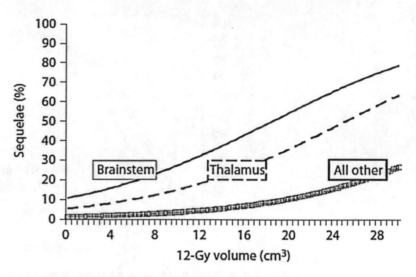

Fig. 5 Neurologic sequelae as a function of dose and location [12]

dose-staged SRS are more heterogeneous and range anywhere from 5 Gy × 6 to 10 Gy × 2 with interfraction time intervals ranging from days to weeks. A recent systematic review of the literature by Sheehan et al. suggests that both techniques result in similarly low rates of symptomatic radiation injury (<15%), but a nearly twofold obliteration rate was noted in the volume-staged group vs the dose-staged group (47.5% vs 22.8%).

Target Delineation

Due to the characteristic steep dose gradients achieved with Gamma Knife and other modern treatment delivery platforms, even small errors in target contouring can result in treatment failure. The treatment target is the nidus of the AVM, not the feeding arteries or draining vessels. Digital subtraction angiography (DSA) has traditionally been the gold standard for defining the AVM nidus because of its superior spatial resolution and dynamic temporal demonstration. However, DSA yields only two-dimensional information potentially resulting in over- and underestimations of the lesion due to complex three-dimensional variations in nidus configuration. Many centers now incorporate CT or MR angiography to provide a 3D view of the nidus.

Complication Management

Immediately following radiosurgery, the brain tissue is affected by disruption of the blood-brain barrier, vasogenic edema, and release of inflammatory cytokines. Clinically, this can manifest as worsening of pre-existing symptoms, headaches, neurologic deficits, and seizures. Although not entirely evidence based, many physicians will prescribe a short prophylactic course of corticosteroids and/or anticonvulsant with SRS. Overall, approximately 10% of patients will experience delayed neurological injury that is symptomatic, occurring anywhere from weeks to years after treatment.

The exact mechanism of this injury is unknown, but pathologic examination has revealed edema, reactive gliosis, cyst formation, blood vessel disruption/dilation with hemorrhage, and even coagulative necrosis in the irradiated tissue. Although surveillance MRI can reveal new regions of contrast enhancement or T2 signal intensity, treatment should only be initiated for symptomatic radiation injury. The mainstay of treatment has been high-dose corticosteroids. However, chronic use of corticosteroids is plagued with complications including weight gain, diabetes mellitus, immunosuppression, secondary infection, and osteopenia. Therefore, if symptoms persist after 1 month of use, next line options should be entertained. These can include therapy with pentoxifylline and vitamin E, bevacizumab, or hyperbaric oxygen. Surgical excision of the affected region should be reserved for medically refractory cases. Permanent neurologic deficits after SRS are rare (<2%).

Response Assessment

Standard follow-up protocol should include a neurological examination and neuroimaging (MRI preferred) every 6 months for 2 years and annually thereafter. When the MRI suggests nidus obliteration by loss of vascular flow voids on T2-weighted imaging, a confirmatory cerebral angiogram or MRA is recommended. Retreatment with SRS can be considered in select cases if obliteration is not successful after 3–4 years.

4 Conclusion

Patients with brain AVMs require careful treatment selection based on lesional angioarchitecture and patient factors for good outcomes. The decision to treat with surgery, radiation, endovascular approach, or a combination of the three is complex and should be made with a multidisciplinary approach focused on minimizing neurological sequelae and preventing future rupture or deterioration.

References

1. Gross BA, Du R (2013) Natural history of cerebral arteriovenous malformations: a meta-analysis. J Neurosurg 118(2):437–443
2. Asif K, Leschke J, Lazzaro MA (2013) Cerebral arteriovenous malformation diagnosis and management. Semin Neurol 33(5):468–475
3. Mohr JP, Kejda-Scharler J, Pile-Spellman J (2013) Diagnosis and treatment of arteriovenous malformations. Curr Neurol Neurosci Rep 13(2):324
4. Spetzler RF, Martin NA (1986) A proposed grading system for arteriovenous malformations. J Neurosurg 65(4):476–483
5. Lawton MT (2003) Spetzler-Martin Grade III arteriovenous malformations: surgical results and a modification of the grading scale. Neurosurgery 52(4):740–748. discussion 748–749
6. Lawton MT et al (2010) A supplementary grading scale for selecting patients with brain arteriovenous malformations for surgery. Neurosurgery 66(4):702–713. discussion 713

7. Fleetwood IG, Steinberg GK (2002) Arterio-venous malformations. Lancet 359 (9309):863–873

8. Russin J, Spetzler R (2014) Commentary: the ARUBA trial. Neurosurgery 75(1):E96–E97

9. Darsaut TE et al (2015) Treatment of brain AVMs (TOBAS): study protocol for a pragmatic randomized controlled trial. Trials 16:497

10. Pasqualin A et al (1991) Treatment of cerebral arteriovenous malformations with a combination of preoperative embolization and surgery. Neurosurgery 29(3):358–368

11. Rangel-Castilla L, Spetzler RF, Nakaji P (2015) Normal perfusion pressure breakthrough theory: a reappraisal after 35 years. Neurosurg Rev 38(3):399–404. discussion 404–405

12. Niranjan A, Kano H, Lunsford LD (2013) Gamma knife radiosurgery for brain vascular malformations. In: Kano H, Niranjan A, Lunsford LD (eds) Progress in neurological surgery. Karger, Basel, pp 49–57

Chapter 13

Managing Complex Dural Arteriovenous Fistulas: Integrating Endovascular and Surgical Approaches

Ryan C. Turner and SoHyun Boo

Abstract

Dural arteriovenous fistula (dAVF) treatment continues to evolve with improvements in endovascular embolization techniques. Transvenous endovascular embolization techniques in particular remain favored due to favorable conditions for identifying the fistula and associated draining veins. Anatomical challenges associated with endovascular navigation for treatment of dAVF can be overcome in many instances with direct access techniques in which a burr hole or craniotomy is planned over a venous sinus responsible for draining the fistula. The sheath can then be inserted directly into the arterialized venous structure, eliminating many of the challenges of navigating for transvenous approaches. Herein we demonstrate the direct access technique for successful endovascular dAVF treatment.

Key words Dural arteriovenous fistula (dAVF), Direct access, Transvenous

1 Introduction

Dural arteriovenous fistulas (dAVFs) are abnormal connections between meningeal arteries and dural venous sinuses, meningeal veins, or cortical veins that result in direct arteriovenous shunting. Dural AVFs account for approximately 10–15% of all intracranial vascular malformations [1, 2]. Clinically, dAVFs may present with pulsatile tinnitus, headache, cranial nerve deficits, and spontaneous intracranial hemorrhage. Intracranial hemorrhage has been reported in up to 35–42% of dAVFs and is thought to be secondary to the rupture of congested cortical veins from retrograde venous drainage [2–4].

The management of dural arteriovenous fistulas has evolved in recent years from a primarily surgical lesion to one that can now be managed utilizing a multitude of techniques including:

- Conventional open surgery.
- Endovascular embolization.

Fawaz Al-Mufti and Krishna Amuluru (eds.), *Cerebrovascular Disorders*, Neuromethods, vol. 170,
https://doi.org/10.1007/978-1-0716-1530-0_13, © Springer Science+Business Media, LLC, part of Springer Nature 2021

- Stereotactic radiosurgery.
- Some combination of the above.

Historically, surgery has been considered the gold standard in the management of dAVFs. However, endovascular management is playing a seemingly larger role with improvement of microcatheters and embolization materials. The heightened role of endovascular management has introduced new challenges such as identification of access routes for embolization and how to overcome fistula accessibility challenges as is often the case when performing repetitive embolization procedures.

Endovascular treatment of dAVFs involves understanding the arterial supply, venous drainage, and assessing the best access route (s) to the fistula connection. In this chapter, we provide a brief introduction on the pathophysiology of dAVFs, explore common clinical presentations and diagnostic tools utilized for dAVFs, and explore the most recent treatment strategies.

2   Pathophysiology and Clinical Presentation

2.1   Pathophysiology

Representing approximately 15% of all intracranial vascular lesions, dural arteriovenous fistulas are comprised of abnormal vascular connections between dural arteries and dural (or cortical) venous systems [5, 6]. These vascular lesions are seen across the spectrum of ages and genders, although women over 40 years old represent the most commonly afflicted population. Males are more likely to suffer from aggressive dAVFs that present with significant neurological symptoms and/or hemorrhage [1].

While advances in diagnosing and treating dAVFs have been made, the elucidation of the pathophysiology remains in its infancy. Many studies have demonstrated an association of dAVF with thrombosis of a dural sinus or venous thrombosis as well as with hypercoagulable states, such as hyper-homocysteinemia, factor V Leiden, and deficiencies of antithrombin, protein C, and protein S [1, 7–13]. Notably, dAVFs are associated with flow-related changes, particularly cortical venous hypertension, that may cause dural sinus thrombosis or occlusion, further complicating elucidation of dAVF pathophysiology [1].

2.2   Clinical Presentations

The clinical presentation of dAVF is varied and generally related to [2, 14]:

- Location.
- Arterial supply.
- Degree of shunting present.

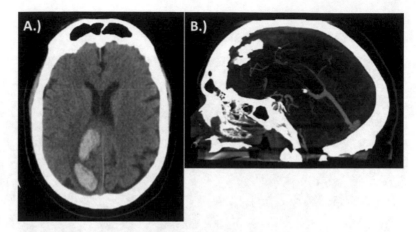

Fig. 1 (**a**) Non-contrast CT in the axial plane demonstrating right occipital intraparenchymal hemorrhage secondary to ruptured dural arteriovenous fistula. (**b**) Intracranial CTA in the sagittal plane demonstrating dural arteriovenous fistula

- Venous drainage characteristics (vessel and flow direction) which ultimately lead to hemorrhage risk.

The most severe and perhaps classic presentation of dAVF is severe headache and neurological deficit secondary to intracerebral hemorrhage (ICH) (Fig. 1). Dural AVFs may often lead to ICH due to one of two causes: (1) the high-pressure arterial flow directly into thin-walled leptomeningeal veins and (2) cortical venous congestion leading to hemorrhagic transformation [2]. More difficult to diagnose and subtle presentations of dAVF described in the literature include seizures, cranial nerve palsies, thalamic or cortical dementia, focal cortical deficits, Parkinsonism, cerebellar dysfunction, dysmetria, aphasia or dysphasia, pulsatile tinnitus, visual deterioration, proptosis, chemosis, myelopathy, and quadriparesis [2, 3, 9, 14–18]. Similarly, generalized symptoms such as headache, nausea, and vomiting are also possible [2, 3, 9, 14–18]. Symptoms may be related to changes in blood flow not only within the fistula but also within the surrounding brain, particularly in cases with cortical venous hypertension and cortical ischemia [2, 19, 20].

3 Management Strategies

3.1 Diagnostic Approach

The classification systems most often utilized for risk stratification are those from Borden et al. [3] and Cognard et al. [4] (Tables 1 and 2).

Diagnosis generally occurs following extensive radiographic imaging that often starts with a non-contrast head computed tomography (CT). While certainly useful for identifying common sequelae of dAVFs such as hemorrhage and cerebral edema,

Table 1
Borden classification of dural AVFs [3]

Category	Description
Type 1: Supply	Meningeal arteries
Drainage	Directly into dural venous sinus/meningeal vein
Presentation	Pulsatile tinnitus, cranial bruit, cranial nerve deficits, or radiculopathy
Clinical course	Often benign, high rate of spontaneous remission
Type 2: Supply	Meningeal arteries
Drainage	Directly into dural venous sinus/meningeal vein with retrograde flow into cortical veins
Presentation	Pulsatile tinnitus, cranial bruit, focal neurological deficits, seizure, hemorrhage, elevated ICP
Clinical course	Progressive neurological deficit or hemorrhage
Type 3: Supply	Meningeal arteries
Drainage	Directly into cortical veins; not through dural venous sinus
Presentation	Focal neurological deficit, seizure, hemorrhage, elevated ICP, myelopathy
Clinical course	Progressive neurological deficit or hemorrhage

Table 2
Cognard classification of dural AVFs [4]

Type	Venous drainage
1	Antegrade dural sinus drainage, *without* cortical venous reflux
2a	Retrograde dural sinus drainage, *without* cortical venous reflux
2b	Antegrade dural sinus drainage, *with* cortical venous reflux
2a + b	Retrograde dural sinus drainage, *with* cortical venous reflux
3	Direct cortical venous drainage, *without* venous ectasia
4	Direct cortical venous drainage, *with* venous ectasia
5	Direct cortical venous drainage, *with* spinal perimedullary drainage

non-contrast CT otherwise plays little role in the diagnosis and study of dAVFs [1, 2]. Magnetic resonance imaging (MRI) may provide further insight into dAVF, particularly in more aggressive dAVF (higher grade) through the demonstration of prominent vasculature, flow voids, microhemorrhages, and cerebral edema. When clinical suspicion and/or radiographic evidence indicates the possible presence of a dAVF, catheter-based cerebral angiography remains the definitive study (gold standard) for visualization and diagnosis secondary to enhanced spatial and temporal resolution [1, 2]. Notably, time-resolved CT angiography and MR angiography are gaining increased acceptance and playing a larger role in the diagnosis of dAVF, as indicated by a multitude of recent studies, but these techniques still lag behind classic digital subtraction angiography to some degree [2].

3.2 Treatment Approach

Management paradigms for dAVF include observation, endovascular treatment, direct surgical resection, and/or any combination of these options (Figs. 2, 3, 4, and 5). The decision to intervene is generally based upon the classification of the lesion. It should be noted that dAVFs are dynamic and can change with time from one classification to another due to development of venous stenosis, thrombosis, or arterial recruitment. This dynamic nature often times reveals itself even after treatment, if a fistula is not completely obliterated. The general algorithm we have followed and we believe is consistent with that reported in the literature is the following:

- Borden 1 or Cognard 1 and 2a.

  Low-grade lesions thought to often be asymptomatic and/or unlikely to result in neurological deficits given the low risk of hemorrhage. Generally managed conservatively
- Borden 2 and 3 or Cognard 2b and greater.

  Higher-grade dAVF, characterized by cortical or deep venous involvement. Generally, associated with higher risk of hemorrhage and neurologic sequelae and, therefore, warrant treatment.

Endovascular treatment can occur via either trans-arterial or transvenous embolization. Trans-arterial embolization techniques, while historically popular, are notorious for high failure rates secondary to recruitment of collateral flow, similar to how surgical arterial ligation has been associated with formation of other arterial anastamotic routes and fistula revascularization [21–23]. Direct surgical resection, while perhaps offering the greatest permanence, is not without morbidity and mortality, primarily related to blood loss and infarction, often venous in nature [21–24]. Given the risk of treatment failure with trans-arterial embolization and the risk of open surgical resection, transvenous embolization has become a preferred technique within the literature. The vast array of transvenous embolization approaches include the transfemoral IPS route, anterior approaches (via the superficial temporal vein [25, 26], ophthalmic vein [25, 26], and facial vein [27]), and other miscellaneous access routes such as via the superior petrosal sinus [28], contralateral cavernous sinus, basilar plexus, and cortical venous approaches. Notably, the extensive endovascular approaches

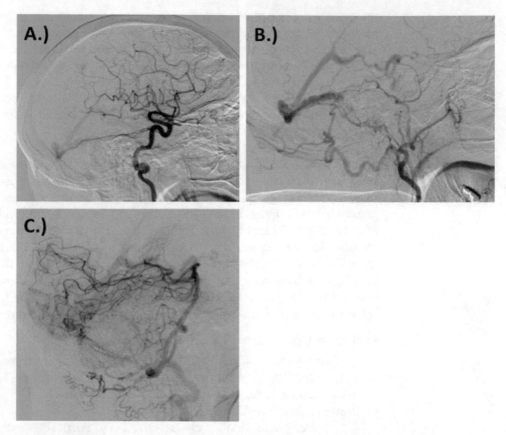

Fig. 2 Digital subtraction angiography (DSA) with (**a**) right internal carotid artery, (**b**) right external carotid artery, and (**c**) right vertebral artery injections demonstrating a Borden 2 and Cognard 2a + b dural arteriovenous fistula with numerous arterial feeders arising from meningeal arterial branches and transosseous occipital and vertebral artery branches. These feeders all appeared to shunt into the right transverse sinus with thrombosis of the right sigmoid sinus. Retrograde venous flow through the straight sinus and into the deep venous system and subsequent cortical veins makes this a high-grade lesion

described within the literature best signify and represent the increasing role of endovascular management for dAVF as endovascular management generally represents the preferred approach.

In our experience, the transvenous approach is preferred based on the likelihood of achieving a definitive cure. Transvenous approaches allow for clear identification of the fistula and draining veins, whereas trans-arterial approaches are less likely to demonstrate the true fistula and carry the risk of incomplete treatment of the fistula through individual feeding arteries. Similarly, occlusion of arterial feeders without occlusion of the fistula may result in recruitment of additional arterial supply and propagation of the dAVF. Therefore, general consensus within the literature favors a transvenous approach for the treatment of dAVF when attempting to achieve a definitive cure via endovascular techniques.

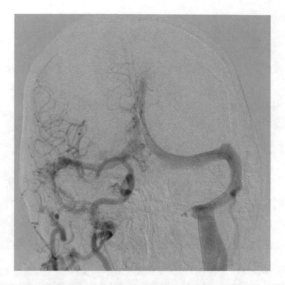

Fig. 3 Attempted transvenous approach via the left internal jugular vein for embolization of the dural arteriovenous fistula, which was unsuccessful given the inability to cross the torcula to gain access to the fistula

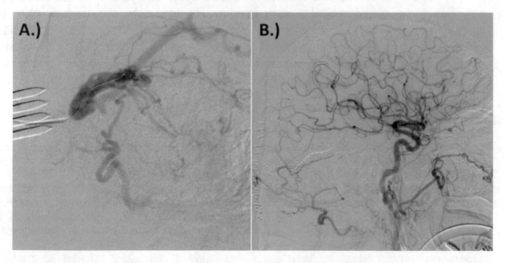

Fig. 4 (**a**) "Direct stick" approach for further transvenous embolization following burr hole craniectomy over the right transverse sinus; imaging demonstrates the access route directly at site overlying sinus. (**b**) Significant reduction in dural arteriovenous fistula following transvenous coil embolization of the right transverse sinus

4 Conclusion

The management of dural arteriovenous fistulas has evolved over time, particularly with the significant advances made in endovascular management secondary to improvements in embolization techniques. While surgical management was considered the gold

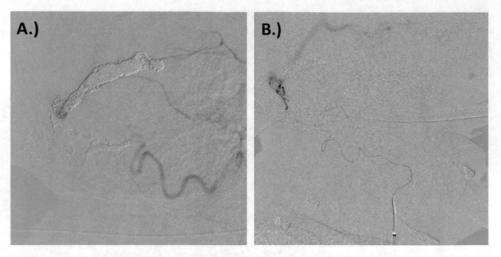

Fig. 5 (a) Follow-up right external carotid artery DSA of residual supply of fistula, which shows Onyx cast embolization of residual arterial supply, as seen on (b) super-selective angiogram of external carotid supply. Residual fistula supply was subsequently treated with stereotactic radiosurgery

standard historically, current treatment paradigms generally favor endovascular management. Endovascular techniques vary and include trans-arterial, transvenous, and direct access approaches. We provide an illustrative case implementing all of these approaches, demonstrating the need for various strategies in handling these dynamic lesions.

References

1. Miller TR, Gandhi D (2015) Intracranial dural arteriovenous fistulae: clinical presentation and management strategies. Stroke 46:2017–2025

2. Reynolds MR, Lanzino G, Zipfel GJ (2017) Intracranial dural arteriovenous fistulae. Stroke 48:1424–1431

3. Borden JA, Wu JK, Shucart WA (1995) A proposed classification for spinal and cranial dural arteriovenous fistulous malformations and implications for treatment. J Neurosurg 82:166–179

4. Cognard C, Gobin YP, Pierot L, Bailly AL, Houdart E, Casasco A, Chiras J, Merland JJ (1995) Cerebral dural arteriovenous fistulas: clinical and angiographic correlation with a revised classification of venous drainage. Radiology 194:671–680

5. Caplan JM, Kaminsky I, Gailloud P, Huang J (2015) A single burr hole approach for direct transverse sinus cannulation for the treatment of a dural arteriovenous fistula. J Neurointerv Surg 7:e5

6. Newton TH, Cronqvist S (1969) Involvement of dural arteries in intracranial arteriovenous malformations. Radiology 93:1071–1078

7. Chou CH, Lin JC, Hsueh CJ, Peng GS (2008) A thrombophilic patient with a dural arteriovenous fistula presenting sensory aphasia and complicated with an acute pulmonary embolism. Neurologist 14:327–329

8. Izumi T, Miyachi S, Hattori K, Iizuka H, Nakane Y, Yoshida J (2007) Thrombophilic abnormalities among patients with cranial dural arteriovenous fistulas. Neurosurgery 61:262–268; discussion 268–269

9. Kim MS, Han DH, Kwon OK, Oh CW, Han MH (2002) Clinical characteristics of dural arteriovenous fistula. J Clin Neurosci 9:147–155

10. Kraus JA, Stuper BK, Berlit P (1998) Association of resistance to activated protein C and dural arteriovenous fistulas. J Neurol 245:731–733

11. Kraus JA, Stuper BK, Muller J, Nahser HC, Klockgether T, Berlit P, Harbrecht U (2002)

Molecular analysis of thrombophilic risk factors in patients with dural arteriovenous fistulas. J Neurol 249:680–682

12. Safavi-Abbasi S, Di Rocco F, Nakaji P, Feigl GC, Gharabaghi A, Samii M, Valavanis A, Samii A (2008) Thrombophilia due to factor V and Factor II mutations and formation of a dural arteriovenous fistula: case report and review of a rare entity. Skull Base 18:135–143

13. Saito A, Takahashi N, Furuno Y, Kamiyama H, Nishimura S, Midorikawa H, Nishijima M (2008) Multiple isolated sinus dural arteriovenous fistulas associated with antithrombin III deficiency—case report. Neurol Med Chir (Tokyo) 48:455–459

14. Gandhi D, Chen J, Pearl M, Huang J, Gemmete JJ, Kathuria S (2012) Intracranial dural arteriovenous fistulas: classification, imaging findings, and treatment. AJNR Am J Neuroradiol 33:1007–1013

15. Hirono N, Yamadori A, Komiyama M (1993) Dural arteriovenous fistula: a cause of hypoperfusion-induced intellectual impairment. Eur Neurol 33:5–8

16. Holekamp TF, Mollman ME, Murphy RK, Kolar GR, Kramer NM, Derdeyn CP, Moran CJ, Perrin RJ, Rich KM, Lanzino G, Zipfel GJ (2016) Dural arteriovenous fistula-induced thalamic dementia: report of 4 cases. J Neurosurg 124:1752–1765

17. Hurst RW, Bagley LJ, Galetta S, Glosser G, Lieberman AP, Trojanowski J, Sinson G, Stecker M, Zager E, Raps EC, Flamm ES (1998) Dementia resulting from dural arteriovenous fistulas: the pathologic findings of venous hypertensive encephalopathy. AJNR Am J Neuroradiol 19:1267–1273

18. Wachter D, Hans F, Psychogios MN, Knauth M, Rohde V (2011) Microsurgery can cure most intracranial dural arteriovenous fistulae of the sinus and non-sinus type. Neurosurg Rev 34:337–345; discussion 345

19. Iwama T, Hashimoto N, Takagi Y, Tanaka M, Yamamoto S, Nishi S, Hayashida K (1997) Hemodynamic and metabolic disturbances in patients with intracranial dural arteriovenous fistulas: positron emission tomography evaluation before and after treatment. J Neurosurg 86:806–811

20. Kuroda S, Furukawa K, Shiga T, Ushikoshi S, Katoh C, Aoki T, Ishikawa T, Houkin K, Tamaki N, Iwasaki Y (2004) Pretreatment and posttreatment evaluation of hemodynamic and metabolic parameters in intracranial dural arteriovenous fistulae with cortical venous reflux. Neurosurgery 54:585–591; discussion 591-582

21. Barnwell SL, Halbach VV, Higashida RT, Hieshima G, Wilson CB (1989) Complex dural arteriovenous fistulas. Results of combined endovascular and neurosurgical treatment in 16 patients. J Neurosurg 71:352–358

22. Mullan S (1994) Reflections upon the nature and management of intracranial and intraspinal vascular malformations and fistulae. J Neurosurg 80:606–616

23. Selvarajah E, Boet R, Laing A (2005) Combined surgical and endovascular treatment of a posterior fossa dural arteriovenous fistula. J Clin Neurosci 12:723–725

24. Sundt TM Jr, Piepgras DG (1983) The surgical approach to arteriovenous malformations of the lateral and sigmoid dural sinuses. J Neurosurg 59:32–39

25. Klisch J, Huppertz HJ, Spetzger U, Hetzel A, Seeger W, Schumacher M (2003) Transvenous treatment of carotid cavernous and dural arteriovenous fistulae: results for 31 patients and review of the literature. Neurosurgery 53:836–856; discussion 856-837

26. White JB, Layton KF, Evans AJ, Tong FC, Jensen ME, Kallmes DF, Dion JE, Cloft HJ (2007) Transorbital puncture for the treatment of cavernous sinus dural arteriovenous fistulas. AJNR Am J Neuroradiol 28:1415–1417

27. Bink A, Goller K, Luchtenberg M, Neumann-Haefelin T, Dutzmann S, Zanella F, Berkefeld J, du Mesnil de Rochemont R (2010) Long-term outcome after coil embolization of cavernous sinus arteriovenous fistulas. AJNR Am J Neuroradiol 31:1216–1221

28. Ishihara H, Ishihara S, Kanazawa R, Kohyama S, Yamane F, Ogawa M, Sato A, Tanahashi N (2007) Transarterial NBCA embolization with transvenous partial outflow obstruction for superior petrosal sinus dural arteriovenous fistula: a case report. No Shinkei Geka 35:1157–1162

Chapter 14

Carotid Cavernous Fistulas

Mohammad El-Ghanem, Muhammad Niazi, and Kevin Cockroft

Abstract

Carotid cavernous fistulas (CCFs) are pathologic connections between the cavernous segment of the carotid artery and the cavernous venous sinus. It involves shunting blood from the internal carotid artery (ICA) and/or the external carotid artery (ECA) to the cavernous sinus (CS) (Hamby, J Neurosurg 21:859–866, 1964). The majority of the clinical signs observed in patients with CCFs are the result of hemodynamic changes in the cavernous sinus leading to increased pressure and engorgement of the draining veins (Korkmazer, World J Radiol 5:143–155, 2013). Treatment options for CCFs have evolved over the years. Starting with ligation of the common carotid artery which subsequently progressed to ICA ligation, more options in CCF ttreatment, have evloved with time. Advancement in neurointerventional technology, more specifically the invention of balloon catheters, has introduced a new era of neuroendovascular treatment modalities of CCFs.

Key words Carotid cavernous fistula, Transarterial, Transvenous, Neuroendovascular, Cavernous sinus, Carotid artery

1 Introduction

The condition of pulsating exophthalmos was first highlighted in 1809 by Benjamin Travers [1]. In 1835, a French autopsy report from Baron showed an abnormal connection between the ICA and the CS in a case of a ruptured intracavernous ICA aneurysm [2]. Later in 1870, it was shown that the portion of the ICA passing through the CS was likely to disrupt under high pressure. This was confirmed later on by autopsy [3, 4].

Carotid cavernous fistulas (CCFs) are intracranial vascular anomalies, which involve shunting blood from the internal carotid artery (ICA) and/or the external carotid artery (ECA) to the cavernous sinus (CS) [5]. The majority of the clinical signs observed in patients with CCFs are the result of hemodynamic changes in the cavernous sinus leading to increased pressure and engorgement of the draining veins [6]. MRI and catheter-based angiography are the methods of choice for diagnosis, and endovascular techniques are the mainstay of treatment [7].

Fawaz Al-Mufti and Krishna Amuluru (eds.), *Cerebrovascular Disorders*, Neuromethods, vol. 170,
https://doi.org/10.1007/978-1-0716-1530-0_14, © Springer Science+Business Media, LLC, part of Springer Nature 2021

Treatment for CCFs has evolved over the years. The first documented surgical treatment of a CCF was reported in 1809 when Benjamin Travers performed a ligation of the common carotid artery [1]. Over the next hundred years, common carotid, and then eventually ICA, ligation was the treatment of choice for CCF with strikingly high morbidity and mortality [8, 9]. With the invention of balloon catheters, a new era of endovascular neurosurgery was introduced, which had implications on the treatment of CCFs. In 1974, Serbinenko and colleagues were the first to successfully treat CCF using a detachable balloon while preserving the ICA [10, 11]. Over the years, the rapid evolution of endovascular neurosurgery technology has made lasting impacts on the treatment approach to CCFs. Currently, transarterial and transvenous embolization with detachable metallic coils or liquid embolic agents such as n-butyl cyanoacrylate (Trufill, Cordis Neurovascular, Miami Lakes, FL) or ethylene-vinyl alcohol copolymer (Onyx LES, Medtronic, Minneapolis, MN) are the primary treatment approach for CCF in the United States.

2 Pathophysiology and Clinical Presentation

2.1 Pathophysiology

A CCF permits pressurized arterial blood to be relayed directly into the CS and in turn into the draining veins, leading to significant venous hypertension. As a consequence, the reversed venous drainage of the CCFs can cause reversal of flow toward the ophthalmic venous system anteriorly; the superior petrosal sinus, the inferior petrosal sinus, or the basilar plexus posteriorly; the sphenoparietal sinus laterally; the intercavernous sinus contralaterally; and the pterygoid plexus via the vein of the foramen rotundum and the vein of the foramen ovale inferiorly. Drainage is often multidirectional [12, 13].

Several classification schemes have evolved for categorizing CCFs. The frequently used methods include classification according to:

1. Angiographic arterial architecture (direct vs. indirect).

2. Etiology (traumatic vs. spontaneous).

3. Hemodynamic features (high vs. low flow).

Perhaps the most popular of these is the angiographic classification system. This method defines the angio-architecture of the lesion and provides a basis for planning an appropriate therapeutic strategy. Barrow et al. provided a detailed anatomical classification which categorizes CCFs into four distinct types based on their arterial supply (types A–D) [14]:

Type A (direct) fistulas (Fig. 1): are direct communications between the ICA and the CS. Although flow rates in type A fistulas are variable and depend on the size of the ostium and venous

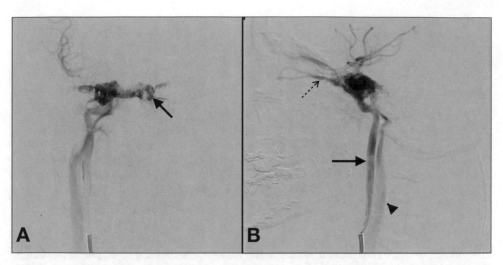

Fig. 1 AP (**a**) and lateral (**b**) right internal carotid artery angiography. (**a**) showing type A (direct) CCF, with direct communication between the ICA and CS arrow. (**b**) decreasing opacification of the ipsilateral intracranial arterial circulation "steal phenomena) and retrograde filling of the ophthalmic vein (dotted arrow) and early opacification of the ipsilateral jugular vein (arrow head)

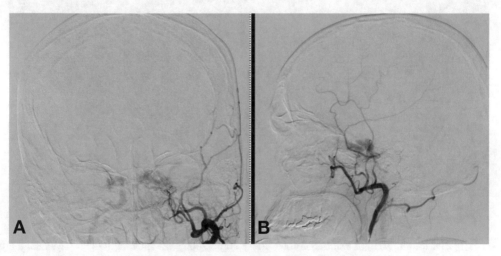

Fig. 2 Indirect CCF. AP (**a**) and lateral (**b**) left external carotid artery angiography showing external carotid artery feeders though the internal maxillary artery, with opacification of the ipsilateral as well as contralateral CS

drainage, they are typically much higher than in other types of fistulas. Type A fistulas usually range from 1 to 5 mm in size (average = 3 mm) [15]. These fistulas can be either traumatic or spontaneous. Complete steal, which is defined as the complete absence of filling of the ICA above the fistula, occurs in 5% of patients at diagnosis. The complete steal phenomenon deserves special attention because it confirms that the CCF is of high flow, that the rent in the ICA is large, and that the patient has an excellent collateral flow through the circle of Willis (assuming there are no

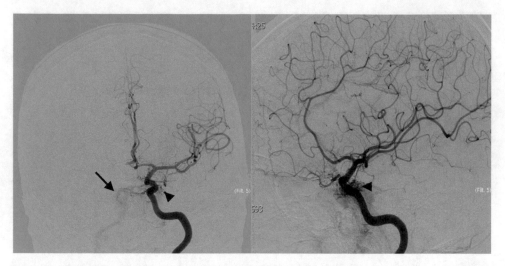

Fig. 3 AP (**a**) and lateral (**b**) left internal carotid artery angiography. (**a**) showing multiple dural ICA branches (arrow head)connected to the CS (CCF type B). Note contralateral CS opacification (arrow)

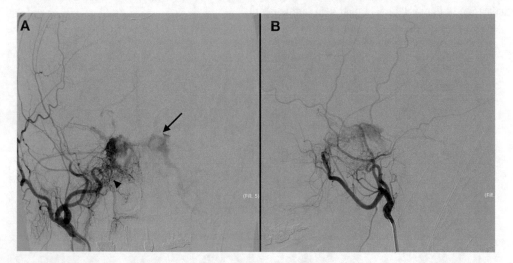

Fig. 4 AP (**a**) and lateral (**b**) right external carotid artery angiography. (**a**) showing multiple dysplastic arterial dural ECA feeders towards a type C CCF (arrow head). Note contralateral CS opacification (arrow)

clinical deficits suggesting cerebral ischemia/infarct) [16]. Direct fistulas are less likely to resolve spontaneously and generally require intervention.

Types B, C, and D (indirect) (Fig. 2): are CCFs supplied by the meningeal arteries of the ICA, the external carotid artery (ECA), or both.

- Type B (Fig. 3): fistulas have dural ICA branches shunting to the CS and are relatively uncommon.

- Type C (Fig. 4): fistulas are supplied solely by the dural branches of the ECA.

– Type D: fistulas have both dural ICA and ECA branches shunting to the CS and are the most common type. The subtype D1 denotes a unilateral fistula, while subtype D2 indicates bilateral arterial supply.

The exact cause of CCFs is still unknown; however, the majority are felt to be spontaneous, with predisposing factors including hypertension, diabetes, pregnancy, straining, atherosclerotic disease, cavernous sinus thrombosis, and collagen vascular disease [17]. Infants presenting with these types of fistulas provide evidence to support a congenital origin in some cases. Trauma is less likely to be associated with indirect CCFs.

Classification of CCFs may also be based on the underlying etiology of the fistula. Traumatic CCFs account for almost 75% of all CCFs and often demonstrate a single communication between the ICA and the CS. Traumatic CCFs are almost always found to be type A direct fistulas. Traumatic disruption of the vessel wall is the most common etiological factor. Blunt and penetrating head trauma, as well as iatrogenic damage (transsphenoidal surgery, glycerol rhizotomy, Fogarty catheter manipulation for carotid angioplasty, etc.) may lead to direct CCFs [12, 18].

Spontaneous fistulas account for approximately 25% of all CCFs, usually have multiple dural feeders, and numerous microfistulas within the CS wall. Spontaneous CCFs may fall into any of the four angiographic categories defined by Barrow et al.

Approximately 20% of type A CCFs are spontaneous and may result from any condition, which weakens the ICA wall. This could be secondary to either a rupture of a cavernous segment aneurysm or a weakened artery from atherosclerosis. Predisposing genetic factors for the development of spontaneous type A CCFs include Ehlers-Danlos syndrome, fibromuscular dysplasia, and pseudoxanthoma elasticum.

Other acquired factors that may predispose patients to the development of dural CCFs include hypertension, diabetes, pregnancy, straining, CS thrombosis, sinusitis, and collagen vascular disease [12].

2.2 Clinical Presentation

Clinical manifestations of CCFs are variable and related to the flow dynamics across the fistula and related venous drainage. Symptoms and signs that may present in patients with CCF are summarized in Table 1. Other factors affecting clinical presentation are related to the size of the fistula, location, venous drainage, and the presence of arterial/venous collaterals [18]. A CCF's flow dynamics may change over the course of the disease even without intervention, and exacerbation/remission of symptoms may occur. White-eye syndrome is an example of such dynamic change; reflecting a remission of eye congestion secondary to spontaneous occlusion of the venous drainage of the eye, causing increased intracranial venous

Table 1
Symptoms and signs of CCFs

Symptoms and signs of CCFs
Red eye (conjunctival chemosis)
Retro-orbital bruit
Proptosis
Decreased visual acuity (vision loss)
Cranial neuropathy (III, IV, VI, and trigeminal)
Headache
Intracranial venous hypertension with secondary intracranial hemorrhage
Epistaxis, otorrhagia

outflow and a potentially higher risk of a hemorrhagic complication. Therefore, careful evaluation and follow-up must be done if patients experience any changes in symptoms and signs.

Clinical presentation of CCF may be based on the angiographic arterial architecture type:

2.2.1 Direct Fistula

Direct CCFs are classically described to present with sudden onset (due to high-flow nature) of the following triad: exophthalmos, red eye "conjunctiva congestions," and bruit. Although the complete triad is the most common presentation [19], additional symptoms may be found including cranial neuropathy leading to diplopia or trigeminal nerve dysfunction, headache, vision loss that might be permanent, intracranial hemorrhage (secondary to venous hypertension), otorrhagia, and epistaxis. Functional ipsilateral ICA occlusion secondary to flow diversion across the fistula may precipitate ipsilateral ischemic events.

Vision loss is one of the most serious consequences of CCF and warrants immediate intervention. This complication results from increased venous pressure with subsequent elevation in the intraocular pressure compromising retinal perfusion. In some cases, depending on the degree and duration of retinal damage, visual acuity may not improve even after closure of fistula.

2.2.2 Indirect Fistula

Although overlap in the clinical presentations of the two fistula types often occurs, the symptoms in indirect CCFs usually emerge slowly overtime secondary to the low-flow state across the fistula. For the same reason, an orbital bruit may not present in this type. In majority of cases, slowly progressive glaucoma, proptosis, and conjunctival congestion may be the main presenting features. In this type, a relapsing and remitting pattern with changes in signs and symptoms may occur secondary to changes in flow dynamics across the fistulization site.

3 Management Strategies: Diagnostic Approach

3.1 Diagnostic Approach

Although cerebral angiography is considered the gold standard imaging modality in the diagnosis of CCFs, noninvasive cerebrovascular imaging (computed tomography (CT) scanning, magnetic resonance imaging (MRI), or CT/MR angiography) is often used for the initial work-up. However, importantly, normal noninvasive imaging studies do not necessarily exclude the diagnosis of CCF.

- Computed tomography (CT): Non-contrast CT scan of the head and orbit can be useful in patients with CCFs. CT may demonstrate radiographic proptosis, dilation of orbital veins, and enlargement of the affected cavernous sinus. In the case of trauma, careful examination may reveal an underlying fracture. In some cases, engorgement of cortical veins and signs of elevated intracranial pressure may be evident.

- Magnetic resonance imaging (MRI): Findings may overlap with the CT findings described above. Prominent flow-void signals may be seen in and around the cavernous sinus. Other findings include cerebral or orbital edema and leptomeningeal or cortical venous dilation secondary to retrograde reflux phenomena.

- CT/MR angiography: Commonly the first-line diagnostic tools used in clinical practice. These studies often have the ability to delineate the various draining veins and arterial supply, as well as providing a view of the underlying brain parenchyma. However, information about flow dynamics cannot be depicted.

- Cerebral angiography: Angiography is considered the gold standard for definite diagnosis, detailed evaluation, and planning treatment. Typically, imaging of the bilateral CCAs, ICAs, and ECAs, as well as the vertebral arteries, is needed for full evaluation [16, 20]. The following potential information can be obtained from the cerebral angiography: type, size, and location of the fistula, associated lesions including aneurysms, presence of complete or partial steal phenomena, assessment of the global cortical arterial circulation and collateral blood flow, venous drainage patterns, and evaluation of the carotid bifurcation before compression therapy.
    - Special considerations: Evaluating the exact location of the fistulous communication can be challenging due to washout of intra-arterial contrast and instantaneous opacification of the cavernous sinus. Using high frame rate imaging (>5 frames/s) and a rapid injection rate may help overcome such problem. Even with these techniques, it may still be difficult to capture the fistulous connection, particularly in high-flow CCFs. The Mehringer-Hieshima maneuver can be used which involves manual compression of the ipsilateral CCA during ICA injection at a slower frame rate. Potentially

this maneuver may slow the rate of opacification of the fistula, thus allowing better evaluation of the fistula site. The same concept can be applied using a double lumen balloon catheter in the ipsilateral ICA using a slow frame and injection rate. Furthermore, ipsilateral CCA compression during vertebral artery injection, known as the Heuber maneuver, may help visualize the fistula through a patent posterior communicating artery [17].

– Assessing tolerance to ICA occlusion is a key step in the initial cerebral angiographic evaluation. Performing a balloon test occlusion is currently the accepted technique for this evaluation.

4   Management Strategies: Treatment Approach

The main goal of treatment is to achieve complete fistula closure while preserving the parent blood vessel. In multiple series, both goals were achieved in 80% of the fistulas [21–23]. In some cases, it is impossible to achieve these two goals, taking into consideration the risk of vision loss and increasing risk of catastrophic intracranial bleeding. In these situations, it may be necessary to sacrifice the cavernous carotid artery.

The treatment modalities include conservative management, which consists of medical treatment of symptoms and manual compression therapy, surgical management, stereotactic radiosurgery, and endovascular closure via a transarterial or transvenous route.

4.1 Conservative Management

Conservative management is usually reserved for low-risk CCF patients. This strategy consists of manual compression maneuvers (external carotid-jugular manual compression of the ipsilateral side several times a week for 4–6 weeks) and treatment of the patient's symptoms while allowing time for possible spontaneous fistula closure. Spontaneous resolution has been reported in low-flow fistulas [17], secondary to thrombosis of the involved segment of the cavernous sinus [16]. In many series, complete CCF occlusion, using this maneuver, has been reported to be 30% in patients with indirect CCFs. However, the technique is usually ineffective in high-flow CCFs [17, 24, 25].

Compression therapy aims to transiently reduce the amount of shunting by decreasing arterial flow and at the same time increasing the outlet venous pressure by applying pressure on the jugular veins as well, thus helping spontaneous thrombosis within the fistula [26]. When the decision is made to use conservative treatment, the patient should be followed closely with detailed eye examination (visual acuity and intraocular pressure monitoring). Signs of progressive visual decline, optic disk swelling, and refractory elevation of intraocular pressure are all warnings for emergency endovascular intervention.

- Special consideration.
  - Careful cervical carotid bifurcation evaluation is recommended to assess for any intraluminal stenosis before initiating compression treatment.
  - The patient performs the compression using their contralateral hand so that any associated symptomatic cerebral ischemia will result in a loss of compressive force.
  - Manual compression therapy is contraindicated in the following circumstances: symptomatic bradycardia with carotid compression, significant cortical venous drainage (potentially increasing the risk of venous infarction or hemorrhage during therapy), ipsilateral significant common carotid atherosclerotic stenosis, ulcerative plaque, or history of cerebral ischemia [12, 17].

4.2 Surgical Management

An open surgical approach may be necessary when every endovascular route has been exhausted or when endovascular therapy results in venous outflow obstruction or a redirecting of flow with concomitant clinical deterioration. Direct embolizations using balloons, coils, or liquid adhesives all are available options. Intraoperative angiography can be used for temporary ICA occlusion to control intraoperative bleeding or to assess fistula obliteration. Discussion of the surgical techniques for treating CCFs is beyond the scope of this book.

4.3 Radiosurgical Management

Stereotactic radiosurgery has developed as an alternative treatment option. Radiosurgery may be used either alone or in combination with other modalities. However, the lag between treatment and clinical effect is a significant drawback. A detailed discussion of the radiosurgical methods for treating CCFs is beyond the scope of this book.

4.4 Endovascular Management

The development of detachable balloons led to a major change in the treatment of choice from surgical exposure to an endovascular approach. Furthermore, with recent advances in endovascular technology, the endovascular approach has evolved as the primary treatment option in the majority of CCF cases.

Although the clinical manifestations of direct and indirect CCFs often overlap, their natural history and the choice of endovascular treatment modality may differ significantly. The choice of endovascular method of treatment is made according to the size, type, and anatomy of the fistula, as well as operator/institutional preferences.

Using different techniques, endovascular fistula obliteration can be accomplished by either transarterial or transvenous approach. For direct CCFs, the goal of fistula obliteration can be accomplished in most cases by using a transarterial approach, either with detachable balloons (currently not commonly used) or embolization with coils or other embolic materials, or using a flow-diverting stent across the fistulous connection. Rarely, in the case of large defect, the ICA may need to be sacrificed [17]. On the other hand, in low-flow, indirect CCFs, the goal of treatment is to interrupt the fistulous communications and decrease pressure in the cavernous sinus. This can be achieved by transarterial embolization of the arterial branches supplying the fistula or, more frequently, by occluding the ipsilateral cavernous sinus using a transvenous approach. In the following section, we will discuss the various endovascular options for the treatment of direct versus indirect CCFs.

4.4.1 Endovascular Treatment: Direct CCFs

The most efficient way to approach the cavernous sinus in a direct fistula is through the arterial route. In cases where the transarterial approach is not possible or has failed, transvenous navigation to the fistulous site may be possible. For simplicity, all treatment modalities discussed below are described using the transarterial route. The transvenous approach will be discussed in a later section.

- Detachable Balloons

  Although no longer widely used, the detachable balloon catheter was one of the first endovascular devices used in the treatment of direct CCFs. In the late 1970s, it was accepted as the endovascular treatment of choice for direct CCFs. Currently there are no detachable balloon catheters approved for intracranial use in the United States, and the detachable balloon technique has largely been supplanted by embolization with coils or liquid agents.

  The main advantage of balloon occlusion of a CCF is the capability to obliterate the fistula in the venous side immediately while preserving the ICA. Although it is relatively quick and straightforward to perform, technical difficulties included balloon migration, cranial nerve palsy secondary to cavernous sinus thrombosis or mechanical pressure caused by the balloon, and/or balloon rupture.

- Microcoil Embolization

  Given the unavailability of detachable balloons, transarterial embolization with coils or other embolic material is now the mainstay of endovascular treatment for high-flow direct CCFs [17]. Using the same technique used in aneurysmal embolization, embolization of CCFs can be achieved with detachable platinum coils and liquid embolic agents such as n-butyl cyanoacrylate (n-BCA) and ethylene-vinyl alcohol copolymer (EVOH) [12].

The advantages of coils as embolizing agents include widespread availability, rapid deployment, opacity, thrombogenicity, and compatibility with various microcatheters. Advances in coil technology have led to extra features, for example, using detachable microcoils allowed interventionalists to create complex coil constructs with the ability of repositioning and removing the coil within the fistula.

- Technical consideration.
Coiling is performed through the fistula hole.

A balloon-tipped microcatheter can be used to protect the parent vessel. The balloon is inflated over the fistula hole on the arterial side, while coiling is performed through the microcatheter placed in the venous side (Fig. 5).

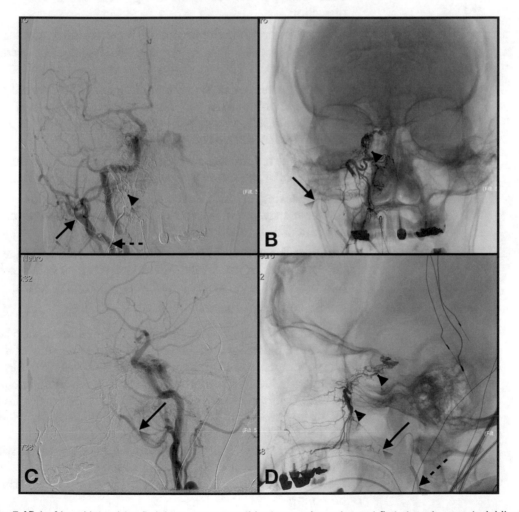

Fig. 5 AP (**a**, **b**) and lateral (**c**, **d**) right common carotid artery angiography post first stage trans-arterial liquid embolic emoblization of type D2 CCF. Note: Micro catheter in the internal maxillary artery (arrow), onyx cast (arrow head) and guid catheter (dotted arrow) positioned in the external carotid artery

An alternative way to protect the parent vessel is by using an adjuvant stent (a stent is placed in the parent vessel, and the microcatheter is placed through the stent toward the fistula opening). If a stent is to be used, pre-procedure antiplatelet therapy is mandatory with continuation for several weeks, which might be challenging in trauma patients [27].

Early case series related successful fistula closure using transarterial coil embolization frequently used other embolic agents in combination [28]; however, more recently complete obliteration using coils only has been reported in up to 75% of the cases in one case series [19]. This improvement is mainly due to advancements in coil technology.

- Complications and special considerations.
  - ICA compromise by protruding coil mass and ICA dissection.
  - Using this method, the fistula will be obliterated gradually and slowly, which may represent a disadvantage in certain circumstances.
  - As in any other treatment modality, there is a potential for incomplete obliteration and loss of transarterial access to the fistula; as a result, a second step procedure using the transvenous approach will be needed [15].
  - Cranial neuropathy can emerge post treatment either due to thrombosis of the sinus or mass effect from a large coil mass.
  - Loose packing of the fistula and the adjacent venous compartment may result in residual flow through the coil mass and potential fistula recurrence.
  - Dense packing to avoid flow within the coil mass and potential recurrence of the fistula can lead to potential coil herniation or compression of the adjacent carotid artery lumen.
- Other Embolization Materials (Fig. 6)

(a) n-Butyl Cyanoacrylate (n-BCA): n-butyl cyanoacrylate (Trufill, Cordis Neurovascular, Miami Lakes, FL) is one of the liquid embolic agents that polymerizes when exposed to blood and subsequently induces thrombosis. It is mainly used in cerebral arteriovenous malformation (AVM) and dural arteriovenous fistulas (dAVF); however, there have been some reports of its use in direct CCFs [27, 29].

Commonly, n-butyl cyanoacrylate is used in conjunction with other treatment modalities (i.e., coils) in which complete obliteration of the fistula was not achieved and sacrificing the carotid artery is not an option.

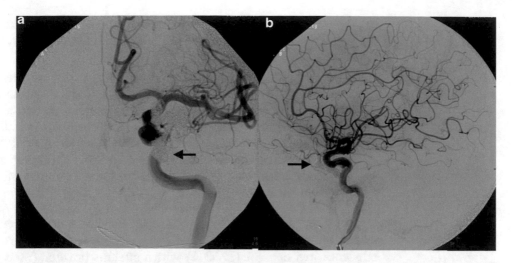

Fig. 6 Successful post trans-venous coil embolization of type C CCF: AP (**a**) and lateral (**b**) left internal carotid artery angiography. Note not residual filling of the cavernous sinuses with a coil mass in the left CS (arrow)

- Technical considerations.
    - General anesthesia may be beneficial to prevent potential patient movement, which can lead to improper visualization or arterial reflux of the material. However, conscious sedation method can be used as well.
    - The microcatheter is flushed with 5% dextrose to prevent polymerization of the n-BCA within the catheter.
    - The material is delivered under flow arrest technique (using a nondetachable balloon at the tip of the microcatheter inflated in the ICA) mainly to prevent reflex of n-BCA into the arterial circulation via the parent vessel.
    - Under continuous fluoroscopy (using a roadmap or a plain mask), careful observation for any material flow into venous component or unsuspected reflux into the ICA is essential.

In a series of 18 patients with CCFs who failed balloon closure, repeat embolization using n-butyl cyanoacrylate resulted in 89% angiographic cure with preservation of parent vessel. The rest of the patients (11%) achieved spontaneous obliteration after initial angiographic imaging showed residual filling [29].

- Complication and special considerations.
    - n-BCA leakage into the carotid artery can occlude the vessel or cause embolic ischemic phenomena.
    - Passage of n-butyl cyanoacrylate into the venous system can cause pulmonary or cerebral venous complications.
    - In CCFs with associated ICA dissection adjacent to the fistula opening, inflation of the balloon at the fistula hole is not possible. Instead the balloon can be transiently inflated distal

to the fistula opening at the level of the ophthalmic artery origin while infusing the n-butyl cyanoacrylate to protect the brain circulation [30].

– Potentially, n-butyl cyanoacrylate can bond to microcatheters and protective balloons. To avoid that, careful visualization during delivery of the material is very important, and slow delivery of n-butyl cyanoacrylate under roadmap is highly recommended. Stopping injection once reflux around the microcatheter tip is observed and quickly removing the microcatheter once reflux is observed are also important to prevent such a complication.

(b) *Ethylene-Vinyl Copolymer (Onyx)*: Onyx is a non-adhesive material in comparison to n-butyl cyanoacrylate It is a mixture of ethylene-vinyl alcohol (EVOH) with tantalum powder (used to provide opacification), dissolved in dimethyl sulfoxide (DMSO).

Compared to n-butyl cyanoacrylate Onyx is cohesive, does not bind to microcatheters, provides more controllable infusion, and does not adhere to the vessel wall. Onyx in direct CCFs was first used as an adjunct method after placement of microcoils into the venous component of the fistula to reduce the flow [31].

• Technical considerations.
  – It is essential to use a DMSO-compatible microcatheter.
  – The microcatheter is flushed with DMSO to avoid crystallization of Onyx within the catheter.
  – Onyx is infused slowly into the fistula in the venous side.
  – Given Onyx characteristics and the requirement of slow infusion, continuous monitoring by roadmap during the infusion process is highly recommended.
  – Although Onyx provides more controllable injection compared to n-butyl cyanoacrylate a nondetachable balloon catheter placed over the fistula opening may still be used to protect the parent vessel and stop flow across the fistula.
  – Intermittent balloon deflation technique with contrast injection through the guide may be used to evaluate the effectiveness of the embolization and to allow cerebral perfusion through the parent vessel.
  – After achieving the goal of fistula obliteration, the microcatheter is withdrawn into the ICA with the balloon inflated to avoid recoil of the Onyx mass into the lumen.
• Complications and special considerations.
  – Embolic infarcts may occur secondary to Onyx reflux into the parent vessel.

- Transient bradycardia and hypotension during Onyx injection, likely secondary to trigeminal stretching by rapid infusion of Onyx, may also develop [10].
- The recommended slow infusion of Onyx under continuous fluoroscopy carries an increased radiation exposure risk.
- Due to the density of Onyx, visualizing any adjacent residual fistula or pseudoaneurysm formation may be difficult.

• Flow Diverters

Flow diverters were developed to treat wide-neck cerebral aneurysms. Pipeline embolization device (PED) is the first flow diverting stent used in neuroendovascular field, other flow diverting stents were apporved by the time this chapter was published. PED is a self-expanding, cylindrical scaffolding device designed to divert the flow away from the aneurysmal sac and promote thrombosis while maintaining flow in the parent vessel and adjacent side branches. Iancu et al. and others have described treatment of CCFs using the PED with adequate closure rates [32, 33].

• Special considerations.

- Pretreatment with dual antiplatelet drugs is necessary and should be continued for several months to ensure stent patency and prevent thromboembolic complications.
- More than one device may be needed to treat a single fistula.
- The data concerning long-term effectiveness, safety, and rate of recurrence are lacking.
- Use of flow diverters for the treatment of CCFs is considered off-label at this time.

• Parent Artery Occlusion

Arterial sacrifice may be obligatory as a life-saving emergency treatment when parent vessel preservation is not feasible due to one of the following reasons: extensive trauma to the vessel wall, active hemorrhage, and a rapidly expanding hematoma of the soft tissues.

Prior to endovascular arterial sacrifice, assessment of the collateral flow and patient's ability to tolerate ICA occlusion is paramount. Balloon test occlusion (BTO) is recommended to ensure distal perfusion from collaterals. BTO can be waived in cases of complete steal presenting without any ischemic symptoms; however, careful evaluation of the quality of collateral flow through the circle of Willis should be performed (Fig. 7) [15, 17].

Endovascular ICA sacrifice can be performed using different devices (coil, balloon, and vascular plug). Coil embolization to achieve occlusion is performed using a distal-to-proximal

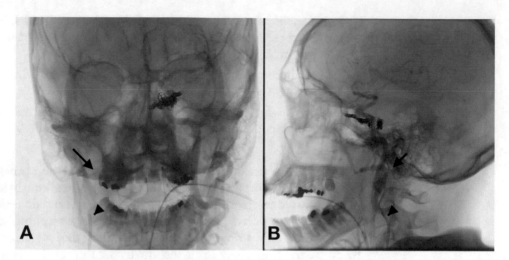

Fig. 7 Trans-venous coil embolization of an indirect CCF: AP (**a**) and lateral (**b**) Using the posterior approach through the inferior petrosal sinus. Note the micro-catheter & micro-wire (arrow). Note the guide catheter in the right internal jugular (arrow head)

approach to prevent the retrograde arterial filling of the fistula from the supraclinoid ICA [12, 17].

It is important to perform fistula entrapment during ICA sacrifice. This can be done with the aid of two balloons, which are positioned proximal and distal to the fistula. Deployment of a vascular plug is an alternative method, but due to poor navigability in the distal ICA, these devices tend to be limited to occlusions below the base of the skull [34].

- Covered Stents

Recent advancement in endovascular technology and specifically the introduction of self-expanding neurovascular stents and polytetrafluoroethylene covered stents have created additional alternatives to parent vessel sacrifice and are particularly important in patients who have failed balloon test occlusion [12, 17]. The method can be used after embolization using coils or liquid embolic materials [27].

- Technical considerations.

  - Covered stents are not approved for intracranial applications in the United States, so their use is considered off-label.

  - Evaluating tolerance for parent artery occlusion (ICA) plays a very important role in this treatment modality and should be performed preoperatively.

  - Detailed measurement of the parent vessel intended to be stented is essential for selection of the correct stent size and appropriate deployment.

  - As with bare stents, the patient must be on an antiplatelet regimen prior to and after treatment. (This step is problematic in trauma patients who usually have higher risk of bleeding in setting of other serious injuries.)

- Complications and special considerations.
    - Covered stent grafts have limited longitudinal flexibility, making it difficult to navigate through tortuous vasculature.
    - Due to irritation caused by the stiffness of the stents, peri-procedural vasospasm may be encountered.
    - Other potential complications include endoleak, covering of vital perforators, and potential injury to the parent vessel receiving the stent (dissection and rupture).

Although covered stent grafts offer a promising endovascular technique, their usage is still limited, and long-term safety date is lacking.

4.4.2 Endovascular Treatment: Indirect CCFs

As discussed earlier, spontaneous resolution of an indirect fistula is not uncommon, with reported rates of spontaneous remission up to 70% [14, 35].

A. Transvenous embolization approach: Transvenous embolization is an alternative technique for direct CCFs that cannot be treated by a transarterial route. This technique is also the preferred treatment for indirect CCFs because of lower ischemic risk, simplicity, and higher success rates. In a systematic review published in 1997, the overall success rate of transvenous embolization was 78% (Fig. 8) and 62% for transarterial approaches [36].

- Technique and special considerations.
    - Femoral vein access is established, and the desired sheath size is selected (depending on the catheters to be used) [due to the large size of the femoral vein and low risk of puncture site bleeding, even larger sheaths are usually well tolerated].

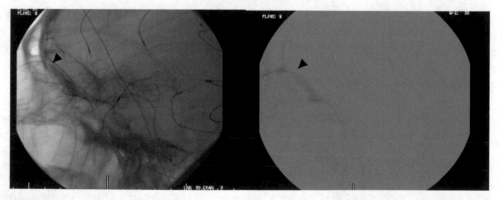

Fig. 8 Trans-venous embolization of Tpy-D CCF using the superior ophthalmic vein to access the cavernous sinus. Note the microcatheter and micro-wire (arrow head) in the SOV at the roof of the orbit, with its distal end in the cavernous sinus.

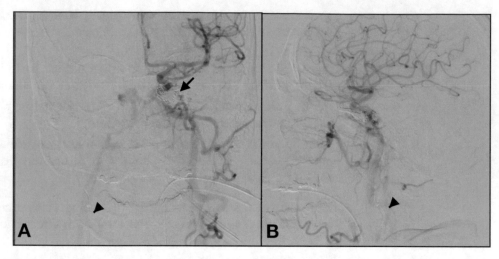

Fig. 9 Stage-1 post trans-venous coil embolization of type D CCF: AP (**a**) and lateral (**b**) left common carotid artery angiography. Note persistent filling of bilateral CS with a coil mass in the right CS (arrow). Note the guide catheter in the right internal jugular (arrow head)

- Delicate navigation through the venous structures is crucial to avoid injury to the vessel wall.

- Several different routes to the cavernous sinus may be used; the most commonly used are the inferior petrosal sinus (IPS) "posterior approach" (Fig. 9) and the superior ophthalmic vein (SOV) "anterior approach".

- Although posterior approach via IPS is more feasible, accessibility may become difficult as the disease progresses, leading to venous hypertension and potential sinus thrombosis.

- Anterior approach via SOV can be achieved directly through a surgical exposure or through the facial vein either percutaneously using ultrasound guidance or after surgical exposure.

- Less commonly used approaches are through the lateral pterygoid plexus, the superior petrosal sinus, the cortical veins, the inferior ophthalmic vein, and the contralateral IPS or even directly through the cavernous sinus puncture via the supraorbital fissure [12, 17, 20].

- It is important that embolization only be performed if the catheter tip is located within the sinus to avoid obstruction of draining veins and leaving a patent fistula, which might redirect flow to other veins and result in the loss of access to the sinus in the future.

After catheterization of the cavernous sinus, various embolic materials can be used either alone or in combination. Microcoils are usually the embolic agent of choice because they pass easily through a microcatheter without displacing it (Fig. 10). If residual filling is

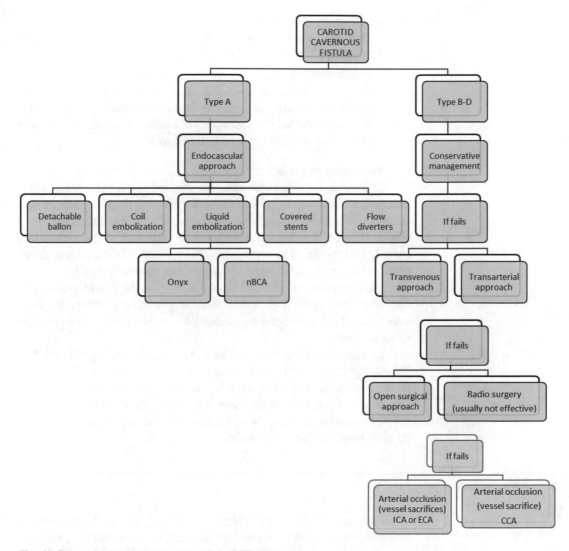

Fig. 10 Propsed management apporach in CCF treatment

noted after packing (Fig. 9), n-BCA or Onyx may be used. Alternatively, a multistage treatment approach can be used.

(c) Transarterial embolization: Low-flow (indirect) CCFs generally are smaller in size and have complex anatomy and multiple arterial feeders. All these characteristics make a transarterial approach a secondary choice of treatment. If used, a multistage strategy may be necessary. In these situations, a transarterial approach is mainly used to reduce the flow prior to transvenous occlusion for high-flow indirect CCFs and as a feasible alternative after failure of transvenous embolization [15, 17].

- Technical considerations.
  - This technique involves distal catheterization of the small meningeal branches supplying the fistula.
  - Superselective microcatheter angiograms are performed. Once the microcatheter tip is as close as possible to the fistulous communication, the liquid embolic agent of choice is injected under fluoroscopic visualization (other agents such as coils or particles can be used as well).

Postoperative period:

Transient symptom worsening, "paradoxical worsening phenomenon" may occur post CCF obliteration, due to propagation of thrombus throughout the cavernous sinus and extending into the SOV. Although concerning to the patient, these symptoms usually resolve spontaneously over time. A brief course of corticosteroids to reduce associated inflammation may help [24].

On the other hand, severe progressive ocular manifestations in the early postoperative period may suggest recurrence. In such cases, cerebral angiography should be performed with possible retreatment.

In cases where a covered stent was used, stent-graft patency follow-up is highly recommended due to the lack of long-term data. Furthermore, patients who were treated by parent artery occlusion will also benefit from long-term follow-up to monitor for the possible development of flow-related aneurysm secondary to alterations in hemodynamics.

5   Conclusion

With advances in catheter design, embolic agents, and fluoroscopic imaging equipment, neurointerventional endovascular techniques have become the preferred treatment modality for CCFs with favorable long-term outcomes. The endovascular approach should be tailored to each case. Multiple factors play role in choosing the best modality to achieve fistula obliteration. With increasing knowledge of novel endovascular techniques, such as flow-diverting devices, potentially higher success rates can be achieved while preserving ICA patency.

References

1. Travers B (1811) A case of Aneurism by anastomosis in the orbit, cured by the ligature of the common carotid artery. Med Chir Trans 2:1–420.1

2. Baron (1835) A case of ruptured internal carotid aneurysm in the cavernous sinus. Bull Soc Anat Paris 10:178

3. Delens É (1870) De la communication de la Carotide interne et du sinus caverneux (Anévrysme artério-veineux). : A. Delahaye

4. Dandy WE, The treatment of carotid cavernous arteriovenous aneurysms (1935) Ann Surg 102(5):916–926

5. Hamby WB, CAROTID-CAVERNOUS FISTULA (1964) Report of 32 surgically treated cases and suggestions for definitive operation. J Neurosurg 21:859–866

6. Korkmazer B et al (2013) Endovascular treatment of carotid cavernous sinus fistula: a systematic review. World J Radiol 5(4):143–155

7. Lang M et al (2016) A brief history of carotid-cavernous fistula. J Neurosurg:1–7

8. Murray FW (1904) XI. The treatment of pulsating exophthalmos. Ann Surg 39 (3):421–432.1

9. Locke CE (1924) Intracranial arterio-venous aneurism or pulsating exophthalmos. Ann Surg 80(2):272–285

10. Zenteno M et al (2010) Management of direct carotid-cavernous sinus fistulas with the use of ethylene-vinyl alcohol (Onyx) only: preliminary results. J Neurosurg 112(3):595–602

11. Serbinenko FA (1974) Balloon catheterization and occlusion of major cerebral vessels. J Neurosurg 41(2):125–145

12. Tjoumakaris SI, Jabbour PM, Rosenwasser RH (2009) Neuroendovascular management of carotid cavernous fistulae. Neurosurg Clin N Am 20(4):447–452

13. Coskun O et al (2000) Carotid-cavernous fistulas: diagnosis with spiral CT angiography. AJNR Am J Neuroradiol 21(4):712–716

14. Barrow DL et al (1985) Classification and treatment of spontaneous carotid-cavernous sinus fistulas. J Neurosurg 62(2):248–256

15. Ringer AJ, Salud L, Tomsick TA (2005) Carotid cavernous fistulas: anatomy, classification, and treatment. Neurosurg Clin N Am 16 (2):279–295. viii

16. Debrun GM (1995) Angiographic workup of a carotid cavernous sinus fistula (CCF) or what information does the interventionalist need for treatment? Surg Neurol 44(1):75–79

17. Gemmete JJ, Ansari SA, Gandhi DM (2009) Endovascular techniques for treatment of carotid-cavernous fistula. J Neuroophthalmol 29(1):62–71

18. Connors JJ, Wojak J (1999) Strategies and practical techniques. Interv Neuroradiol:215–226

19. Gupta AK et al (2006) Endovascular treatment of direct carotid cavernous fistulae: a pictorial review. Neuroradiology 48(11):831–839

20. Meyers PM et al (2002) Dural carotid cavernous fistula: definitive endovascular management and long-term follow-up. Am J Ophthalmol 134(1):85–92

21. Barry RC et al (2011) Interventional treatment of carotid cavernous fistula. J Clin Neurosci 18 (8):1072–1079

22. Ng PP et al (2003) Endovascular strategies for carotid cavernous and intracerebral dural arteriovenous fistulas. Neurosurg Focus 15(4): Ecp1

23. Higashida RT et al (1989) Interventional neurovascular treatment of traumatic carotid and vertebral artery lesions: results in 234 cases. AJR Am J Roentgenol 153(3):577–582

24. Komiyama M et al (1999) Brachial plexus and supraclavicular nerve injury caused by manual carotid compression for spontaneous carotid-cavernous sinus fistula. Surg Neurol 52 (3):306–309

25. Ellis JA et al (2012) Carotid-cavernous fistulas. Neurosurg Focus 32(5):E9

26. McConnell KA et al (2009) Neuroendovascular management of dural arteriovenous malformations. Neurosurg Clin N Am 20 (4):431–439

27. Moron FE et al (2005) Endovascular treatment of high-flow carotid cavernous fistulas by stent-assisted coil placement. AJNR Am J Neuroradiol 26(6):1399–1404

28. Halbach VV et al (1991) Transarterial platinum coil embolization of carotid-cavernous fistulas. AJNR Am J Neuroradiol 12(3):429–433

29. Luo CB et al (2006) Transarterial balloon-assisted n-butyl-2-cyanoacrylate embolization of direct carotid cavernous fistulas. AJNR Am J Neuroradiol 27(7):1535–1540

30. Troffkin NA, Given CA 2nd (2007) Combined transarterial N-butyl cyanoacrylate and coil embolization of direct carotid-cavernous fistulas. Report of two cases. J Neurosurg 106 (5):903–906

31. Baccin CE, Compos C, Abicalaf R et al (2005) Traumatic carotid-cavernous fistula: endovascular treatment with Onyx and coils. Interv Neuroradiol 4(11):363–367

32. Iancu D et al (2015) Flow diversion in the treatment of carotid injury and carotid-cavernous fistula after transsphenoidal surgery. Interv Neuroradiol 21(3):346–350

33. Amuluru K et al (2016) Direct carotid-cavernous fistula: a complication of, and treatment with, flow diversion. Interv Neuroradiol 22(5):569–576

34. Ross IB, Buciuc R (2007) The vascular plug: a new device for parent artery occlusion. AJNR Am J Neuroradiol 28(2):385–386

35. Sasaki H et al (1988) Long-term observations in cases with spontaneous carotid-cavernous fistulas. Acta Neurochir 90(3–4):117–120

36. Lucas CP et al (1997) Treatment for intracranial dural arteriovenous malformations: a meta-analysis from the English language literature. Neurosurgery 40(6):1119–1130. discussion 1130–1132

Part V

Cerebrovascular Stroke

Chapter 15

Thrombotic Strokes

Brian Mac Grory, Nasir Fakhri, and Shadi Yaghi

Abstract

In this chapter, we will describe the management strategy in the large artery extracranial atherosclerotic disease of the neck, dissection of the extracranial arteries of the head and neck, as well as small vessel disease. The main mechanism of stroke in patients who have extracranial atherosclerosis and/or dissection is artery-to-artery embolism, occasionally associated with hemodynamic disturbances. Although these mechanisms are also important in patients with intracranial atherosclerosis, branch occlusion and in situ thrombotic occlusion play a relatively more important role in these patients. Small vessel disease/cerebral small vessel disease describes pathology due to occlusion or stenosis of small, penetrating arteries of the brain. We discuss the pathophysiology, clinical presentations, diagnostic workups, and current treatment recommendations. Any treatment strategy should be based on the correct understanding of the stroke mechanism in individual patients.

Key words Dissection, Ischemic stroke, Pathogenesis, Stroke mechanism, Transient ischemic attack, Cerebrovascular diseases, Cerebral small vessel disease, Lacunar stroke, Atherosclerosis, Carotid stenosis, CEA, NASCET, CREST, SPACE, CAVATAS, EVA-3S, ICSS, CASANOVA, VA-COOP, ACAS, ACST

---

1 Introduction

"Thrombosis" describes a compromise of blood flow due to local occlusion of a blood vessel (i.e., an occlusive process that does not arise remotely which would be described as an "embolus"). A thrombotic stroke refers to infarct of brain parenchyma as a result of thrombosis. In this chapter, we will describe the management strategy in the large artery extracranial atherosclerotic disease of the neck, dissection of the extracranial arteries of the head and neck, as well as small vessel disease.

Thrombotic stroke arising from intracranial atherosclerotic disease will be discussed in a separate, dedicated chapter.

Fawaz Al-Mufti and Krishna Amuluru (eds.), *Cerebrovascular Disorders*, Neuromethods, vol. 170,
https://doi.org/10.1007/978-1-0716-1530-0_15, © Springer Science+Business Media, LLC, part of Springer Nature 2021

2   Large Artery Extracranial Atherosclerotic Disease of the Neck

2.1   Introduction

Disease of the large arteries of the neck includes disease affecting the common carotid arteries, internal carotid arteries, and vertebral arteries. There are many disease processes affecting these large arteries including atherosclerosis, arterial dissection, and fibromuscular dysplasia. Overwhelmingly, the most common cause of large artery disease is atherosclerotic stenosis/occlusion, and this is the area to which we will devote the majority of this discussion. Arterial dissection will be discussed in a separate section to follow.

2.2   Pathophysiology and Clinical Presentation

Atherosclerosis of the carotid artery system tends to occur preferentially at sites where there is turbulent blood flow—a factor that increases stress on the arterial intima and set in place the cascade of molecular events that lead to atherosclerotic plaque formation. Within the carotid system, the most common site for this to take place is at the bifurcation of the common carotid artery. In the vertebral system, the most common site for atherosclerotic disease is at the origin of the vertebral artery from the subclavian artery. While Atherosclerosis also preferentially occurs at the proximal basilar artery, this represents intracranial atherosclerosis and is discussed in a separate chapter. The chief risk factor for atherosclerosis of the large arteries of the neck is smoking; however, other cardiovascular risk factors also contribute including hypertension, hyperlipidemia, diabetes mellitus, obesity, and physical inactivity [1].

Atherosclerosis of the large arteries of the neck increases the liability of distal tissue infarct by three chief mechanisms:

1. Atherothrombosis: Thrombus forms on a plaque and—if this is sufficiently large—causes occlusion of the vessel and compromised distal blood flow. Within this category, one also considers hemorrhage into a plaque which can cause a sudden occlusion of a large artery. In theory, a person with an intact circle of Willis should be able to compensate for even three large vessels being occluded with collateral circulation originating from the one remaining patent neck vessel. However, in practice, many people have an incomplete circle of Willis or atherosclerosis affecting multiple vessels simultaneously.

2. Atheroembolism: Thrombus can form on the surface of an atherosclerotic plaque and migrate distally causing a distal vessel occlusion. Additionally, cholesterol fragments or calcified plaque can also break off from vessel wall plaque and itself embolize distally. From the internal carotid artery, the most common site of distal embolism is the ipsilateral middle cerebral artery (MCA). However, it is possible for emboli from the internal carotid artery to lodge in the ipsilateral anterior

cerebral artery (ACA) or even ipsilateral posterior cerebral artery (PCA) (even without a fetal posterior cerebral artery).

3. Critically decreased perfusion: In the case of—say—internal carotid stenosis, it may be the case that under normal physiology, the lumen is wide enough to permit satisfactory blood flow to the brain. However, conditions that decrease blood pressure (such as dehydration, acute blood loss, or sepsis) may expose this compromised blood vessel flow and lead to infarct in the watershed areas of the brain (the MCA/PCA and MCA/ACA territory junctions).

2.3 Management Strategies

2.3.1 Diagnostic Approach

Large artery disease of the head and neck is considered based on (1) the symptoms that a patient presents with as well as (2) signs on physical examination and is identified with the use of (3) imaging modalities.

1. The symptoms attributed to large artery disease overlap with those from any other mechanism of stroke, with some caveats. In the carotid system, the most common site for embolism is the ipsilateral middle cerebral artery, and therefore the resulting symptoms include contralateral motor weakness, contralateral sensory loss, contralateral visual loss, and gaze deviation to the side of the stroke as well as language dysfunction (in the case of a dominant hemisphere stroke) and visuospatial neglect (in the case of non-dominant hemisphere stroke). One particular phenomenon that is associated with internal carotid artery stenosis is transient retinal ischemia occurring in what is known as amaurosis fugax. This occurs when the thrombus or plaque migrates from the internal carotid artery to the ophthalmic artery (the first branch of the internal carotid artery). Another characteristic phenomenon of internal carotid artery stenosis is the so-called limb-shaking TIA which occurs when there is a decreased perfusion to the watershed region of the brain. In strokes arising from large artery disease of the vertebral arteries, the symptoms are referable to the posterior fossa and include diplopia, dysphagia, facial sensory loss, facial droop, visual loss, and ataxia.

2. There are no physical examination findings that reliably can be used to diagnose large artery atherosclerosis in the head and neck. The classically taught "carotid bruit" is neither sensitive nor specific for internal carotid artery stenosis [2].

3. Imaging.
   (a) MRI of the brain: Strokes of embolic origin (irrespective of whether they originate in the venous system, left heart, aorta, or neck vessels) have a predilection to the cortex and the watershed regions of the brain. This pattern seen on MRI is therefore suggestive of a stroke of embolic

origin which may be reflective of large artery atherosclerotic disease (but equally could represent cardioembolic or aortoembolic disease).

(b) CTA of the neck: A CTA of the neck with contrast is highly sensitive for the detection of internal carotid artery and vertebral artery stenosis. In addition to being able to detect large vessel stenosis, it is also useful for examining the aortic arch anatomy and vessel depth and orientation and (when combined with a CTA of the head) for looking for tandem stenosis or occlusion including ipsilateral intracranial stenosis.

(c) Carotid duplex ultrasonography: Despite its name, this test also assesses flow through the vertebral arteries. Duplex ultrasonography includes both B-mode (grayscale) imaging of the soft tissues of the neck and arteries and color Doppler imaging which provides dynamic imaging of blood flow patterns in the neck. It is a sensitive, versatile, and minimal-risk modality, and indeed some surgeons are prepared to perform carotid endarterectomy with only this imaging modality.

(d) Cerebral angiography: Angiography is considered the gold standard investigation for characterizing vertebral and carotid stenosis. It permits both diagnosis of carotid stenosis and immediate treatment via stenting. However, in contrast to the above non-invasive investigations, angiography is an invasive procedure with a small, but non-negligible risk of neurologic sequelae of 0.14–0.5% [3]. Angiography was the method of choice for enrollment into the NASCET trial [4]. When determining the extent of carotid artery stenosis, the NASCET investigators used the following formula:

$$\frac{(\text{Diameter of normal artery distal to the bulb}) - (\text{Diameter of maximally stenosed artery})}{(\text{Diameter of normal artery distal to the carotid bulb})}$$

In routine practice, this formula is used to calculate the degree of vessel stenosis but is applied to CTA, MRA, or duplex ultrasonography.

2.3.2 Treatment Approach

- Urgent identification of large vessel stenosis and aggressive management are always warranted. Almost all patients presenting to emergency care with symptoms suggestive of an acute ischemic stroke undergo vascular imaging. Therefore, in the majority of our patients, large artery disease can be ruled in or ruled out early in their hospital presentation. Asymptomatic large artery stenosis can also be diagnosed if it is identified during evaluation of the contralateral internal carotid or vertebral artery or if imaging is ordered because of a bruit heard in the neck.

Table 1
Trials examining CEA for symptomatic carotid stenosis

	VA-COOP group [5]	NASCET [4]	Barnett et al. [6]	ECST [7]
Year	1991	1991	1998	1998
No. of subjects	189	659	2226	3024
Inclusion criteria	>50% stenosis	70–99% stenosis	<70% stenosis	Any ICA stenosis
Intervention	CEA vs. medical management			
Primary outcome	Stroke, TIA, or perioperative mortality	Ipsilateral stroke/any stroke/death	Ipsilateral stroke	Stroke or mortality
Mean follow-up	11.9 months	18 months	5 years	6.1 years
Result	ARR of 11.7% for CEA ($p < 0.011$)	ARR of 17% for CEA ($p < 0.001$) over 2 years	ARR of 6.5% over 5 years if 50–69% stenosis. No benefit in <50%	CEA beneficial for 80–99% stenosis

Table 2
Trials examining CAS for symptomatic carotid stenosis

	SPACE [8]	EVA-3S [9]	ICSS [10]
Year	2008	2008	2015
No. of subjects	1214	527	1710
Inclusion criteria	>70% stenosis	>60% stenosis	>50% stenosis
Intervention	CEA vs. CAS		
Primary outcome	Stroke/mortality/MI in 30 days OR mortality/ipsilateral stroke in 1 year	Perioperative stroke or mortality	Mortality/disabling stroke in any territory
Mean follow-up	2 years	4 years	4.2 years
Result	Equivalent rates of recurrent ipsilateral ischemic stroke	Equivalent rates of recurrent ipsilateral ischemic stroke. Higher risk of perioperative stroke/MI with stenting	Equivalent rates of recurrent ipsilateral ischemic stroke

Table 3

Trials examining CAS for both symptomatic and asymptomatic carotid stenoses

	CAVATAS [11, 12]	SAPPHIRE [13]	CREST [14]
Year	2001	2004	2010
No. of subjects	504	334	2502
Inclusion criteria	Any carotid stenosis requiring treatment	>50% stenosis if symptomatic and >80% stenosis if asymptomatic	>50% stenosis on angiography if symptomatic and >60% if asymptomatic[a]
% symptomatic	62	29	53
% asymptomatic	38	71	47
Intervention	CAS vs. CEA		
Primary outcome	Stroke/TIA	Mortality/ipsilateral stroke in 1 year or perioperative stroke/MI/death	Stroke in 4 years or perioperative stroke/MI/death
Mean follow-up	5[b]	1 year	2.5 years
Result	Risk of ipsilateral stroke was 11.3% in CAS group and 8.6% in CEA group. Non-significant difference.	12.2% risk of stroke at 1 year in CAS group and 20.2% in CEA group. Non-significant difference. Less revascularization procedures performed in CAS group	7.2% risk of stroke at 4 years in CAS group and 6.8% in CEA group

[a]Also included symptomatic patients with >70% stenosis by CUS and those with 50–69% stenosis on CUS if a follow-up CTA or MRA demonstrated >70% stenosis *and* asymptomatic patients with >70% stenosis by CUS and those with 50–69% stenosis on CUS if follow-up CTA or MRA demonstrated >80% stenosis
[b]Median follow-up

- Hyperacute management of a stroke due to large vessel disease with tissue plasminogen activator (tPA) administration and/or intra-arterial therapy is no different than that for any other stroke subtype. The only caveat is that high-grade vessel stenosis can pose technical problems in achieving arterial access distal to the stenosis when performing an angiogram.

- The question when evaluating a patient with large artery stenosis is what is the optimal management to reduce their risk of future stroke? Note that this can represent either primary (pertaining to an asymptomatic carotid artery) or secondary (in a symptomatic carotid artery) stroke prevention.

- A "symptomatic carotid artery" describes a stenosed artery ipsilateral to which there has been a stroke or TIA with a given arbitrary period of time (by convention, 6 months prior to evaluation). There has been a wealth of clinical data pertaining

Table 4
Trials examining CEA for asymptomatic carotid stenosis

	CASANOVA [16]	VA-COOP [17]	ACAS [18]	ACST [19]
Year	1991	1993	1995	2004
No. of subjects	410	444	1662	3120
Inclusion criteria	50–90% stenosis. Included patients with bilateral stenosis	50–99% stenosis	60–99% stenosis	60–99% stenosis
Intervention	CEA vs. delayed CEA (or unilateral CEA if bilateral stenosis)	CEA vs. medical management		
Primary outcome	Stroke or perioperative mortality	Recurrent TIA or stroke	Ipsilateral stroke or perioperative stroke/mortality	Stroke or perioperative stroke/MI
Mean follow-up	3 years	4 years	2.7 years	3.4 years
Result	No difference in stroke or mortality between two groups	RR of 0.38 for stroke/TIA	RR of 0.53 for primary endpoint	6.4% risk of primary endpoint in treatment group vs. 11.8% in control group

Table 5
Trials examining CAS for asymptomatic carotid stenosis

	ACT-1 [20]
Year	2016
No. of subjects	1453
Inclusion criteria	70–99% stenosis and at low-to-moderate risk of perioperative complications
Intervention	CAS vs. CEA (enrolled in 3:1 ratio)
Primary outcome	Stroke in 1 year following procedure or perioperative stroke/death/MI
Mean follow-up	2.8 years
Result	3.4% risk of stroke at 1 year in CAS group and 3.8% in CEA group. CAS non-inferior to CEA. No statistically significant difference in perioperative complications

to the surgical treatment of symptomatic carotid artery stenosis over the past three decades [1]. Options for treatment of symptomatic carotid artery stenosis include medical management (which encompasses risk factor modification and anti-platelet therapy), carotid endarterectomy, or carotid artery angioplasty with/without stenting. The major clinical trials that addressed the topic of carotid endarterectomy (CEA) for treatment of symptomatic carotid stenosis are presented in Table 1 and carotid artery stenting (CAS) are presented in Table 2. Trials including both symptomatic and asymptomatic subjects are presented in Table 3.

In light of the above evidence:

- All patients presenting for management of symptoms suspicious for stroke/TIA obtain urgent vessel imaging with CTA of the head/neck or with carotid duplex ultrasonography.

- If there is evidence of stenotic disease ipsilateral to the brain territory to which original symptoms were referable, intervention should be considered to include either carotid endarterectomy or carotid artery stenting.

- Definitive surgical management is offered to patients with >70% stenosis of the internal carotid artery.

- Patients with stenosis of 50–69% are only offered surgical management if they are of very low operative risk and they understand the benefits/risks of undertaking treatment.

- Patients with stenosis of <50% are not offered surgical management.

- Stroke due to large artery stenosis has a very high risk of early recurrence, and therefore surgical management should be performed as soon as possible after the index event. A meta-analysis of NASCET and ECST suggested that the most benefit is achieved when surgical management is performed within the first 2 weeks after a stroke [15].

- Special circumstances:
  - Large infarcts: The benefits of early surgery must be counterbalanced with the risks of anesthesia in a person who has suffered a recent stroke. Therefore, in the case of a large infarct, surgery will usually be delayed by 10–14 days to allow the infarct to mature and for cerebral autoregulation to be restored (in order that, theoretically, the risk of blood pressure volatility during induction of anesthesia is lower).
  - Total carotid occlusion: Patients with total carotid occlusion are not intervened upon surgically. In patients with an ipsilateral total carotid occlusion by a single imaging modality (carotid ultrasound or CTA of the neck), conventional angiography to ensure that there is not a trickle of flow through the carotid which would make them a candidate for surgical intervention may be necessary.

- No salvageable tissue: If a patient has suffered a very large hemispheric stroke and there is very little tissue to be saved in the territory of the affected vessel, surgical management is usually withheld.

- Limited life expectancy: There is a high upfront risk of stroke in the perioperative period during both carotid endarterectomy and carotid artery stenting (e.g., 2.3% and 4.1%, respectively, in CREST [14]). Due to the relatively high risk of early recurrence in patients with a symptomatic carotid, there may still be a net benefit for intervention even in patients with a limited life expectancy.

- The decision to pursue carotid endarterectomy as opposed to carotid artery stenting rests on addressing individual patient factors. The evidence for the two modalities in the treatment of carotid artery stenosis does not make a compelling case for one procedure over the other, with respect to risk of recurrent ipsilateral stroke.

- Technical factors that inform the decision include plaque location and length and the presence or absence of multiple plaques because these factors dictate whether or not CEA or CAS is feasible. For instance, a plaque that is located distally in the internal carotid artery may not be accessible through a neck dissection. By contrast, a plaque that occurs at an area of tortuosity within the vessel may not be amenable to stenting.

- Stenting is typically pursued if a patient has complex cardiopulmonary issues that make general anesthesia a risk. CAS can be performed without general anesthesia and without intubation.

- Stenting necessitates the administration of dual anti-platelet therapy for a period of 30 days after placement and lifelong single anti-platelet therapy thereafter. In the case of a patient with a high risk of hemorrhage, stenting would be therefore relatively contraindicated.

- The major clinical trials that addressed the topic of carotid endarterectomy (CEA) for treatment of asymptomatic carotid stenosis are presented in Table 4 and carotid artery stenting (CAS) are presented in Table 5.

- Surgical management of asymptomatic carotid stenosis is more controversial than that for symptomatic carotid stenosis. CEA has been shown to reduce the risk of future stroke in patients with carotid stenosis. However, the magnitude of effect is very small (1% absolute risk reduction per year in the ACAS trial [18], for instance). Therefore, the procedure could theoretically benefit people with a very long life expectancy, assuming a very low operative risk and assuming that they were undergoing surgery in a high-volume center with a low complication rate. The one further caveat is that the trials on asymptomatic carotid stenosis did not include a medical arm with aggressive risk factor control. For example, statins were not widely used for stroke secondary prevention in the mid-1990s when these trials were being performed. It is probable that this marginal benefit of CEA for asymptomatic carotid stenosis would be obviated with administration of anti-platelets and a statin as well as blood pressure control and control of serum glucose. Indeed, a recent meta-analysis suggested that rates of recurrent strokes in patients with ipsilateral carotid stenosis who have aggressive medical management are as low as in patients from historic trials who have undergone carotid endarterectomy [21]. Ongoing studies, such as the CREST-2 study comparing carotid endarterectomy, carotid artery stenting, and medical treatment, may help to clarify this controversial topic.

3 Dissection of the Extracranial Arteries of the Neck

3.1 Introduction

An arterial dissection describes a separation of the layers of the arterial wall with an ensuing collection of blood between those layers [22]. Dissection of the wall of an artery creates two lumens—a false lumen in which blood aggregates and a true lumen which is the original lumen of the vessel. The arterial wall is divided into three layers, the tunica intima, tunica media, and tunica adventitia, any of which can be involved in dissection.

3.2 Pathophysiology and Clinical Presentation

In the case of the great vessels in the neck, a dissection involving the intima can cause symptoms only by acting as a nidus for thrombosis: Thrombus forms on the surface of the dissection or in the false lumen and then causes arterial occlusion or (more likely) embolizes distally causing cerebral or retinal ischemia. A dissection involving the media and adventitia can—in addition to the above—cause symptoms due to true or false aneurysm formation and ensuing mass effect. A dissection that pierces all three layers can cause a carotid or vertebral artery blowout in the neck or—once the vessel has tracked intradurally—can cause a subarachnoid hemorrhage [23]. Thus, the three mechanisms for disease in arterial dissections are:

- Local symptoms: Due to hemorrhage or mass effect. In the neck, mass effect due to aneurysm formation can manifest with cranial neuropathies.

- Hypoperfusion: A dissection may cause stenosis or occlusion of a vessel which can cause distal hypoperfusion and stroke by that mechanism.

- Embolism: Thrombus forming on the surface of a dissection may migrate distally.

Table 6
Risk factors and precipitating factors for cervical artery dissection

Risk factors	Precipitating factors
Marfan's syndrome	Blunt force trauma to the head/neck
Ehlers-Danlos syndrome	Yoga
Fibromuscular dysplasia	Chiropractic manipulation
Osteogenesis imperfecta	Sexual intercourse
Autosomal dominant polycystic kidney disease	Roller coaster rides
Reversible cerebral vasoconstriction syndrome	Contact sports (e.g., boxing)
Homocystinuria	Sports requiring reaching/neck manipulation (volleyball, basketball)
MTHFR 677TT genotype	Coughing
	Childbirth

Dissection is thought of as a more likely cause of stroke in a young person; however, dissection can occur in a person of any age [22]. They are more common in those with connective tissue disorders. Multiple precipitating factors have also been described and include activities that entail sudden, sharp movements of the neck which, by extension, place stress on arteries as they course through the neck [24]. Risk factors and precipitating factors are summarized in Table 6. Thus far, the only genetic association that has been definitively linked to arterial dissection in the neck is the MTHFR 677TT genotype [25], but most genetic studies are underpowered.

The natural history of cervical artery dissections is that they improve over time though there is usually some residual luminal irregularity that can be detected with sensitive imaging modalities [26].

Common symptoms associated with arterial dissection of the neck are headache, neck pain, dissection and ipsilateral facial and retro-orbital pain. With respect to vertebral artery dissection, pain typically occurs in the posterior neck, occiput, and post-auricular area. Headache is usually not severe in patients with arterial dissection, however thunderclap headache (sudden-onset, worst headache of life) can result if the dissection extends in intracranially.

Signs of arterial dissection in the neck include (1) signs of ischemic stroke as a result of the dissection and (2) signs due to local effect including mass effect from pseudoaneurysm formation and/or disruption of structures adjacent to the vessel wall. The canonical examination finding in internal carotid artery dissection is

Horner's syndrome. Horner's syndrome describes ptosis and myosis which are indicative of sympathetic chain dysfunction because of the involvement of third-order sympathetic fibers that course adjacent to the internal carotid artery (the third component of a Horner's syndrome—anhidrosis/absence of perspiration—does not occur in cervical carotid dissection because the sympathetic fibers that subserve perspiration in the face travel with the external carotid artery as opposed to the internal carotid artery). Mass effect from aneurysm formation in the neck can also cause lower cranial neuropathies manifesting as tongue deviation, shoulder droop, or dysphagia/dysarthria.

3.3 Management Strategies

3.3.1 Diagnostic Approach

Arterial dissections of the neck are visualized on either CTA of the neck with iodinated contrast, on MRA of the neck with (or without) gadolinium (which has approximately a 90% sensitivity for detection of cervical arterial dissection [26]), or even on carotid artery duplex ultrasonography. As with most vascular lesions of the head/neck, conventional angiography remains the gold standard despite the known procedural risks. Imaging features of dissection can include vessel luminal irregularity, vessel stenosis, vessel occlusion, the presence of a false lumen, or an aneurysm. In the case of acute dissection, the definitive imaging modality is T1-weighted MRI with fat suppression, which permits the visualization of acute intramural hematoma as a curvilinear area of signal intensity ("crescent sign").

3.3.2 Treatment Approach and Outcomes

The hyperacute management of a stroke due to arterial dissection is the same as the management for any other stroke subtype. Of particular note, tPA is not contraindicated in patients who have a known cervical arterial dissection [27]. The main management question after diagnosis of cervical artery dissection concerns the optimal strategy for either primary or secondary prevention of acute ischemic stroke as a result of the dissection. More specifically, is the optimal management strategy anti-platelet therapy (using, e.g., aspirin or clopidogrel) or anticoagulation (using, e.g., coumadin, heparin, or a target-specific oral anticoagulant (TSOAC))? [28] Trials comparing treatment strategies are limited by two chief factors: (1) the fact that dissection is an uncommon cause of stroke relative to, say, large artery atherosclerotic disease, atrial fibrillation, or small vessel disease and (2) there is a low rate of recurrent stroke associated with dissection such that it is tremendously difficult to power a trial to detect a difference in treatment outcomes.

The CADISS trial [29] was a randomized controlled trial which attempted to address superiority of either anti-platelet or anticoagulation treatment. The investigators enrolled 250 participants at 46 sites in the UK and Australia. Patients enrolled in the trial were those with either carotid or vertebral artery dissection and whose symptoms had started within 7 days. They were randomized to receive either anti-platelet treatment or anticoagulation treatment at the discretion of the treating physician and followed for 3 months. The mean time to enrollment in this trial was 3.65 days. This trial showed no statistically significant difference in the two arms, but the trial was hampered by the very low rate of stroke after the initiation of treatment (3/126 of patients in the anti-platelet group and 1/1214 patients in the anticoagulation group). These results were echoed in a non-randomized arm of the trial [30] and two meta-analyses of retrospective data [30, 31].

The decision as to whether or not to employ anticoagulation or anti-platelet therapy is based, therefore, on individual clinician preference. Those who advocate for anti-platelet therapy usually evoke the low incidence of recurrent stroke irrespective of the treatment choice and the risks of anticoagulation which include potential enlargement of the intramural hematoma or subarachnoid hemorrhage if the dissection were to propagate intracranially. Proponents of anticoagulation for treatment of cervical artery dissections evoke the front-loaded risk of stroke after initial diagnosis of arterial dissection [32] (which might suggest benefit to a very short course of anticoagulation) and the absence of evidence that anticoagulation causes enlargement of dissections. There is some limited clinical evidence that TSOACs are equally as efficacious as coumadin or warfarin in the secondary prevention of stroke in cervical artery dissection [33].

4 Small Vessel Disease

4.1 Introduction

Small vessel disease/cerebral small vessel disease describes pathology due to occlusion or stenosis of small, penetrating arteries of the brain. Dysfunction of these arteries leads to increased permeability through the blood-brain barrier, impaired cerebral autoregulation, and reduced cerebral blood flow [34]. There are some notable locations within the brain where blood supply is delivered through tiny, perforating arteries that arise directly from large parent vessels. These locations include (a) the corona radiata, internal capsule, and basal ganglia (which are supplied by lenticulostriate branches off the middle cerebral artery), (b) the thalamus (which is supplied by a network of perforators off the posterior cerebral artery), and (c) the pons (which receives its arterial supply from pontine perforators arising from the basilar artery).Small vessel disease can manifest as either a lacunar stroke (defined as a stroke that appears less than 1.5 cm in size that arises in one of the above territories) or as chronic, progressive cognitive impairment, culminating in vascular dementia. It is also possible for a stroke arising from small vessel disease to be entirely silent, i.e., not cause any symptoms.

4.2 Pathophysiology and Clinical Presentation

The arterial supply of the corona radiata, internal capsule, basal ganglia, thalamus, and pons is notable for two chief reasons:

1. The arteries supplying them are prone to atherosclerosis due to hypertension. This is because they are exposed to relatively high pressure as blood is transmitted directly from the parent lumen to a much smaller lumen. This is in contrast to the blood supply of the lobe of the brain and cortex where arterial supply is derived from vessels that become gradually smaller through a dendritic pattern. These arteries arise from the parent vessel in a perpendicular fashion which also creates shear stress at the origin of the artery. The effect of hypertension in causing a liability to atherosclerosis is compounded by hyperlipidemia, diabetes, and smoking, all of which increase the liability to small vessel disease.

2. There is almost no redundant or collateral circulation in these areas. Occlusion of a single small vessel is likely to cause an infarct. In contrast, occlusion of a similarly sized distal branch of the middle cerebral artery may cause no infarct because the area of the cortex it supplies also receives blood flow from leptomeningeal collaterals.

These arteries occlude through three main mechanisms:

1. Atherosclerosis: Atherosclerosis describes thickening of the wall of an artery due to inflammation of the vessel wall and ensuing atheromatous plaque formation. Plaque can form in the small vessel itself and cause occlusion or can occur in the wall of the parent vessel (e.g., the MCA) and then occlude the origin of the branch vessel.

2. Lipohyalinosis: Lipohyalinosis describes segmental narrowing of an arterial wall induced by endothelial damage due to hypertension, but without frank plaque formation [35].

3. Microemboli: It is speculated that some cerebral small vessel disease is due to microscopic emboli from the heart, aorta, or great vessels that lodge in small, perforating arteries [36].

There are several hereditary forms of small vessel disease in which the disease phenotype is seen at an accelerated rate: These include cerebral autosomal dominant arteriopathy with subcortical infarcts and leukoencephalopathy (CADASIL), Fabry's disease, and retinal vasculopathy with cerebral leukodystrophy [37]. Individually, hereditary small vessel disease syndromes are rare, and collectively they account for no more than 5% of instances of cerebral small vessel disease [37].

When cerebral small vessel disease is clinically overt, it either manifests as cognitive impairment (with vascular dementia at the extreme end of the spectrum) or as an ischemic stroke. There is an overlap between the symptoms of an ischemic stroke due to small

Table 7
The five classically described lacunar stroke syndromes

Syndrome	Localization
Pure motor hemiparesis	Corona radiata, pons, or internal capsule
Pure sensory syndrome	Ventroposterolateral nucleus of the thalamus, internal capsule, corona radiata
Ataxic hemiparesis	Internal capsule, pons, corona radiata
Dysarthria-clumsy hand syndrome	Pons, internal capsule, basal ganglia, thalamus
Sensorimotor syndrome	Pons or thalamus/internal capsule

vessel disease and an ischemic stroke due to other subtypes. However, a stroke due to small vessel disease does not have any symptoms/signs referable to the cortex: aphasia, agraphia, alexia, acalculia, visuospatial neglect, or gaze preference. Although physically small, strokes due to small vessel disease may affect highly concentrated areas of brain parenchyma such that a "small" lesion causes an apparently "large" syndrome, e.g., a small internal capsule stroke causing a complete contralateral hemiparesis. The five classically described lacunar stroke syndromes are summarized in Table 7.

Another characteristic feature of strokes arising from small vessel disease is the phenomenon of worsening. This may be because a small perforating artery is critically stenosed such that even small variations in blood pressure are enough to compromise flow through the artery. It can also be related to an active atherosclerotic plaque at the main artery or an embolus from a distant source. It can also give rise to the phenomenon of the "stuttering lacune" where symptoms can abate and relapse over a period of minutes or hours.

4.3 Management Strategies

4.3.1 Diagnostic Approach

The mainstay of investigation when stroke due to cerebral small vessel disease is suspected is magnetic resonance imaging (MRI). Diffuse-weighted imaging (DWI) sequences identify acute tissue ischemia. Fluid-attenuated inversion recovery (FLAIR) sequences identify chronic tissue loss and old (greater than 6 h) infarcts. Typically, the brain of a person afflicted with cerebral small vessel disease will demonstrate FLAIR hyperintensity in a confluent fashion in the thalamus, pons, corona radiata, internal capsule, and basal ganglia. In general, the more advanced the disease, the more this FLAIR hyperintensity will tend to infiltrate toward the cortex. There may also be evidence of lacunes, i.e., areas of frank tissue loss which are ovoid in shape and less than 1.5 cm in size. More advanced MRI sequences are currently being developed which permit quantification of small vessel disease burden and microvascular dysfunction [38].

4.3.2 Treatment Approach and Outcomes

The mainstay of treatment for secondary stroke prevention in patients with cerebral small vessel disease is aggressive control of hypertension, hyperlipidemia and diabetes, smoking cessation, and the usage of anti-platelet therapy [39]. The most rational anti-platelet choice is either aspirin alone or clopidogrel alone. The combination of the two drugs does not reduce the risk of future stroke, and in those patients who have suffered a lacunar stroke, the combination increases the risk of intracranial hemorrhage and death [40]. In patients with stuttering symptoms, however, the short-term use of aspirin and clopidogrel combined may reduce the risk of decline and help improve outcome without significant major bleeding complications [41]. Long-term blood pressure reduction reduces the risk of future stroke, irrespective of the agent used [42]. However, blood pressure should not be lowered in the first 7 days after an ischemic stroke as this has a trend to cause harm [43]. In general, our practice is not to reduce systolic blood pressure below 220 mmHg in the first 24 h unless there is evidence of end organ damage as a result of hypertension (hypertensive encephalopathy, headache, acute kidney injury, or visual impairment) and not below 180 mmHg in the first 7 days. The SPARCL trial [44] demonstrated that high-dose atorvastatin was effective in reducing the risk of future stroke, and a retrospective subgroup analysis of the trial confirmed this effect in patients with small vessel disease [45]. It is not known whether or not this effect is primarily mediated through a reduction in serum cholesterol or some other property of statins.

Thrombolysis with intravenous tPA is effective for the treatment of stroke syndromes that are subsequently found to be due to small vessel disease. However, the presence of small vessel disease is an independent risk factor for post-tPA hemorrhage [46] with the risk of post-tPA hemorrhage doubled in the presence of white matter lesions on CT of the brain [47]. However, tPA still represents the standard of care in the acute management of an ischemic stroke, including in those for whom small vessel disease is felt to be the underlying mechanism.

References

1. Grotta JC (2013) Clinical practice. Carotid stenosis. N Engl J Med 369:1143–1150

2. Ratchford EV, Jin Z, Di Tullio MR, Salameh MJ, Homma S, Gan R et al (2009) Carotid bruit for detection of hemodynamically significant carotid stenosis: the northern manhattan study. Neurol Res 31:748–752

3. Lawson MF, Velat GJ, Fargen KM, Mocco J, Hoh BL (2011) Interventional neurovascular disease: avoidance and management of complications and review of the current literature. J Neurosurg Sci 55:233–242

4. Barnett HJM, Taylor DW, Haynes RB, Sackett DL, Peerless SJ, Ferguson GG et al (1991)

Beneficial effect of carotid endarterectomy in symptomatic patients with high-grade carotid stenosis. N Engl J Med 325:445–453

5. Mayberg MR, Wilson SE, Yatsu F, Weiss DG, Messina L, Hershey LA et al (1991) Carotid endarterectomy and prevention of cerebral ischemia in symptomatic carotid stenosis. Veterans affairs cooperative studies program 309 trialist group. JAMA 266:3289–3294

6. Barnett HJ, Taylor DW, Eliasziw M, Fox AJ, Ferguson GG, Haynes RB et al (1998) Benefit of carotid endarterectomy in patients with symptomatic moderate or severe stenosis. North american symptomatic carotid endarterectomy trial collaborators. N Engl J Med 339:1415–1425

7. (1998) Randomised trial of endarterectomy for recently symptomatic carotid stenosis: final results of the mrc european carotid surgery trial (ecst). Lancet 351:1379–1387

8. Eckstein HH, Ringleb P, Allenberg JR, Berger J, Fraedrich G, Hacke W et al (2008) Results of the stent-protected angioplasty versus carotid endarterectomy (space) study to treat symptomatic stenoses at 2 years: a multinational, prospective, randomised trial. Lancet Neurol 7:893–902

9. Mas JL, Trinquart L, Leys D, Albucher JF, Rousseau H, Viguier A et al (2008) Endarterectomy versus angioplasty in patients with symptomatic severe carotid stenosis (eva-3s) trial: results up to 4 years from a randomised, multicentre trial. Lancet Neurol 7:885–892

10. Bonati LH, Dobson J, Featherstone RL, Ederle J, van der Worp HB, de Borst GJ et al (2015) Long-term outcomes after stenting versus endarterectomy for treatment of symptomatic carotid stenosis: the international carotid stenting study (icss) randomised trial. Lancet 385:529–538

11. Ederle J, Bonati LH, Dobson J, Featherstone RL, Gaines PA, Beard JD et al (2009) Endovascular treatment with angioplasty or stenting versus endarterectomy in patients with carotid artery stenosis in the carotid and vertebral artery transluminal angioplasty study (cavatas): long-term follow-up of a randomised trial. Lancet Neurol 8:898–907

12. (2001) Endovascular versus surgical treatment in patients with carotid stenosis in the carotid and vertebral artery transluminal angioplasty study (cavatas): a randomised trial. Lancet 357:1729–1737

13. Yadav JS, Wholey MH, Kuntz RE, Fayad P, Katzen BT, Mishkel GJ et al (2004) Protected carotid-artery stenting versus endarterectomy in high-risk patients. N Engl J Med 351:1493–1501

14. Brott TG, Hobson RW, Howard G, Roubin GS, Clark WM, Brooks W et al (2010) Stenting versus endarterectomy for treatment of carotid-artery stenosis. N Engl J Med 363:11–23

15. Rothwell PM, Goldstein LB (2004) Carotid endarterectomy for asymptomatic carotid stenosis: asymptomatic carotid surgery trial. Stroke 35:2425–2427

16. (1991) Carotid surgery versus medical therapy in asymptomatic carotid stenosis. The casanova study group. Stroke 22:1229–1235

17. Hobson RW, Weiss DG, Fields WS, Goldstone J, Moore WS, Towne JB et al (1993) Efficacy of carotid endarterectomy for asymptomatic carotid stenosis. The veterans affairs cooperative study group. N Engl J Med 328:221–227

18. (1995) Endarterectomy for asymptomatic carotid artery stenosis. Executive committee for the asymptomatic carotid atherosclerosis study. JAMA 273:1421–1428

19. Halliday A, Mansfield A, Marro J, Peto C, Peto R, Potter J et al (2004) Prevention of disabling and fatal strokes by successful carotid endarterectomy in patients without recent neurological symptoms: randomised controlled trial. Lancet 363:1491–1502

20. Rosenfield K, Matsumura JS, Chaturvedi S, Riles T, Ansel GM, Metzger DC et al (2016) Randomized trial of stent versus surgery for asymptomatic carotid stenosis. N Engl J Med 374:1011–1020

21. Abbott AL (2009) Medical (nonsurgical) intervention alone is now best for prevention of stroke associated with asymptomatic severe carotid stenosis: results of a systematic review and analysis. Stroke 40:e573–e583

22. Caplan LR (2008) Dissections of brain-supplying arteries. Nat Clin Pract Neurol 4:34–42

23. Yaghi S, Kamel H, Elkind MSV (2017) Atrial cardiopathy: a mechanism of cryptogenic stroke. Expert Rev Cardiovasc Ther 15 (8):591–599

24. Engelter ST, Grond-Ginsbach C, Metso TM, Metso AJ, Kloss M, Debette S et al (2013) Cervical artery dissection: trauma and other potential mechanical trigger events. Neurology 80:1950–1957

25. Debette S, Markus HS (2009) The genetics of cervical artery dissection: a systematic review. Stroke 40:e459–e466

26. Kasner SE, Hankins LL, Bratina P, Morgenstern LB (1997) Magnetic resonance angiography demonstrates vascular healing of carotid

and vertebral artery dissections. Stroke 28:1993–1997

27. Zinkstok SM, Vergouwen MD, Engelter ST, Lyrer PA, Bonati LH, Arnold M et al (2011) Safety and functional outcome of thrombolysis in dissection-related ischemic stroke: a meta-analysis of individual patient data. Stroke 42:2515–2520

28. Yaghi S, Maalouf N, Keyrouz SG (2012) Cervical artery dissection: risk factors, treatment, and outcome; a 5-year experience from a tertiary care center. Int J Neurosci 122:40–44

29. Markus HS, Hayter E, Levi C, Feldman A, Venables G, Norris J et al (2015) Antiplatelet treatment compared with anticoagulation treatment for cervical artery dissection (cadiss): a randomised trial. Lancet Neurol 14:361–367

30. Kennedy F, Lanfranconi S, Hicks C, Reid J, Gompertz P, Price C et al (2012) Antiplatelets vs anticoagulation for dissection: cadiss non-randomized arm and meta-analysis. Neurology 79:686–689

31. Chowdhury MM, Sabbagh CN, Jackson D, Coughlin PA, Ghosh J (2015) Antithrombotic treatment for acute extracranial carotid artery dissections: a meta-analysis. Eur J Vasc Endovasc Surg 50:148–156

32. Biousse V, D'Anglejan-Chatillon J, Touboul PJ, Amarenco P, Bousser MG (1995) Time course of symptoms in extracranial carotid artery dissections. A series of 80 patients. Stroke 26:235–239

33. Caprio FZ, Bernstein RA, Alberts MJ, Curran Y, Bergman D, Korutz AW et al (2014) Efficacy and safety of novel oral anticoagulants in patients with cervical artery dissections. Cerebrovasc Dis 38:247–253

34. Tan R, Traylor M, Rutten-Jacobs L, Markus H (2017) New insights into mechanisms of small vessel disease stroke from genetics. Clin Sci (Lond) 131:515–531

35. Fisher CM (1979) Capsular infarcts: the underlying vascular lesions. Arch Neurol 36:65–73

36. Futrell N (2004) Lacunar infarction: embolism is the key. Stroke 35:1778–1779

37. Søndergaard CB, Nielsen JE, Hansen CK, Christensen H (2017) Hereditary cerebral small vessel disease and stroke. Clin Neurol Neurosurg 155:45–57

38. Blair GW, Hernandez MV, Thrippleton MJ, Doubal FN, Wardlaw JM (2017) Advanced neuroimaging of cerebral small vessel disease. Curr Treat Options Cardiovasc Med 19:56

39. Mok V, Kim JS (2015) Prevention and management of cerebral small vessel disease. J Stroke 17:111–122

40. Benavente OR, Hart RG, McClure LA, Szychowski JM, Coffey CS, Pearce LA et al (2012) Effects of clopidogrel added to aspirin in patients with recent lacunar stroke. N Engl J Med 367:817–825

41. Marsh EB, Llinas RH (2014) Stuttering lacunes: an acute role for clopidogrel? J Neurol Transl Neurosci 2(1):1035

42. Rashid P, Leonardi-Bee J, Bath P (2003) Blood pressure reduction and secondary prevention of stroke and other vascular events: a systematic review. Stroke 34:2741–2748

43. Sandset EC, Bath PM, Boysen G, Jatuzis D, Kõrv J, Lüders S et al (2011) The angiotensin-receptor blocker candesartan for treatment of acute stroke (scast): a randomised, placebo-controlled, double-blind trial. Lancet 377:741–750

44. Amarenco P, Bogousslavsky J, Callahan A, Goldstein LB, Hennerici M, Rudolph AE et al (2006) High-dose atorvastatin after stroke or transient ischemic attack. N Engl J Med 355:549–559

45. Amarenco P, Benavente O, Goldstein LB, Callahan A, Sillesen H, Hennerici MG et al (2009) Results of the stroke prevention by aggressive reduction in cholesterol levels (sparcl) trial by stroke subtypes. Stroke 40:1405–1409

46. Pantoni L, Fierini F, Poggesi A (2014) Thrombolysis in acute stroke patients with cerebral small vessel disease. Cerebrovasc Dis 37:5–13

47. Curtze S, Haapaniemi E, Melkas S, Mustanoja S, Putaala J, Sairanen T et al (2015) White matter lesions double the risk of post-thrombolytic intracerebral hemorrhage. Stroke 46:2149–2155

Chapter 16

Intracranial Atherosclerotic Diseases (ICAD)

Sammy Pishanidar, Mais N. Al-Kawaz, and Alexander E. Merkler

Abstract

Intracranial atherosclerotic disease (ICAD) is a common etiology of stroke, accounting for up to 20% of all strokes worldwide. The risk of developing ICAD increases with age and risk factors including smoking, hypertension, and dyslipidemia. Stroke mechanisms secondary to ICAD vary and include hypoperfusion, artery-to-artery embolism, and plaque extension. Compared to other stroke mechanism, the risk of recurrent stroke among patients with ICAD is high, despite aggressive medical management. Further research evaluating novel therapies is necessary as surgical approaches including stenting have to date not shown benefit. This chapter aims to discuss the pathophysiology and epidemiology of ICAD. In addition, the efficacy of different management and treatment approaches is discussed based on recent clinical evidence.

Key words Intracranial atherosclerosis, Ischemic stroke, Plaque, Stenosis, Risk factors, Prevention, Treatment

1 Introduction

Intracranial atherosclerotic disease (ICAD) refers to the buildup of plaque within the intracranial vasculature of the brain. ICAD was recently recognized as the most common cause of stroke worldwide accounting 20–53% of all ischemic strokes, depending on race, gender, and age [1]. This chapter will discuss the epidemiology, pathophysiology, risk factors, diagnosis, and treatment considerations of ICAD.

2 Epidemiology and Pattern of Disease

ICAD typically develops in the second decade of life and usually becomes symptomatic between ages 60 and 80 years [1]. Certain subgroups appear particularly susceptible to ICAD, including Asians, African Americans, and Hispanics. For example, when compared to Americans and Europeans, Asians develop ICAD earlier in life with more extensive vessel involvement [2].

Fawaz Al-Mufti and Krishna Amuluru (eds.), *Cerebrovascular Disorders*, Neuromethods, vol. 170,
https://doi.org/10.1007/978-1-0716-1530-0_16, © Springer Science+Business Media, LLC, part of Springer Nature 2021

Plaque accumulation within the major intracranial arteries including the internal carotid artery, middle cerebral artery, posterior cerebral artery, basilar artery, and vertebral artery is typical of ICAD [3]. ICAD is often most severe in the petro-cavernous segment of the internal carotid artery (ICA), basilar artery (BA), vertebral arteries, the supraclinoid segment of ICA, the M2 segment of middle cerebral artery (MCA), and the A2 segment of the anterior cerebral artery (ACA) [4]. The communicating arteries and the cerebellar arteries are often spared from atherosclerotic changes.

ICAD often occurs later in life compared to extracranial atherosclerosis within the cervical internal carotid artery. This is thought to be secondary to loss of autoregulation and cerebral blood flow control mechanism later in life. Cerebral blood flow control and autoregulation are generally affected by microvascular disease secondary to diabetes mellitus, obesity, or hypertension—all comorbidities that typically occur with advanced age. Disruption of these control mechanisms results in pressure passive cerebral blood flow with transfer of systemic high blood pressure to intracranial vasculature. This effect exposes intracranial vessels to be more susceptible to fibrosis, plaque formation, and subsequently ICAD [1].

3   Pathophysiology and Clinical Presentation

The initial lesion that precedes the formation of an atherosclerotic plaque is thought to consist of a focal increase of lipoprotein within the intima resulting in formation of fatty streaks. These lipoprotein particles favor oxidative modification and trigger local formation of pro-inflammatory cytokines. Cytokines produced secondary to the inflammatory response augment expression of adhesion molecules and leukocyte receptors expressed on arterial endothelial cells, which can recruit leukocytes to the evolving atheroma. Once adherent, the leukocytes migrate into the intima. Mononuclear phagocytes ingest lipids forming lipid-laden foam cells. Apoptotic foam cells result in formation of a lipid core necrotic center of the plaque. The mononuclear phagocytes within the fatty streak also signal smooth cell migration and proliferation from the media. These smooth cells produce extracellular matrix forming a fibro-fatty plaque. Rupture of the fibrous cap can result in thrombosis. Subsequent healing and fibrosis result in a fibro-proliferative process increasing the size of the lesion and resulting in a more critical stenosis, hypoperfusion, and ischemia [5].

Age is one of the most important risk factors for ICAD that correlates with both disease prevalence and severity [6]. Race also appears to be an important risk factor, as Asians, African Americans, and Hispanics are significantly more prone to ICAD than

Caucasians. In a multi-ethnic cohort study assessing determinants of ICAD-related strokes, both Hispanics and African Americans were found to have a higher incidence of ICAD when compared to Caucasians [2].

The incidence of ICAD is similar between men and women though the progression of the disease is different. The risk in men appears to be highest in the fourth and fifth decade of life. Women are often affected later in life, during the fourth to sixth decade, with certain races having a higher propensity for ICAD in women beyond the sixth decade. Several studies have postulated that discrepancies in the course of ICAD and ICAD-related risk factors between genders are related to estrogen levels. Estrogen is thought to play a key factor in endothelium-dependent maintenance of vascular tone. Indeed, several studies have found that estrogen may enhance nitric oxide leading to a relaxation of the vascular tone and subsequently slowing the progression of atherosclerosis [7].

No clear genetic risk factors have been shown to contribute to ICAD, but a few susceptibility genes have been identified. Ring finger protein 213 gene polymorphism (RNF213), which was initially identified as a susceptibility gene for Moyamoya disease in East Asians, was found to correlate with onset of ICAD at an early age in Korea [8]. A recent genome-wide association study identified two loci including 9p21.3 and 11p11.2 that were significantly associated with calcification volume in intracranial ICA [9]. Genetic polymorphisms in pathways affecting lipid and homocystcine metabolism were found to be associated with concurrent ICAD and extracranial atherosclerosis stenosis (ECAS) in Asian populations. Phosphodiesterase 4D genetic variants were found to be associated with ICAD-related strokes in South Indian populations [10].

It has been suggested that the circle of Willis serves some compensatory mechanism in the event of severe stenosis or occlusion of ICA. Given this contribution, it has been hypothesized that different anatomic variations within the circle of Willis could explain some racial discrepancies in the progression of ICAD noted between populations [11].

Other non-modifiable risk factors have been reported but still require more longitudinal studies to confirm association with the disease. The WASID trial demonstrated that more extensive collaterals decreased the risk of recurrent strokes in severe stenotic lesions but were associated with an increased recurrent risk in milder stenotic lesions [12]. Three-vessel CAD has been reported to correlate with severe cervical and ICAD [13]. Duration of smoking was found to be a predictor of ICAD, while smoking status and alcohol intake were a predictor of ICAS only in men [14]. Elevated homocysteine level was also associated with severity of ICAD [15].

Hypertension is an independent risk factor for ICAD that has been shown to correlate strongly with severity of disease [6]. Similarly, DM is also associated with ICAD and disease severity. In a multivariate analysis of the Northern Manhattan Study, DM was found to confer a significantly higher risk for ICAD-related stroke compared with non-atherosclerotic-related strokes [2]. In a population of asymptomatic Hispanic patients, DM was found to correlate with moderate to severe ICAD [6]. Metabolic syndrome is also a factor that shows a stronger association with ICAD-related strokes in all distributions when compared to ECAS-related strokes in the Northern Manhattan Study [2].

Dyslipidemia is another important risk factor for ICAD; the ratio of apolipoprotein B/apolipoprotein A-1 strongly correlates with the development of ICAD [16]. A study investigating ICAD in a symptomatic Chinese population showed that lower HDL-C levels specifically correlated with incidence of the ICAD subtype of stroke [17]. However, this association was not seen in the Northern Manhattan Study where hypercholesterolemia similarly increased the risk of ICAD, ECAS, and small-vessel disease [2].

ICAD is one of the most common causes of stroke worldwide [18]. Randomized trial found a 15% increased risk of ischemic strokes in patients with ICAS > 50% [19]. Multiple hypothesized stroke mechanisms have been reported for ICAD including: hypoperfusion, artery-to-artery embolism, and plaque extension over small penetrating artery ostia (also known as branch atheromatous disease) resulting in lacunar strokes. A combination of the above mechanisms can also occur where hypoperfusion can prevent resolution of a distal embolus. Imaging may aid in differentiating between the above stroke mechanisms. A watershed infarct between vascular territories is typically indicative of hypoperfusion, whereas a wedge-shaped territorial infarct is generally due to an artery-to-artery embolism.

ICAD has one of the highest rates of stroke recurrences. A 2-year follow-up study showed an ischemic stroke recurrence rate as high as 38.2% in the same vascular territory despite optimal medical therapy [20], making it an important public health epidemic. Predictors of recurrent stroke in the same vascular area in the WASID study included severe stenotic lesions ($\geq$70%), baseline severe neurologic deficits, and female sex [21]. Progression of stenosis is associated with a 20% incidence of ischemic stroke as noted on follow-up of MCA stenotic lesions with transcranial Dopplers (TCDs) [22].

4 Diagnostic Considerations

Although catheter angiography is considered the gold standard method for diagnosis, noninvasive imaging is often adequate. Computed tomography angiogram (CTA) is frequently the diagnostic imaging test of choice; CTA has a positive predictive value (PPV) of 46.7% and a negative predictive value (NPV) of 73% for a catheter angiography-proven stenosis of 77–90%. The SONIA trial showed that transcranial Dopplers (TCDs) have a PPV of 36% but a

NPV of 85%. Magnetic resonance angiogram (MRA) had a 59% PPV and a 91% NPV. Overall, noninvasive tests including TCD, MRA, and CTA can fairly reliably exclude ICAD, but catheter angiography remains the gold standard [23]. Catheter angiography can further characterize occlusion from pseudo-occlusion in severe stenosis, identify collaterals, and look for other possible etiologies including Moyamoya disease, vasculitis, intracranial dissection, etc. However, catheter angiography is associated with a 2.6% risk of neurologic complications and 0.14% risk of ischemic stroke with permanent disability [24], is less readily available than noninvasive modalities, and should be performed with caution in older patients, patients with diabetes, and chronic kidney disease. Therefore, ICAD is often discovered or excluded based on the noninvasive neuroimaging.

Over the past few years, high-resolution vessel wall imaging (HR-VWI) has gained popularity as an important tool to differentiate ICAD from other intracranial vasculopathies, which often pose a diagnostic challenge. Typically, plaque appears as eccentric wall thickening resulting in an intermediate to high T1 signal in the plaque wall with heterogeneous signal otherwise. High T1 signal on fat-saturated imaging represents intra-plaque hemorrhage. On T2 sequences, an atherosclerotic plaque usually demonstrates a thin juxtaluminal hyperintense band with an underlying hypointense core. This juxtaluminal band represents a fibrous cap that can differentiate atherosclerosis from vasculopathy including vasculitis or RCVS (reversible cerebral vasoconstriction syndrome). In addition, recent work suggests that as compared to asymptomatic plaques, the vast majority of symptomatic plaques will enhance on post-contrast MR imaging [25].

Differential diagnosis usually includes Moyamoya disease, intracranial dissection, vasculopathies such as vasculitis (primary or secondary to infectious etiologies), and reversible cerebral vasoconstriction syndrome, and fibromuscular dysplasia. Certain clinical and imaging characteristics may help further guide physicians to differentiate between these pathologies; characteristics are listed in Table 1.

5 Management Strategies

Several key landmark studies have assessed the best management for patients with ICAD. Three trials include Warfarin versus Aspirin for Symptomatic Intracranial Disease (WASID), Stenting versus Aggressive Medical Management for Preventing Recurrent Strokes in Intracranial Stenosis (SAMMPRIS), and the Vitesse Intracranial Stent Study for Ischemic Stroke Therapy (VISSIT).

Table 1
Differential diagnosis for intracranial atherosclerosis with identifying clinical and imaging features

Diagnosis	Clinical features	Appearance on catheter angiography	Appearance on HR-VWI
Dissection	Usually presents as headache, motor leg/arm weakness, dysarthria/aphasia, and vertigo. Can present as infarction, SAH, or both	String sign on DSA, tapered or flame-shaped occlusion, intimal flap, dissecting aneurysm, distal pouch	High T1 signal intramural hemorrhage. Enhancement of vessel wall
Reversible cerebral vasoconstriction syndrome	Sudden thunderclap HA, no neurologic deficits	Multifocal segment constriction with reversibility of angiographic stenosis within 12 weeks of findings	Mild, multifocal, vessel wall thickening with no or minimal concentric enhancement
Fibromuscular dysplasia	HTN (renal vessel involvement), dizziness, TIA, tinnitus, stroke	Multifocal "string-of-beads" pattern most commonly; can be a unifocal short or long concentric stenosis	
Moyamoya disease	Children present more commonly with ischemic events, while adults present with hemorrhagic complications more often	Bilateral stenosis affecting the distal ICA and proximal Circle of Willis, prominent basal collaterals	Concentric enhancement of distal ICA/MCA if enhancement is present. Thinning of stenotic segments is seen due to negative remodeling
Primary angiitis of the central nervous system	Subacute onset with multiple infarcts in different vascular supply areas, HA, cognitive decline	String-of-beads in the distal and smaller vessels	Smooth, concentric, homogeneous enhancement in a majority of cases due to wall thickening
Vasculitis secondary to infectious etiology (CMV, VZV, HIV)	Subacute onset with multiple infarcts in different vascular supply areas, HA, cognitive decline	String-of-beads in the distal and smaller vessels	Smooth, concentric, homogeneous enhancement in a majority of cases due to wall thickening

Warfarin versus Aspirin for Symptomatic Intracranial Disease (WASID) was a NIH/NINDS-funded randomized, double-blinded, multicenter clinical trial published in 2005 that compared anticoagulation to aspirin for the prevention of ischemic stroke in patients with ICAD. Patients who had an angiographically confirmed stenosis of 50–99% in a major intracranial artery, resulting in either a TIA or a non-disabling stroke, were randomly assigned to treatment with either warfarin (INR goal of 2.0 to 3.0) or aspirin (1300 mg/day). The study found that oral anticoagulation was not superior to aspirin and, in fact, was stopped early as concern for the patients' safety in the warfarin arm arose [26]. In conclusion, based on WASID, antiplatelet therapy is preferred over warfarin for patients with symptomatic intracranial arterial stenosis of 50–99% given the increased risk of adverse events without any benefit in preventing ischemic stroke.

Given the high recurrent risk of stroke in patients with ICAD despite medical therapy and the recent success of interventional cardiology with coronary stent placement for myocardial infarction, a new hope of stenting open the intracranial vasculature arose. Stenting versus Aggressive Medical Management for Preventing Recurrent Strokes in Intracranial Stenosis (SAMMPRIS) was published in 2011. SAMMPRIS was a randomized, multicenter trial comparing intracranial stenting versus maximal medical management in patients with ICAD. Patients with intracranial stenosis of 70–99% who had a TIA or ischemic stroke within 30 days of enrollment were randomized to treatment with percutaneous transluminal angioplasty and stenting (PTAS) with the use of the Wingspan Stent System plus aggressive medical therapy versus aggressive medical therapy alone. The primary outcome was recurrent stroke or death within 30 days and stroke in the qualifying artery beyond 30 days.

Trial enrollment was stopped prematurely after recruitment of only 451 of the planned 764 patients as the 30-day stroke or death rate was 14.7% (nonfatal stroke, 12.5%; fatal stroke, 2.2%) in the PTAS group versus 5.8% (nonfatal stroke, 5.3%; non-stroke-related death, 0.4%) ($P = 0.002$). At 1 year, the trend continued; the rate of stroke or death was higher in the PTAS group as compared to the medical management group (19.7% vs 12.6%, $P = 0.0009$). In the stenting arm, the major cause of early ischemic strokes was attributed to occlusion of a perforating vessel [19].

The SAMMPRIS trial defined aggressive medical therapy to include aspirin 325 mg daily, clopidogrel 75 mg daily for 90 days, and intensive risk factor modification and to target a blood pressure < 140/90 mmHg or <130/80 in diabetics and a low-density lipoprotein (LDL) <70 mg/dL (<1.81 mmol/L).

In conclusion, SAMMPRIS demonstrated that angioplasty and stenting were associated with an increased risk of recurrent stroke and death, compared to aggressive medical management alone. For patients with a TIA or stroke secondary to symptomatic ICAD of 70–99%, aggressive medical management is superior to PTAS with the use of the Wingspan Stent System. Not only was the risk of early stroke after PTAS higher than expected, but also the risk of stroke in the aggressive medical therapy arm was also lower than expected compared to data from the WASID trial.

Extrapolating from SAMMPRIS, one can support the use of dual antiplatelet therapy for 90 days in the context of ischemic stroke or TIA attributed to intracranial stenosis of 70–99%. Furthermore, as SAMMPRIS used aspirin 325 mg/day, which is higher than the typical dose used for secondary stroke prevention, it is unclear if similar results would have yielded using a lower dose of 81 mg/day.

The Vitesse Intracranial Stent Study for Ischemic Stroke Therapy (VISSIT) was an international, multicenter, 1:1 randomized parallel group trial published in the *Journal of the American Medical Association* (JAMA) in 2015, which confirmed the results of SAMMPRIS. In this trial, a balloon-expandable stent was employed for patients with symptomatic intracranial arterial stenosis and was found to be harmful compared to medical therapy alone with respect to recurrent stroke or TIA in the same territory [27]. Similar to SAMMPRIS, medical therapy consisted of aspirin 81–325 mg daily*, clopidogrel 75 mg daily for 90 days, intensive risk factor modification, and a target blood pressure < 140/90 mmHg* and an LDL goal <100 mg/dL.*

112 patients with symptomatic intracranial stenosis of 70–99% in the ICA, MCA, or vertebral or basilar artery in the previous 30 days were randomized to receive balloon-expandable stent plus medical therapy ($n = 59$) or medical therapy alone ($n = 53$). TIAs were defined as lasting 10 min to 24 h. The primary outcome measure was a composite of stroke in the same vascular territory within 12 months of randomization or a TIA in the same vascular territory day 2 through month 12 of randomization. The sponsor stopped enrollment prematurely after negative results from SAMMPRIS prompted an early evaluation of outcomes. This interim analysis of the trial suggested futility after 112 patients of the planned 250 patients were enrolled [27].

The primary safety endpoint, a composite of stroke, death, or intracranial hemorrhage within 30 days, was significantly higher in the stent group (24.1%) versus the medical group (9.4%) ($P = 0.05$). Intracranial hemorrhage within 30 days occurred only in the stenting group (8.6%) and not in the medical group ($P = 0.06$). At 12 months, the rate of the primary outcome measure was significantly higher in the stent group (36.2%) versus the medical group (15.1%) ($P = 0.02$). Furthermore, the modified Rankin Scale (mRS), used to measure baseline disability, revealed more patients in the stent group (24.1%) versus the medical group (11.3%) ($P = 0.09$) had worsening in their mRS at 12 months [27].

Several limitations to the study question the validity of these results. As enrollment was halted, treatment groups were not well matched for demographics resulting in the trial population of ~70%

---

*Of note SAMMPRIS used aspirin 325 mg daily, had more aggressive LDL targets of <70 mg/dL, and had more aggressive BP goals of <130/80 for diabetics.

white males. Second, smoking, a major risk factor for intracranial atherosclerosis, was not assessed in about half the patient population. Third, the study was unable to be blinded, as it was procedure based, similar to SAMMPRIS, raising the possibility of bias. In addition, authors cite the operators may not have been experienced enough with the device and that the device was technically limited stating experience and a newer-generation device may be superior in the future.

Among patients with symptomatic intracranial arterial stenosis of 70–99%, the use of a balloon-expandable stent was associated with an increased 30-day risk of any stroke, increased 12-month risk of stroke or TIA in the same vascular territory, and increased risk of intracranial hemorrhage versus medical therapy. Therefore, the use of a balloon-expandable stent for patients with symptomatic intracranial stenosis is not supported.

Both SAMMPRIS and VISSIT clearly indicate angioplasty and stenting of high-grade symptomatic intracranial stenosis are inferior to best medical management. Despite these randomized trials, management of patients with recurrent stroke on optimal medical therapy remains uncertain. Some experts still suggest stenting as a last resort for patients who have multiple ischemic events despite optimal medical management; however, a subgroup analysis of the SAMMPRIS trial of patients with prior ischemic stroke in the territory of the symptomatic intracranial artery showed no benefit for stenting [28, 29]. Results of the abovementioned randomized clinical trials are summarized in Table 2.

Of note, given certain misgivings regarding patient selection and site selection in SAMMPRIS and VISSIT, future studies are underway to evaluate specific subpopulations that may still benefit from intracranial stenting rather than medical management. The China Angioplasty and Stenting for Symptomatic Intracranial Severe Stenosis (CASSISS) trial is an ongoing, government-funded, prospective, multicenter, randomized trial recruiting patients with recent TIA or stroke caused by 70–99% stenosis of a major intracranial artery. The trial will exclude patients with previous stroke related to perforator ischemia, and only high-volume centers with a proven track record will enroll patients. This proposed trial aims to address certain deficiencies of SAMMPRIS including patient and participating center selection. Investigators hope this study will allow for a critical reappraisal of the role of intracranial stenting for selected patients in high-volume centers [30].

A key limitation to a couple of these studies is that they only looked at patients with intracranial atherosclerosis in the anterior circulation only (other than VISSIT, which was prematurely stopped and does not have good power). Is this generalizable to the posterior circulation as well? There is clinical equipoise, and further studies need to be done to confidently determine best management.

Table 2
Major randomized clinical trials affecting ICAD management summarizing interventions, results, and conclusions

Trial	Intervention	Results	Conclusion
WASID	Aspirin ($N = 280$) Warfarin ($N = 289$)	• No difference between aspirin and warfarin for the primary outcome measure[a] • Major hemorrhage significantly lower in aspirin group[b] • Aspirin associated with lower rate of death[b] • Trial stopped early	Aspirin is preferred over warfarin for ICAD
SAMMP RIS	Medical management ($N = 227$) PTAS + medical management ($N = 224$)	• Higher rate of primary outcome (stroke or death) in PTAS group • Trial enrollment stopped early	Aggressive medical management alone is superior to PTAS for ICAD
VISSIT	Medical group ($N = 53$) Stent group ($N = 58$)	• Stent group had a higher 30-day risk of any stroke • Stent group had a higher 12-month risk of stroke or TIA in the same vascular territory • Intracranial hemorrhage within 30 days only occurred in the stent group • More patients in stent group had worsening of their mRS • Trial enrollment stopped after results of SAMMPRIS published	Balloon-expandable stent was harmful compared to medical therapy alone with respect to recurrent stroke/TIA in the same territory in patients with ICAD

[a]Primary outcome measure: combined ischemic stroke, brain hemorrhage, or death from a vascular cause other than stroke
[b]Primary safety endpoint: a composite of stroke, death, or intracranial hemorrhage

In addition, little evidence with head-to-head trials comparing dual antiplatelet to aspirin or clopidogrel, in the setting of intracranial stenosis, has been published but is underway. The recent *Recurrent Ischemic Lesions After Acute Atherothrombotic Stroke: Clopidogrel Plus Aspirin Versus Aspirin Alone (COMPRESS)* trial, published in *Stroke* in 2016, raises the concern that clopidogrel plus aspirin might not be superior to aspirin alone for preventing new ischemic lesions and vascular events caused by ICAD [31]. Further investigation needs to be completed to help answer these questions.

6   Treatment Approach

Given the data that is available, how do we manage a patient with an ischemic infarct or TIA secondary to symptomatic intracranial atherosclerosis of 50–99%?

Current guidelines recommend maximal medical therapy consisting of antiplatelet and antihypertensive agents, statins, and lifestyle modification. For a newly diagnosed ischemic infarct or TIA secondary to symptomatic intracranial stenosis, the following is recommended:

- Dual antiplatelet therapy with aspirin 81–325 mg daily plus clopidogrel 75 mg daily for 90 days followed by long-term antiplatelet therapy with aspirin 81–325 mg daily.

- For patients with hypertension, we suggest a BP goal of <140/90 (or < 130/80 for diabetics). Blood pressure should be lowered cautiously; however, the optimal blood pressure has never been examined systematically.

- Independent of baseline LDL, high-intensity statin therapy with atorvastatin 80 mg daily or rosuvastatin 20–40 mg daily with a long-term goal of LDL < 70.

- Lifestyle modification including smoking cessation, alcohol moderation, exercise, weight reduction if overweight, and a Mediterranean diet.

- Aggressive risk factor management of hypertension, hyperlipidemia, diabetes, and smoking. Post hoc analysis from WASID trial revealed that patients who had poorly controlled blood pressure or elevated cholesterol during follow-up had a significantly higher rate of stroke, myocardial infarction, and death versus patients with good control of their risk factors [32]. Also, in the aggressive medical arm of SAMMPRIS, the lower than expected rate of stroke is quite possible a result of the intensive risk factor control.

- Although interventional therapies such as angioplasty and stenting exist and are feasible, the SAMMPRIS and VISSIT randomized trials showed that for patients with symptomatic intracranial large artery stenosis, angioplasty and stenting had worse outcomes than those who received medical therapy [19, 27]. Furthermore, the American Heart Association/American Stroke Association (AHA/ASA) has stated the Wingspan Stent System, used in the SAMMPRIS trial, is not recommended for patients with stroke or TIA attributed to intracranial artery stenosis [33].

- Intracranial bypass surgery was considered a therapeutic option for patients in the past; however, the extracranial-intracranial (EC-IC) bypass study of 1985 demonstrated that EC-IC bypass was not only ineffective in patients with distal carotid stenosis but hazardous in patients with MCA stenosis, so it is not routinely recommended [34].

- Multiple retrospective studies looking at patients who were medically treated for 50–99% stenosis of a major intracranial artery suggest the annual risk of stroke is of ~3–15% in any vascular territory and 2–11% in the territory of the stenotic artery [35, 36]. Similar rates are found in several prospective studies including the GESICA study and the WASID trial where the ischemic stroke rate in the territory of the stenosis in the treatment groups was 11–12% at 1 year [20, 26]. As suggested in the SAMMPRIS and VISSIT trials, these recurrent stroke rates may in fact be lower with the implementation of intensive medical therapy as outlined above.

- Several investigators have looked at what parameters make a patient "high risk" for recurrent ischemic strokes or TIA. In the medical arm of the SAMMPRIS trial, high-risk features included the presence of an old infarct in the territory of the stenosis, presentation with stroke, and absence of statin use at trial entry [37]. Earlier retrospective studies based on the WASID trial suggested patients with vertebrobasilar disease to be high risk [38]. However, the prospective WASID trial did not support that claim and found no increased stroke risk associated with vertebrobasilar disease (HR 1.05, 95% CI 0.66–1.68) or use of anti-thrombolytics at the time of the ischemic event [21]. Furthermore, the WASID trial found a greater frequency of stroke or vascular death in women compared to men (28.4 versus 16.6%; HR 1.58, 85% CI 1.1–2.9) [39].

- It is still unclear how to best care for patients with asymptomatic ICAD. It is reasonable to extrapolate from the current recommendations for symptomatic ICAD. If risk factors are not well controlled, given the pathophysiology and natural course of the disease, asymptomatic ICAD can progress to becoming symptomatic. Therefore, we recommend a similar approach of risk factor management of hypertension, hyperlipidemia, diabetes, and smoking. Lifestyle modification consisting of smoking cessation (if applicable), alcohol moderation, exercise, weight reduction (if overweight), and a Mediterranean diet is recommended regardless of presence or absence of ICAD. Especially if multiple vascular risk factors are present, it is reasonable to consider antiplatelet monotherapy in patients with asymptomatic ICAD for primary stroke prevention following a consideration of risks and benefits.

- Management of a patient who has failed medical therapy who presents with a recurrent stroke remains uncertain. Current guidelines suggest that stenting is not recommended in patients with a first time event secondary to ICAD; however, it is unclear how to manage patients with recurrent strokes despite maximal medical therapy. There is much debate in the next step of management in these patients. Patients with symptomatic ICAD who fail antithrombotic therapy have extremely high rates of recurrent stroke/TIA or death most notable within a few months after failure of standard medical therapy [40].

- At this point, there is clinical equipoise to consider stenting, angioplasty, bypass surgery, or even anticoagulation for such patients. It is a dilemma that has yet to be answered. Given the high risk of recurrent vascular events following failed medical management, further testing of alternative treatment strategies such as those mentioned above is warranted.

7 Conclusion

ICAD has proven to be an important cause of ischemic strokes worldwide. Understanding the epidemiology, pathophysiology, and risk factors in certain patient populations will help identify and diagnose these individuals. Several studies have given insight and guided our practice in how to manage patients with ICAD; however, there is still much to be learned.

References

1. Ritz K et al (2014) Cause and mechanisms of intracranial atherosclerosis. Circulation 130 (16):1407–1414
2. Rincon F et al (2009) Incidence and risk factors of intracranial atherosclerotic stroke: the Northern Manhattan Stroke Study. Cerebrovasc Dis 28(1):65–71
3. Akins PT et al (1998) Natural history of stenosis from intracranial atherosclerosis by serial angiography. Stroke 29(2):433–438
4. Alkan O et al (2009) Intracranial cerebral artery stenosis with associated coronary artery and extracranial carotid artery stenosis in Turkish patients. Eur J Radiol 71(3):450–455
5. Hansson GK, Libby P (2006) The immune response in atherosclerosis: a double-edged sword. Nat Rev Immunol 6(7):508–519
6. Lopez-Cancio E et al (2012) The Barcelona-Asymptomatic Intracranial Atherosclerosis (AsIA) study: prevalence and risk factors. Atherosclerosis 221(1):221–225
7. Kublickiene K, Luksha L (2008) Gender and the endothelium. Pharmacol Rep 60(1):49–60
8. Bang OY et al (2016) A polymorphism in RNF213 is a susceptibility gene for intracranial atherosclerosis. PLoS One 11(6):e0156607
9. Adams HH et al (2016) Heritability and genome-wide association analyses of intracranial carotid artery calcification: the Rotterdam study. Stroke 47(4):912–917
10. Munshi A et al (2009) Phosphodiesterase 4D (PDE4D) gene variants and the risk of ischemic stroke in a South Indian population. J Neurol Sci 285(1-2):142–145
11. Eftekhar B et al (2006) Are the distributions of variations of circle of Willis different in different populations?—Results of an anatomical study and review of literature. BMC Neurol 6:22
12. Liebeskind DS et al (2014) Computed tomography angiography in the stroke outcomes and neuroimaging of intracranial atherosclerosis (SONIA) study. Interv Neurol 2(4):153–159
13. Uekita K et al (2003) Cervical and intracranial atherosclerosis and silent brain infarction in Japanese patients with coronary artery disease. Cerebrovasc Dis 16(1):61–68
14. Bos D et al (2012) Intracranial carotid artery atherosclerosis: prevalence and risk factors in the general population. Stroke 43 (7):1878–1884

15. Yoo JH, Chung CS, Kang SS (1998) Relation of plasma homocyst(e)ine to cerebral infarction and cerebral atherosclerosis. Stroke 29(12):2478–2483

16. Park JH et al (2011) High levels of apolipoprotein B/AI ratio are associated with intracranial atherosclerotic stenosis. Stroke 42(11):3040–3046

17. Qian Y et al (2013) Low HDL-C level is associated with the development of intracranial artery stenosis: analysis from the Chinese Intra-Cranial AtheroSclerosis (CICAS) study. PLoS One 8(5):e64395

18. Gorelick PB et al (2008) Large artery intracranial occlusive disease: a large worldwide burden but a relatively neglected frontier. Stroke 39(8):2396–2399

19. Chimowitz MI et al (2011) Stenting versus aggressive medical therapy for intracranial arterial stenosis. N Engl J Med 365(11):993–1003

20. Mazighi M et al (2006) Prospective study of symptomatic atherothrombotic intracranial stenoses: the GESICA study. Neurology 66(8):1187–1191

21. Kasner SE et al (2006) Predictors of ischemic stroke in the territory of a symptomatic intracranial arterial stenosis. Circulation 113(4):555–563

22. Arenillas JF et al (2001) Progression and clinical recurrence of symptomatic middle cerebral artery stenosis: a long-term follow-up transcranial Doppler ultrasound study. Stroke 32(12):2898–2904

23. Feldmann E et al (2007) The stroke outcomes and neuroimaging of intracranial atherosclerosis (SONIA) trial. Neurology 68(24):2099–2106

24. Kaufmann TJ et al (2007) Complications of diagnostic cerebral angiography: evaluation of 19,826 consecutive patients. Radiology 243(3):812–819

25. Qiao Y et al (2014) Intracranial plaque enhancement in patients with cerebrovascular events on high-spatial-resolution MR images. Radiology 271(2):534–542

26. Chimowitz MI et al (2005) Comparison of warfarin and aspirin for symptomatic intracranial arterial stenosis. N Engl J Med 352(13):1305–1316

27. Zaidat OO et al (2015) Effect of a balloon-expandable intracranial stent vs medical therapy on risk of stroke in patients with symptomatic intracranial stenosis: the VISSIT randomized clinical trial. JAMA 313(12):1240–1248

28. Derdeyn CP et al (2014) Aggressive medical treatment with or without stenting in high-risk patients with intracranial artery stenosis (SAMMPRIS): the final results of a randomised trial. Lancet 383(9914):333–341

29. Lutsep HL et al (2015) Does the stenting versus aggressive medical therapy trial support stenting for subgroups with intracranial stenosis? Stroke 46(11):3282–3284

30. Gao P et al (2015) China angioplasty and stenting for symptomatic intracranial severe stenosis (CASSISS): a new, prospective, multicenter, randomized controlled trial in China. Interv Neuroradiol 21(2):196–204

31. Hong KS et al (2016) Recurrent ischemic lesions after acute atherothrombotic stroke: clopidogrel plus aspirin versus aspirin alone. Stroke 47(9):2323–2330

32. Chaturvedi S et al (2007) Risk factor status and vascular events in patients with symptomatic intracranial stenosis. Neurology 69(22):2063–2068

33. Kernan WN et al (2014) Guidelines for the prevention of stroke in patients with stroke and transient ischemic attack: a guideline for healthcare professionals from the American Heart Association/American Stroke Association. Stroke 45(7):2160–2236

34. (1985) Failure of extracranial-intracranial arterial bypass to reduce the risk of ischemic stroke. Results of an international randomized trial. The EC/IC bypass study group. N Engl J Med 313(19):1191–1200

35. Sacco RL et al (1995) Race-ethnicity and determinants of intracranial atherosclerotic cerebral infarction. The Northern Manhattan Stroke Study. Stroke 26(1):14–20

36. Qureshi AI et al (2003) Stroke-free survival and its determinants in patients with symptomatic vertebrobasilar stenosis: a multicenter study. Neurosurgery 52(5):1033–1039; discussion 1039–40

37. Waters MF et al (2016) Factors associated with recurrent ischemic stroke in the medical group of the SAMMPRIS trial. JAMA Neurol 73(3):308–315

38. (1998) Prognosis of patients with symptomatic vertebral or basilar artery stenosis. The Warfarin-Aspirin Symptomatic Intracranial Disease (WASID) study group. Stroke 29(7):1389–1392

39. Williams JE et al (2007) Gender differences in outcomes among patients with symptomatic intracranial arterial stenosis. Stroke 38(7):2055–2062

40. Thijs VN, Albers GW (2000) Symptomatic intracranial atherosclerosis: outcome of patients who fail antithrombotic therapy. Neurology 55(4):490–497

Chapter 17

Embolic Ischemic Stroke

Matthew M. Padrick, Shouri Lahiri, and Konrad Schlick

Abstract

Cardioembolism represents a common source of ischemic stroke. The pathophysiology of emboli can differ significantly, with differing treatments. These can include embolism from atrial fibrillation, congestive heart failure, infective endocarditis, inflammatory endocarditis, aortic valve atheroma, aortic dissection, deep venous thrombosis in the presence of right-to-left shunt, atrial septal defect, and mechanical circulatory support devices. Here we will discuss when to suspect cardioembolism, diagnostic approach, and optimal treatment strategies.

Key words Embolism, Embolus, Cardioembolism, Atrial fibrillation, Congestive heart failure, Endocarditis, Atheroma, Aortic dissection, Paradoxical embolism, Shunt, Atrial septal defect, Mechanical circulatory support device

1 Introduction

The NIH has estimated 60% of all strokes are caused by embolism, with cardiogenic embolism responsible for 14–30% of ischemic strokes. Embolism from any source has been previously reported to account for between 15% and 70%, illustrating the wide gap across the medical community regarding stroke etiology [1]. There have been numerous classification systems created in the effort to effectively diagnose and ultimately choose the most effective secondary stroke prevention method. This chapter will deal specifically with embolic stroke in adults rather than the pediatric population, with a focus on cardiac disease.

2 Pathophysiology and Clinical Presentation

Mural thrombi and platelet aggregates are the materials most commonly embolized. While quite destructive as a solid, these emboli can also be remarkably evanescent. To illustrate this concept, consider that embolic fragments are found in more than 75% of cases

Fawaz Al-Mufti and Krishna Amuluru (eds.), *Cerebrovascular Disorders*, Neuromethods, vol. 170,
https://doi.org/10.1007/978-1-0716-1530-0_17, © Springer Science+Business Media, LLC, part of Springer Nature 2021

within 8 h of symptoms, 40% in 72 h after symptoms, and 15% after 72 h as detected by angiography in ischemic stroke patients [1, 2]. However, clearly this timetable still leaves patients with the potential for considerable loss of brain tissue.

No reliable means have been developed to identify which embolic occlusions will persist and which will disappear. After initial occlusion, a wall of blood forms between embolus and vessel wall that will erode the clot over hours to days. Patients with a cardiac source of emboli experienced recanalization more commonly than those with an arterial source, indicating a different composition between the two [3].

While ischemic stroke has an association with both systolic and diastolic congestive heart failure (CHF), the pathogenesis of thrombus formation differs. In diastolic CHF (with preserved ejection fraction), it is hypothesized that turbulent flow across faulty valves can lead to thrombus formation. Systolic CHF, with left ventricular dysfunction affecting ejection fraction, allows blood stasis in dilated cardiac chambers with poor contractility invoking Virchow's triad. There is also laboratory evidence to suggest a hypercoagulable state associated with systolic HF, with increases in platelet activation, blood viscosity, fibrinogen, von Willebrand factor, and fibrin D-dimer [4]. Data in patients with systolic HF suggest that the risk of stroke or thromboembolism is similar to slightly increased compared to patients with diastolic HF (ACTIVE trials, CHARM-Preserved, I-PRESERVE, and CHARM) [5].

While not proven, it has been hypothesized that cerebral ischemia in patients with isolated atrial septal aneurysm (ASA) without a shunt occurs because of fibrin-platelet particles which adhere to the left atrial side of the aneurysm that can be dislodged by oscillations into arterial circulation.

Aortic arch atherosclerotic disease is a manifestation of systemic atherosclerotic disease and thus has the same risk factors and is more common in older patients and patients with CAD (coronary artery disease). There are two mechanisms for neurovascular insult: thromboembolism and atheroembolism itself. Atheroembolism represents cholesterol crystal embolization and generally occurs when there is instrumentation around the atheroma, although it can happen sporadically.

Thromboembolism arises when atherosclerotic plaque becomes unstable and superimposed thrombi embolize. These thrombi are generally smaller compared to a cardiogenic thrombus and thus can result in ischemia to small- and medium-sized arteries producing less severe deficits on presentation. Atheroembolism produces a shower of emboli, which can lead to systemic embolization elsewhere as well.

Historically, it was taught that onset of symptoms in embolic stroke is generally abrupt and maximal at the onset, as evidenced by the Harvard Cooperative Stroke Registry [6]. In contrast, in situ thrombosis has been thought to be more likely to produce evolving and fluctuating symptoms. However, it has now been demonstrated that regardless of etiology, ischemic stroke can produce both sudden and maximal symptoms at onset, and non-sudden or fluctuating symptoms can occur in a small percentage of documented embolic strokes [7, 8]. Additionally, lacunar infarcts have classically been attributed to microvascular disease (such as lipohyalinosis); however, embolic strokes have been demonstrated to result in classical lacunar syndromes, along with radiographic small vessel infarcts [9]. In a cohort of endocarditis patients, a significant proportion of strokes appeared lacunar, both clinically and radiographically [10]. Embolism is a likely mechanism of many strokes commonly described as lacunar in nature. [11]

The CRYSTAL-AF study demonstrated detection of atrial fibrillation in 8.9% of ischemic stroke patients wearing an implanted cardiac monitor to detect arrhythmias versus 1.4% of patients in the conventional care, indicating the prevalence of occult atrial fibrillation in stroke patients is likely higher than previously recognized [12].

Patients may have accompanying signs and symptoms of congestive heart failure, including but not limited to:

- Shortness of breath.
- Orthopnea.
- Pitting extremity edema.
- Rales on auscultation of lung bases.
- Pulmonary edema on chest X-ray.
- Elevated BNP.

Patients presenting with acute myocardial infarction develop ischemic cardiomyopathy and are at risk of developing intracardiac thrombus, likely via similar pathogenesis associated with cardiac wall motion abnormalities.

Endocarditis can be classified into viral, bacterial, and non-infectious. Bacterial and non-infectious endocarditis give rise to emboli that can cause ischemic stroke, and thus this is where our focus will be. Depending on the specific pathogen, bacterial endocarditis can present in fulminant or subacute fashion. An acute, fulminant presentation of chest pain, profound fatigue, fever, and chills is most likely a staphylococcus pathogen. Subacute presentation—with fatigue, myalgia, weight loss, and dyspnea—is generally a streptococcus or HACEK (*Haemophilus, Aggregatibacter, Cardiobacterium, Eikenella, Kingella*) organism. Risk factors include recent dental procedure, IV drug abuse, mechanical valve, rheumatic heart disease, autoimmune conditions, HIV, as well as a prior endocarditis history. Bacterial vegetations in IV drug abuse patients preferentially form on the tricuspid valve. In a patient with a right-to-left shunt (such as patent foramen ovale), this can result in embolization to the right atrium, onward to the left atrium, and enter the systemic arterial circulation.

More commonly, embolic stroke in the setting of endocarditis arises from aortic and mitral valve lesions in which septic emboli can simply break off from a valve vegetation and enter arterial circulation. These strokes generally appear punctate and clustered but can be spread across multiple vascular territories. These infarcts have a propensity to bleed compared to typical thromboembolism.

A common congenital structural variant of the heart is between the atria, called a patent foramen ovale (PFO), that shunts blood from the right side to the left side of the heart. This can allow a clot formed in the peripheral venous vasculature, usually deep venous thrombosis in major vessels of extremities, to enter the systemic arterial circulation, referred to as a paradoxical embolus. A paradoxical embolus bypasses the lungs where it could otherwise cause a pulmonary embolism. A similar process can happen with a pulmonary shunt (such as an arteriovenous malformation) large enough to allow a clot to pass from venous to arterial circulation. Pulmonary shunts can be detected by clinical clues including:

- Cyanosis.
- Clubbing.
- Unexplained hemoptysis.
- Hypoxemia/dyspnea.
- Hemorrhagic telangiectasia.
- Nodule of chest radiograph.

Atrial septal aneurysm (ASA) is defined as redundant and mobile interatrial septal tissue in the region of the fossa ovalis with phasic excursion of at least 10–15 mm during the cardiorespiratory cycle. They can be classified according to the direction they oscillate, either into the left or right atrium, and according to the motion during the respiratory cycle [13]. ASA has a weak but present association with cryptogenic and presumed embolic stroke. Mattioli et al. reports a 27.7% higher prevalence in patients with cerebral ischemia and normal carotid arteries compared to a control group [14]. ASA was frequently associated with PFO in these patients as well.

Aortic dissection is a surgical emergency that is frequently in the setting of severe back pain. A neurologic exam may prove difficult depending on patient's ability to participate, level of consciousness, and significant pain. Aortic dissection can produce a wide range of symptoms by affecting the outflow of supra-aortal, abdominal, spinal, extremity, and renal vessels. Most often, symptoms are severe chest and/or back pain with hypotension, similar to a myocardial infarction. Neurologic symptoms arise because of occlusion of carotid, vertebral, or spinal arteries or because of profound hypotension and watershed ischemia in the brain and spinal cord.

There is a small population of patients who present with neurologic deficits that can be attributed to painless aortic dissection. Gaul et al. reported 13.9% of type A aortic dissection patients reported no pain and approximately half of pain-free patients presented with neurologic symptoms [15]. Transient symptoms may occur as the result of the dissection propagating up the vessel and causing temporary occlusions. Symptoms are generally right-sided and hemispheric rather than posterior as the carotids will become involved much more readily compared to vertebral arteries. Additional diagnostic clues include involvement of both central and peripheral nerve ischemia, syncope, and seizures.

Fluctuating symptoms can arise as emboli migrate through the neurovascular tree, intermittently blocking blood flow at various junctures along the way. The embolus can lyse entirely, leaving the patient asymptomatic. This may be deemed a transient ischemic attack (TIA), or the patient may still experience a smaller volume of infarcted tissue downstream with associated symptoms commensurate with the tissue affected. The initial interruption of blood flow at the largest most proximal artery would likely induce the largest clinical deficit the patient is to experience at the onset of symptoms.

Historically, recurrent TIAs in the same vascular territory suggest vascular pathology and artery-to-artery embolism. The comprehensive Lausanne Stroke Registry investigators were unable to conclude that previous TIAs ipsilateral to cerebral infarction were more suggestive of atherosclerosis than of embolic heart disease in overall stroke population. However, if vessel imaging reveals $\geq 50\%$ arterial stenosis or occlusion, there is a significant association with previous ipsilateral TIAs [7, 8].

While both the Harvard Cooperative Stroke Registry and the Lausanne Stroke Registry were unable to concretely establish criteria regarding infarct etiology, modern diagnostic clues have emerged. Essentially, a clinician should look not just for the cause of stroke but also the absence of other causes as well.

2.1 Management: Diagnostic Approach

2.1.1 TOAST

The Trial of Org 10172 in Acute Stroke Treatment (TOAST) criteria were published in 1993 to help classify stroke etiology [16, 17]. The TOAST classification denotes five subtypes of ischemic stroke:

1. Large artery atherosclerosis
2. Cardioembolism
3. Small vessel occlusion (lacunar)
4. Stroke of other determined etiology
5. Stroke of undetermined etiology

(a) Complete evaluation

(b) Incomplete evaluation

In devising these criteria, it should be noted that if investigators were to find multiple possible etiologies, the patient would be excluded from classification to any of the above categories.

Since its inception, the TOAST criteria have been used in numerous studies despite their limitations (only fair to moderate interrater agreement). There have been attempts to refine this classification system, including SSS-TOAST, which further delineates the five classifications into "Evident," "Probable," and "Possible." More recently, clinicians devised the Causative Classification System (CCS). The CCS is a computerized algorithm that consists of questionnaire-style classification scheme, with improved interrater reliability across multiple centers [18]. This constant refinement among the neurovascular community illustrates the complexity of ischemic stroke diagnosis.

2.1.2 Imaging

In addition to the clinical history and examination, the modern diagnosis of embolic stroke is heavily dependent on neuroimaging. In addition to demonstrating an infarct in brain parenchyma, the diagnosis of embolism is predicated on demonstration of a lesion (especially a vascular lesion) proximal to the territory infarct.

- In the most typical presentation of a cardioembolic stroke, diffusion-weighted imaging (DWI) demonstrates a wedge-shaped infarct, a plausible cardiac pathologic origin is identified, and no intervening vascular lesion is present.

- Parenchymal imaging is most commonly obtained with CT and MRI modalities.

  – The advantage of CT remains its widespread availability, cheaper cost (as compared with MRI), and rapid acquisition.

  – MRI (DWI in particular) is more sensitive in identifying ischemia in its earliest stages and is more likely to identify smaller infarcts not easily seen on CT scans of the brain.

- Numerous techniques can identify vascular lesions that are a source of embolism; angiography can be performed with CT (CTA), MRI (MRA), as well as conventional catheter angiography (i.e., digital subtraction angiography (DSA)).

  – These are largely "lumenograms," describing the vessel lumen rather than any potential pathology contained within the vessel wall proper.

- MRI with fat suppression techniques can identify pathology within the walls of arteries.

- – In particular, this technique can demonstrate arterial dissection, which may cause local thrombosis and subsequent downstream embolism.
- Carotid duplex scanning can identify lesions in the cervical carotid arteries (especially atheromatous lesions).
  - – The B-mode imaging component of duplex scanning can characterize the morphology of stenosis carotid lesions, while the Doppler waveforms provide information about the velocity of blood flow.

    In combination, this provides information regarding the severity of stenosis. Carotid stenosis and associated atherothromboembolism are a prototypical example of artery-to-artery embolism resulting in hemispheric stroke syndromes.
- Transcranial Doppler (TCD) ultrasound can identify intracranial stenotic lesions, which may themselves lead to downstream embolism.
  - – TCD is highly operator-dependent, but by producing flow velocities, it can provide semiquantitative information regarding the degree of intracranial stenosis, especially in the MCAs where insonation is most reproducible.
  - – Other TCD techniques take advantage of its temporal resolution, allowing for the direct monitoring of microembolic signals (MES), also called high-intensity transient signals (HITS).

    TCD monitoring can observe and quantify HITS in the vessel segments monitored and may provide information regarding stroke risk.
- As cardiac pathology is a prototypical cause for embolic strokes, multiple modalities exist for structural evaluation of the heart.
  - – Most commonly implemented is echocardiography.

    The transthoracic echocardiogram (TTE) is easily obtainable and can demonstrate systolic or diastolic dysfunction, valvular pathology, and even right-to-left shunts when color Doppler is used or agitated saline contrast (visualized as "bubbles") is administered.

    Mural thrombus (especially in the anterior chambers) can be visualized on TTE; this visualization may be enhanced with ultrasound contrast.

    As pathology in the posterior aspects of the heart is common (especially clot in the left atrial appendage, LAA), a transesophageal echocardiographic (TEE) evaluation is sometimes necessary to identify a cardioembolic source. This approach is more sensitive than TTE in identifying clot, spontaneous echocontrast ("pre-clot"), atrial tumors, some valvular lesions, and aortic pathology.

- Less commonly used techniques for identifying cardioembolic sources include CT with contrast (especially for atrial clots) and cardiac MRI (for mural thrombi, especially in the cardiac ventricles).

2.1.3 Diagnostic Clues for Cardioembolism

- Wedge-shaped infarct seen on neuroimaging (both CT and MRI)
- Lack of compelling atherosclerotic disease on vessel imaging (Adams, Goldstein, Hart)
- Stroke territory confined to the cerebral surface territory of a single branch
- Ischemia across multiple vascular territories
- Hemorrhagic conversion [2, 19]
- Elevated brain natriuretic protein (BNP) for up to 72 h [20]

  Common sources of cardiac embolism are:

- Intracardiac thrombus formation in patients with atrial fibrillation
- Intracardiac thrombus formation in patients with anterior myocardial infarction
- Septal wall abnormalities
  - Patent foramen ovale
  - Atrial aneurysm
  - Ventricular aneurysm
- Cardiomyopathies (ischemic and non-ischemic)
  - Systolic congestive heart failure with impaired ejection fraction
  - Diastolic congestive heart failure with preserved ejection fraction
- Aortic arch atheromas
- Valvular heart disease
  - Mitral stenosis
  - Mitral regurgitation
  - Prolapsed mitral valve
- Atrial myxoma
- Marantic endocarditis

2.1.4 Diagnosing Atrial Fibrillation

Atrial fibrillation on arrival is most commonly diagnosed via electrocardiogram (ECG), if the patient has active disease at the time.

- ECG will show irregularly irregular rhythm and lack of p waves.
- In patients with paroxysmal atrial fibrillation in sinus rhythm on arrival, diagnosis may require further investigation.

- The patient may report a sense of racing heartbeat, dizziness, or palpitations.

- ECG may show normal sinus rhythm, but with increased amplitude of p waves suggestive of atrial enlargement.

- In a systematic review and meta-analysis of prospective cohort studies, retrospective cohort study, and case-control studies, p-wave indices were assessed as electrocardiographic markers that can be used to stratify stroke risk [21].

  - Increased p-wave terminal force in lead V1 (PTFV1), the product of the depth (μV) of the terminal portion of the p wave in V1 multiplied by its duration (ms), is an independent predictor of stroke as both continuous variable and categorical variable.

    The agreed-upon threshold for multiple studies has been 5000μV*ms.

  - P-wave duration alone is a significant predictor of ischemic stroke when analyzed as a categorical value but not as a continuous value.

  - Maximum p-wave area was also shown to be a predictor of ischemic stroke.

- Telemetry monitoring is routinely performed during the acute stroke patient's hospitalization.

  - By the time the patient is discharged, if there is no definitive stroke etiology, further cardiac monitoring can be performed if occult arrhythmia is suspected.

- Echocardiography (TEE more reliably compared to TTE) may also show:

  - Atrial enlargement.

  - Occult clot in the left atrial appendage, the most common site for intracardiac thrombus formation.

  - Spontaneous echo contrast (SEC), or "smoke," which is characterized by dynamic "smoke-like" echoes within the left atrium/left atrial appendage. SEC is strongly associated with enlarged left atrium and left atrial stasis. It is frequently found in patients with LA thrombus and is found in many patients with previous embolic events [19].

- Additional diagnostic tools include CT angiography to include the upper chambers of the heart (which can demonstrate a clot if present), as well as cardiac MRI.

- The lack of temporal relationship between periods of electrographic atrial fibrillation and stroke calls into question thrombogenesis as a result of atrial quivering versus atrial wall remodeling as a whole [22].

2.1.5 Diagnosing
Endocarditis

- Diagnosis will largely rely on infectious evaluation, including cardiac auscultation, echocardiography, and blood cultures.
- To stratify neurovascular insult risk, clinicians can perform TCD with emboli monitoring.
- There are no specific guidelines, but the argument can be made to expedite surgery if there are ongoing HITS during emboli monitoring.
- Non-infectious endocarditis arises from autoimmune and cancerous conditions.
  - Libman-Sacks endocarditis is associated with systemic lupus erythematosus, with verrucous vegetations consisting of immune complexes, mononuclear cells, hematoxylin bodies, fibrin, and platelet thrombi.
  - Marantic endocarditis, or non-bacterial thrombotic endocarditis, is associated with underlying malignancy.

2.1.6 Diagnosing Atrial
Septal Abnormalities

- TTE with agitated saline bubble study can detect a right-to-left shunt.
- TEE and TCD with microembolus detection ("TCD bubble study") with similar maneuvers increases detection.
  - For an optimized TCD bubble study, the patient must be able to follow commands and bear down for Valsalva for bubble study, which increases intrathoracic pressures across the septum and increases detection of microembolic signals during the bubble study.
- While TEE can more clearly visualize the presence of PFO size and shape, TCD may be more sensitive for right-to-left shunt detection.
- TCD, however, does not allow for specificity as to what type of shunt is present.
- In the acute setting, if a PFO is detected, the clinician must check all four extremities for deep venous thrombosis that could lead to paradoxical embolism.

2.1.7 Diagnosing Atrial
Septal Aneurysm

- Like a PFO, an ASA can be detected on TTE but is better visualized with TEE.

2.1.8 Diagnosing Aortic
Atheroma

- CTA of the neck, including the aortic arch, can identify an atheroma but produces only a static image.
- Atheromas can be best characterized with TEE from the aortic valve level to the initial curvature of the arch.
  - Echocardiography, compared to CTA and MRA, is dynamic and can identify both mobile elements and heterogeneity of the plaque.
  - Severe disease burden is determined by:

Size $\geq$4 mm.

Heterogeneity of atheroma.

Presence of ulcerations.

Presence of mobile elements.

- Patients with an atheromatous arch are likely to have atherosclerotic disease elsewhere.
  - For this reason, antiplatelet medications are indicated.
  - Depending on the extent of atherosclerotic disease/atheroma size, one can consider monotherapy with aspirin or clopidogrel or dual antiplatelet therapy with both.
    If only mild-moderate disease burden, monotherapy to avoid excessive bleeding risk is reasonable.
  - If there is severe disease present, dual antiplatelet is reasonable.

2.2 Management: Treatment Approach

Regardless of etiology, acute stroke management is heavily time-dependent. Patients must be rapidly interviewed, evaluated, scanned, and treated with recanalization therapies as appropriate to save maximal brain tissue.

- IV tPA remains the standard of care for patients who are eligible.
  - An analysis of stroke subtypes in the NINDS tPA trial demonstrates that all stroke subtypes appear to benefit from IV tPA treatment [3].
- The majority of stroke patients who are treated with endovascular thrombectomy, however, suffer embolic strokes (especially cardioembolism).
  - This is in part because cardioembolic strokes produce more severe deficits.
  - Current guidelines suggest that otherwise eligible patients with a NIH Stroke Scale (NIHSS) score of 6 or greater should be treated with endovascular therapy [23].
  - Further, embolic strokes are most likely to produce a proximal large vessel occlusion that is technically amenable to endovascular treatment.
- Patients with a suspected large vessel occlusion, who present with dramatic symptoms and higher NIHSS scores, should be rapidly evaluated with angiography as they may be candidates for mechanical thrombectomy.
  - If a proximal large vessel occlusion is identified in a patient with hyperacute stroke, recanalization with endovascular treatment should be considered.

2.2.1 Atrial Fibrillation   A stroke patient with newly discovered atrial fibrillation associated with embolus commonly presents with rapid and maximal symptom onset. Given the size of emboli that can be produced behind the left atrial appendage, vessels of all caliber may be affected, including the internal carotid itself. Vessel imaging may reveal little to no atherosclerotic disease, and therefore embolus moves as distal as possible through the vascular tree with little resistance until reaching vessel caliber equal to diameter of embolus.

- The SPAF trials demonstrated anticoagulation (vitamin K antagonists in particular, such as warfarin) as the preferred stroke prevention regimen as compared to antiplatelet regimen for AF-associated stroke.
  - However, in the setting of acute stroke, there is concern for hemorrhagic conversion if anticoagulation is started too early [24].
  - The HAEST trial demonstrated that within the first 2 weeks after an ischemic stroke, the rates of stroke recurrence and hemorrhagic complications were similar regardless of whether low-dose aspirin or anticoagulation was used [25].

    This is consistent with the relatively low short-term stroke rate after an index AF-associated cardioembolic stroke.
  - AF patients with ischemic stroke have higher risk of hemorrhagic conversion compared to lacunar infarcts, and thus the pros and cons of anticoagulation must always be weighed on an individual case-by-case basis.
  - If patients need interruption of anticoagulation for any reason, bridging therapy with low molecular weight heparin may be considered for short-term management.
- For patients diagnosed with atrial fibrillation but without stroke, a clinician can calculate CHA2DS2 score, a composite score that factors age, sex, congestive heart failure, hypertension, stroke/TIA/thromboembolism episodes, vascular disease, and diabetes, to predict stroke risk and optimize primary stroke prevention.
- For any patient with already diagnosed ischemic stroke and AF, anticoagulation is generally necessary, barring any concomitant coagulopathy or bleeding diathesis.
  - Oral anticoagulants for secondary stroke prevention in atrial fibrillation include dabigatran, rivaroxaban, apixaban, and edoxaban (direct oral anticoagulants (DOACs)).
  - Of note, large randomized trials of DOACs have demonstrated these to be either non-inferior or superior to warfarin in stroke prevention. The above named agents have also been

shown to have lower rates of intracranial hemorrhage than warfarin, thus making these the preferred option when patients are eligible.

– Limitations of DOAC use primarily include impaired renal function; dose adjustment may be necessary in the setting of low body weight and advanced age.

In a large, unselected hypertrophic cardiomyopathy group of 900 patients, Maron et al. report shows a 6% prevalence rate and an incidence of 0.8% per year. Expectedly, ischemic stroke was substantially more common in elderly patients and occurred almost exclusively in patients with documented paroxysmal or chronic atrial fibrillation [26].

2.2.2   Congestive Heart Failure

The role of antithrombotic therapy in HF patients in sinus rhythm is not well established.

• Multiple studies, including Warfarin/Aspirin Study in Heart Failure (WASH) trial and the Heart Failure Long-Term Antithrombotic Study (HELAS), were not able to establish significant differences in outcomes between antithrombotic, anticoagulation, and placebo.

• The randomized WARCEF trial suggested a small benefit of warfarin for ischemic stroke prevention in systolic heart failure patients, but this was offset by an increased risk of hemorrhagic complications [27].

• The American College of Cardiology/American Heart Association guidelines currently recommend no antithrombotic agent nor anticoagulation in patients without established coronary artery disease and without evidence of intracardiac thrombus, atrial fibrillation, or prior thromboembolic event with cardioembolic suspicion.

2.2.3   Atrial Septal Abnormalities

Options for secondary prevention of ischemic stroke in the setting of PFO include:

• Antiplatelet therapy.

• Anticoagulation.

• PFO closure.

• Combinations of these.

Until recently, data had been lacking regarding the optimal treatment modality in these patients.

- The CLOSURE I trial (Evaluation of the STARFlex Septal Closure System in Patients with a Stroke and/or Transient Ischemic Attack due to Presumed Paradoxical Embolism through a Patent Foramen Ovale) provided significant evidence regarding the role of PFO closure in stroke prevention [28].
  - Adult patients <60 years old with PFO and cryptogenic stroke or TIA were randomly assigned to PFO closure ($n = 447$) or medical therapy alone ($n = 463$).
    Investigators concluded there was no significant difference between the two, but the study had flaws.

    TIA patients and lacunar strokes were included, with possible misdiagnosis and concomitant small vessel disease obscuring the overall contribution of the PFO.
- The RESPECT trial (closure of a PFO with Amplatzer PFO Occluder compared with medical therapy alone) randomly assigned patients aged 18–60 years with cryptogenic stroke and PFO to medical therapy alone (aspirin, clopidogrel, warfarin, or combination of aspirin/dipyridamole, $n = 480$) vs closure ($n = 499$), excluding TIA, known large vessel vascular disease, small vessel vascular disease, and hypercoagulable state patients.
  - Again, no superiority could be established between PFO closure and medical therapy alone.
  - Recurrent stroke rates are low overall, which limits statistical power.
- The Risk of Paradoxical Embolism (RoPE) Study was designed to identify patients with PFO most at risk for recurrent stroke and thus who would benefit most from PFO closure.
  - Investigators were able to create a 10-point RoPE point score, with higher point totals indicating lack of traditional stroke risk factors and higher likelihood of paradoxical embolus [29].
    No history of hypertension—1 point.

    No history of diabetes—1 point.

    No history of stroke or TIA—1 point.

    Non-smoker—1 point.

    Cortical infarct on imaging—1 point.

    Age 18–29—5 points.

    Age 30–39—4 points.

    Age 40–49—3 points.

    Age 50–59—2 points.

Age 60–69—1 point.

Age ≥ 70—0 points.

- There is now preliminary data from the CLOSE trial and Gore REDUCE (publication pending), presented at the European Stroke Organization Conference 2017, that shows significant benefit from PFO closure compared to antiplatelet therapy.
  - Both trials randomized ischemic stroke patients to PFO closure versus medical therapy alone.
  - With implementation of very strict criteria to exclude patients with other coronary sources, such as atrial fibrillation, underlying coronary disease, or small vessel disease, investigators could further stratify former cryptogenic stroke patients with most likely PFO as stroke etiology.
  - Gore REDUCE demonstrated a relative risk reduction of recurrent stroke of 77% at 2 years.
  - CLOSE showed an absolute risk reduction of recurrent stroke at 5 years of 4.9%.
  - Further, both of these trials generally included patients who were found to have relatively large right-to-left shunts.
  - It should be noted that rates of atrial fibrillation were higher in the PFO closure group than in the medical therapy-only group.

2.2.4   Aortic Atheroma

Optimal medical therapy for prevention of stroke due to arch atheroma remains uncertain.

- Amarenco et al. investigated patients with stroke, TIA, or peripheral embolism.
  - These patients had thoracic aortic plaque, but no other identified source of their symptoms [30].
  - They were randomized to a combination of aspirin and clopidogrel versus warfarin, with the primary outcome being a composite of cardiovascular events.
  - The trial was stopped early due to lack of power, and no differences were seen in the primary outcome.
  - There did, however, appear to be more instances of vascular death in the warfarin arm.
- With coronary artery bypass graft surgery, the incidence of postoperative neurologic sequelae is approximately 2–6%, most of which is due to stroke.
  - Atheroemboli associated with ascending aortic atherosclerosis dislodged intraoperatively is probably the most common cause rather than thromboembolism.

2.2.5 Aortic Dissection   • Emergency surgical management of aortic dissection.

2.2.6 Mechanical Circulatory Support Devices

The management of advanced heart failure sometimes requires the use mechanical circulatory support devices, often as a bridge until heart transplant can be performed. These include:

- Durable ventricular assist devices (VADs).
- Total artificial hearts (TAHs).
- More short-term interventions such as extracorporeal membrane oxygenation (ECMO) and intra-aortic balloon pump (IABP) placement.

While these devices prolong life, they may be accompanied by various complications. Embolic stroke (as well as hemorrhagic stroke) is a common complication in heart failure patients treated with these devices. Technological and technical advances in cardiac devices in the past decade have significantly reduced adverse events of thrombosis, stroke, and bleeding, but they are still significant at 8–11%, 25–40%, and 13–29%, respectively [31].

- International guidelines currently recommend combined antiplatelet therapy with aspirin and anticoagulation with warfarin.
  - This is largely based on protocols mandating these agents in clinical device trials.
  - Those trials, however, were not designed to optimize dosing to balance bleeding and thrombotic complications.

High shear stress through the narrow channels of a device induces hemolysis of red blood cells, which activates a cascade of increased reactive oxygen species, leading to:

- Increased vascular tone.
- Platelet activation.
- Endothelial dysfunction.
- Inflammation.
- Concurrently, free heme, iron, and carbon monoxide will increase carboxyhemoglobin levels.

Both cascades will result in hypofibrinolysis and hypercoagulability, leading to thrombosis. Serum levels of lactate dehydrogenase (LDH) is an easy laboratory test to gauge degree of hemolysis.

- Thromboelastography (TEG) has been used to help titrate antiplatelet and antithrombotic medications to minimize the risk of pump thrombosis, as well as embolic complications.
  - Rather than extrapolating numerous lab values to ascertain clot potential, TEG tests both platelet function and coagulation by assaying several parameters of clot formation dynamically in whole blood.

- While several reports describe the use of TEG to guide medical management of left ventricular assist device (LVAD) patients, no large randomized trials exist that demonstrate the efficacy of this approach.
- In our practice, we incorporate data from TEG, microembolic signals on TCD, as well as clinical events (including neurologic evaluation) to help guide management of these patients.

References

1. Mohr JP, Wolf PA, Grotta JC, Moskowitz MA, Mayberg MR, von Kummer R (2011) Stroke: pathophysiology, diagnosis, and management, vol 5. Elsevier, Inc., Amsterdam

2. Bozzao L, Fantozzi LM, Bastianello S et al (1989) Ischaemic supratentorial stroke: angiographic findings in patients examined in the very early phase. J Neurol 236(6):340–342

3. National Institute of Neurological Disorders and Stroke rt-PA Stroke Study Group (1995) Tissue plasminogen activator for acute ischemic stroke. N Engl J Med 333(24):1581–1587

4. Lip GY, Gibbs CR (1999) Does heart failure confer a hypercoagulable state? Virchow's triad revisited. J Am Coll Cardiol 33(5):1424–1426

5. Zile MR, Gaasch WH, Anand IS et al (2010) Mode of death in patients with heart failure and a preserved ejection fraction: results from the Irbesartan in heart failure with preserved ejection fraction study (I-Preserve) trial. Circulation 121(12):1393–1405

6. Mohr JP, Caplan LR, Melski JW et al (1978) The Harvard cooperative stroke registry: a prospective registry. Neurology 28(8):754–762

7. Bogousslavsky J, Van Melle G, Regli F (1988) The Lausanne stroke registry: analysis of 1,000 consecutive patients with first stroke. Stroke 19 (9):1083–1092

8. Bogousslavsky J, Cachin C, Regli F, Despland PA, Van Melle G, Kappenberger L (1991) Cardiac sources of embolism and cerebral infarction—clinical consequences and vascular concomitants: the Lausanne stroke registry. Neurology 41(6):855–859

9. Cacciatore A, Russo LS Jr (1991) Lacunar infarction as an embolic complication of cardiac and arch angiography. Stroke 22 (12):1603–1605

10. Hart RG, Foster JW, Luther MF, Kanter MC (1990) Stroke in infective endocarditis. Stroke 21(5):695–700

11. Futrell N (2004) Lacunar infarction: embolism is the key. Stroke 35(7):1778–1779

12. Sanna T, Diener HC, Passman RS et al (2014) Cryptogenic stroke and underlying atrial fibrillation. N Engl J Med 370(26):2478–2486

13. Pearson AC, Nagelhout D, Castello R, Gomez CR, Labovitz AJ (1991) Atrial septal aneurysm and stroke: a transesophageal echocardiographic study. J Am Coll Cardiol 18 (5):1223–1229

14. Mattioli AV, Aquilina M, Oldani A, Longhini C, Mattioli G (2001) Atrial septal aneurysm as a cardioembolic source in adult patients with stroke and normal carotid arteries. A multicentre study. Eur Heart J 22 (3):261–268

15. Gaul C, Dietrich W, Friedrich I, Sirch J, Erbguth FJ (2007) Neurological symptoms in type A aortic dissections. Stroke 38(2):292–297

16. Adams HP Jr, Bendixen BH, Kappelle LJ et al (1993) Classification of subtype of acute ischemic stroke. Definitions for use in a multicenter clinical trial. TOAST. Trial of org 10172 in acute stroke treatment. Stroke 24(1):35–41

17. Goldstein LB, Jones MR, Matchar DB et al (2001) Improving the reliability of stroke subgroup classification using the trial of ORG 10172 in acute stroke treatment (TOAST) criteria. Stroke 32(5):1091–1098

18. Radu RA, Terecoasa EO, Bajenaru OA, Tiu C (2017) Etiologic classification of ischemic stroke: where do we stand? Clin Neurol Neurosurg 159:93–106

19. Fagan SM, Chan KL (2000) Transesophageal echocardiography risk factors for stroke in nonvalvular atrial fibrillation. Echocardiography 17 (4):365–372

20. Llombart V, Antolin-Fontes A, Bustamante A et al (2015) B-type natriuretic peptides help in cardioembolic stroke diagnosis: pooled data meta-analysis. Stroke 46(5):1187–1195

21. He J, Tse G, Korantzopoulos P et al (2017) P-wave indices and risk of ischemic stroke: a systematic review and meta-analysis. Stroke 48 (8):2066–2072

22. Brambatti M, Connolly SJ, Gold MR et al (2014) Temporal relationship between subclinical atrial fibrillation and embolic events. Circulation 129(21):2094–2099

23. Powers WJ, Derdeyn CP, Biller J et al (2015) 2015 American Heart Association/American Stroke Association focused update of the 2013 guidelines for the early management of patients with acute ischemic stroke regarding endovascular treatment: a guideline for healthcare professionals from the American Heart Association/American Stroke Association. Stroke 46(10):3020–3035

24. Stroke Prevention in Atrial Fibrillation Study (1991) Final results. Circulation 84 (2):527–539

25. Berge E, Abdelnoor M, Nakstad PH, Sandset PM (2000) Low molecular-weight heparin versus aspirin in patients with acute ischaemic stroke and atrial fibrillation: a double-blind randomised study. HAEST study group. Heparin in acute embolic stroke trial. Lancet 355 (9211):1205–1210

26. Maron BJ, Olivotto I, Bellone P et al (2002) Clinical profile of stroke in 900 patients with hypertrophic cardiomyopathy. J Am Coll Cardiol 39(2):301–307

27. Homma S, Thompson JL, Pullicino PM et al (2012) Warfarin and aspirin in patients with heart failure and sinus rhythm. N Engl J Med 366(20):1859–1869

28. Carroll JD, Saver JL, Thaler DE et al (2013) Closure of patent foramen ovale versus medical therapy after cryptogenic stroke. N Engl J Med 368(12):1092–1100

29. Kent DM, Ruthazer R, Weimar C et al (2013) An index to identify stroke-related vs incidental patent foramen ovale in cryptogenic stroke. Neurology 81(7):619–625

30. Amarenco P, Davis S, Jones EF et al (2014) Clopidogrel plus aspirin versus warfarin in patients with stroke and aortic arch plaques. Stroke 45(5):1248–1257

31. Shah P, Tantry US, Bliden KP, Gurbel PA (2017) Bleeding and thrombosis associated with ventricular assist device therapy. J Heart Lung Transplant 36(11):1164–1173

Chapter 18

Cryptogenic Stroke and Stroke of "Unknown Cause"

Francisco Eduardo Gomez III, Krishna Amuluru, Yuval Elkun, and Fawaz Al-Mufti

Abstract

Cryptogenic ischemic stroke is considered a diagnosis of exclusion wherein no probable cause is identified after a comprehensive diagnostic workup. Cryptogenic stroke can be further classified as non-embolic or embolic. Embolic stroke of undetermined source can be due to paroxysmal atrial fibrillation, minor emboligenic cardiac conditions, atheroembolism, cancer-associated and paradoxical embolism through a patent foramen ovale, or less often a pulmonary fistula. Currently, risk factor control, statins, and antiplatelets are the main therapeutic measures to prevent recurrent stroke. Advances in high-resolution ultrasound or magnetic resonance imaging of extracranial and intracranial vessels and of the heart and prolonged heart rhythm monitoring will be instrumental techniques to identify arterial and cardiac hidden causes of stroke. In this chapter, we discuss the phenomena of cryptogenic stroke, workup and imaging, and proposed management pathways.

Key words Ischemic stroke, Cardioembolism, Atherosclerosis, Dissection, Thrombophilia, Fibromuscular dysplasia, Reversible cerebral vasoconstriction syndrome, Sickle cell disease, Carotid artery stenosis, CADASIL, CARASIL, Fabry's disease, Cardiac monitoring, Cryptogenic, Embolic stroke of undetermined source, Paroxysmal atrial fibrillation, Patent foramen ovale, Undetermined cause

1 Introduction

Cryptogenic ischemic stroke is considered a diagnosis of exclusion wherein no probable cause is identified after a comprehensive diagnostic workup. This entity comprises 20–30% of ischemic events. Cryptogenic strokes are generally less severe in presentation and carry a lower risk of mortality and recurrence than strokes with known risk factors. Some define strokes in patients under 50 years of age with no known risk factors as cryptogenic.

Stroke etiology notably varies by cohort and has changed through time. As technology advances and new conditions are

---

The original version of this chapter was revised. The correction to this chapter is available at https://doi.org/10.1007/978-1-0716-1530-0_23

Fawaz Al-Mufti and Krishna Amuluru (eds.), *Cerebrovascular Disorders*, Neuromethods, vol. 170, https://doi.org/10.1007/978-1-0716-1530-0_18, © Springer Science+Business Media, LLC, part of Springer Nature 2021, Corrected Publication 2021

defined, the incidence of cerebrovascular events termed "crypto-genic" will decrease as causes become more readily identifiable.

Key Points:

- Up to 80–90% of cryptogenic strokes are embolic in mechanism.
- PFO (patent foramen ovale) is overrepresented in cryptogenic stroke patients.
- Paroxysmal fibrillation is an increasingly recognized cause.
- May be caused by occult atherosclerosis consisting of non-stenosing unstable plaques.
- Other rare causes include cervical artery dissections or vasculitides.

Causes by age
18–30: Dissection, thrombophilic conditions, structural heart abnormalities
31–60: Early-onset atherosclerotic, acquired heart structural disease
>60: Atrial fibrillation

Non-lacunar infarcts with no significant proximal stenosis or cardioembolic source are defined as having an embolic source of unknown origin. This mechanism accounts for 80–90% of cryptogenic strokes. Suspected causes include aortic arch or occult atherosclerosis, paroxysmal AFib, mitral calcifications, and patent foramen ovale (PFO).

Transesophageal echocardiography (TEE) is informative in 50–75% of patients with otherwise cryptogenic events. Most notably, occult or low-burden atrial fibrillation has become increasingly recognized. Reportedly only an hour of atrial fibrillation over a 2-year span doubles stroke risk. The CRYSTAL AF study noted that insertable cardiac monitoring device sensitivity for atrial fibrillation increased with time, from 9% at 6 months to 30% at 3 years, and was superior to standard follow-up.

Monogenic and heritable causes of stroke are also becoming increasingly recognized, with a reported rate of 7% in one cohort, and include CADASIL, Fabry's disease, Marfan syndrome, and hereditary cerebral amyloid angiopathy. These conditions should be considered in an advanced stroke workup and may be guided by suggestive symptoms, signs, or family history (Fig. 1). A selection of such conditions will be further discussed below.

2   Patent Foramen Ovale

Numerous studies have demonstrated an association between patent foramen ovale and cryptogenic stroke. Meta-analysis showed a significant association between presence of PFO and cryptogenic stroke in patients under the age of 55 years. While the prevalence of

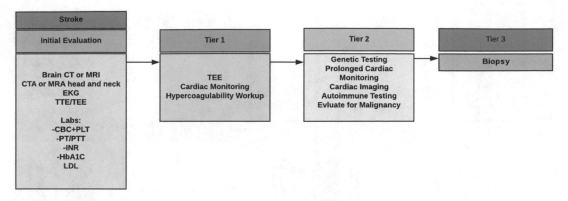

Fig. 1 Workup flowchart for cryptogenic stroke

PFO in adults may be as high as 25%, the prevalence of PFO in patients with cryptogenic strokes may be as high as 63% in some series [1–9].

2.1 Pathophysiology

Patent foramen ovale (PFO) is a remnant of fetal circulation caused by failure of the primum and secundum septa to fuse, creating an atrial septal defect. When the right atrial pressure exceeds the left atrial pressure, a right-to-left shunt develops, and the defect becomes a gateway for venous thromboemboli to enter the arterial circulation. This phenomenon, known as paradoxical embolism, can cause transeint ischemic attacks (TIAs) or ischemic strokes.

2.2 Management

2.2.1 Diagnostic Approach

PFO should be strongly considered in any patient younger than 55 years presenting with acute ischemic stroke or TIA without classical risk factors. Certain imaging patterns may further suggest a PFO [1–9].

- Stroke caused by paradoxical embolism through a PFO often usually appears as a single cortical or multiple small ischemic lesions, frequently in the vertebrobasilar circulation.
  - This differs from the lesion pattern seen in atrial fibrillation, which more often appears as large cortical-subcortical lesions, confluent large lesions with satellites, or lesions involving multiple vascular territories.
- On angiography, the majority of PFO strokes show no apparent vessel occlusion.
- All patients suffering a cryptogenic stroke or TIA, and especially patients with suspected paradoxical embolism causing acute ischemic stroke or TIA, should undergo trans-thoracic echocardiography (TTE) with bubble study initially.
- Similarly, all patients diagnosed with concurrent cryptogenic stroke and PFO should undergo lower extremity Doppler

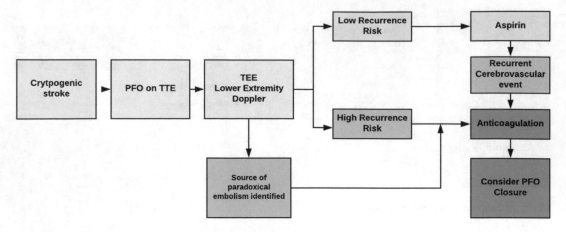

Fig. 2 Management flowchart for PFO (Figure adapted from Wu et al. (2004))

examination to determine possibility or source of a paradoxical embolism.

- Routine TEE in cryptogenic stroke is recommendable, as this modality is superior to TTE for diagnosis and characterization of PFO.

2.2.2 Treatment Approach (Fig. 2)

There is no definitive data or consensus regarding the optimal management strategy for secondary stroke prevention in patients with cryptogenic stroke or TIA and PFO. A meta-analysis of five retrospective studies showed PFO closure to be more efficacious than antithrombotic therapy [1–9].

- Although there is no definitive data in antithrombotics vs anticoagulants, aspirin is recommended routinely.
- Anticoagulation is recommended if a venous source of paradoxical embolism is found.
- PFO closure showed definite benefit for patients with recurrent cardiovascular events.
- Results from the RESPECT trial showed benefit in cryptogenic stroke reduction for the PFO closure group.
- Closure of the PFO should be routinely considered in high-risk patients.

2.2.3 Prognosis of stroke in the setting of PFO

The Lausanne Study showed that while the presenting stroke is often severe, with about half of patients experiencing disabling sequelae, recurrence is generally uncommon. Risk factors as shown in Table 1 are associated with recurrence, although some of these factors (PFO diameter, degree of shunting, and concomitant ASA) have been called into question in the PICCS trial.

Table 1
Features associated with increased risk of initial and recurrent stroke

Features associated with increased risk of initial and recurrent stroke	
Clinical	Echocardiographic
• History of recent migraine	• Large PFO opening
• History of Valsalva maneuver preceding stroke	• Large right-to-left shunting
• Coexisting cause of stroke	• Right-to-left shunting at rest
• Acute pulmonary embolism	• Presence of ASA
• Recurrent strokes or events	• High membrane mobility

3   Hypercoagulable States: Coagulopathies in Ischemic Stroke

Hypercoagulability or thrombophilia is defined as a pathologic tendency toward clotting. These states are more associated with venous thrombosis but have been identified as causative of ischemic stroke in <1% of cases [10–12].

3.1 Pathophysiology and Clinical Presentation

3.1.1 Homocysteinemia

Elevated circulating homocysteine promotes endothelial damage and stroke via various mechanisms, including upregulation of factors V and XII, increased lipid peroxidation with decreased NO production, and induction of vessel smooth muscle cell proliferation. Elevated levels of homocysteine have been found in up to 30% of ischemic stroke patients. Hyperhomocysteinemia may result from inborn errors of metabolism; decreased intake or absorption of metabolic factors like B6, B12, or folate; and depletion. Notably, one prospective study found vitamin supplementation decreased stroke incidence by nearly a quarter.

3.1.2 Protein C and S

Physiologically, thrombin activates protein C which in concert with thrombomodulin and protein S generates activated protein C complex. This complex inhibits activated factors V and VIII. Deficiency in the amount or activity of protein C complex results in thrombophilia. Low levels of protein C are associated with cerebral venous thrombosis (CVT) or ischemic strokes.

3.1.3 Factor V Mutations

Factor V Leiden is the most common form of protein C resistance and hereditary thrombophilias, accounting for about 90% of cases and with an estimated prevalence of nearly 5% in the Caucasian population. In this mutation, ArgΔGlu substitution at position 506 in one the three protein C cleavage sites generates breakdown resistance. It is estimated that thrombosis risk is increased by 80 times in homozygotes. Other less common factor V mutations associated with thromboses include p606GluΔAsp, FV Liverpool, and FV Cambridge [10–12].

3.1.4 Factor II Mutations   Factor II or thrombin plays a role in the common coagulation pathway. 2012 GA variations have been associated with an increase in circulating factor II levels at nearly 30% above normal, generating thrombophilia with both arterial and venous events noted. This entity increases risk of CVT or ischemic strokes.

3.2 Diagnosis of Thrombophilias

- Patients <30 years of age presenting with new or recurrent stroke.
- Ischemic stroke plus personal or familial history of thromboses.
- Abnormal initial coagulation screening tests.
- Initial thrombophilia testing should include:
  - B12.
  - Folate.
  - Homocysteine.
  - Antiphospholipid antibodies.
  - PT-PTT.
  - Russell viper venom test or equivalent.
- If the above findings suggest a protein C resistance, factor V Leiden is the highest yield test, and more esoteric entities like prothrombin mutation or other factor V variations may be considered [10–12].

4   Malignancy and Stroke

Associations between hypercoagulability and malignancies are well known, and embolic events are considered the most common cause of cancer-related strokes. Thrombosis of arteries or veins can also be observed.

4.1 Pathophysiology and Clinical Presentation

Causative mechanisms vary according to the causative neoplasm. Some alter hemostasis, while others increase blood viscosity; yet others involve cardiac valves.

Neoplasms commonly associated with thrombotic events include:

- Colon.
- Gastric.
- Gallbladder.
- Ovarian.
- Pancreatic.
- Myeloproliferative.

Others have been found to cause disruption of hemostasis. Leukemia, multiple myeloma, and pancreatic cancer cause protein S deficiency, while hepatocellular carcinoma, Chronic lymphocytic leukemia (CLL), Acute myelogenous leukemia (AML), and Acute lymphoblastic leukemia (ALL) cause diminished protein C [13–19].

Mucin-producing adenocarcinomas, such as pancreas, colon, breast, lung, and ovary, secrete mucin into the bloodstream, promoting formation of platelet-laden microthrombi. Adenocarcinomas, most commonly pancreatic, are also associated with non-bacterial endocarditis [13–19].

4.2 Management

4.2.1 Diagnostic Approach

Whenever an occult primary is suspected, NCCN guidelines recommend [13–19]:

- Complete anamnesis focusing on prior history of malignancy or biopsies. We recommend this include family history of malignancy.
- Physical examination should include breast, genitourinary, rectal, and pelvic exams.
- Full laboratory workup is warranted, including CBC and chem including calcium, as well as renal and hepatic function tests.
- Initial imaging evaluation should include CT of the chest, abdomen, and pelvis.

Imaging and laboratory findings are helpful in distinguishing cancer-related stroke from other stroke mechanisms.

- MRI with DWI.
    - DWI demonstrating multiple lesions involving multiple arterial territories are more frequently observed in patients without conventional stroke mechanisms.
    - Recent studies have demonstrated that undiagnosed cancer should be considered in patients exhibiting multiple infarcts on DWI.
- D--Dimer.
    - Laboratory findings suggesting coagulopathy may predict possible cancer-related stroke mechanisms.
    - Most cancer patients without common stroke mechanisms have elevated D-dimer levels.

4.2.2 Treatment Approach

Prevention of recurrent thromboembolism is essential to management of malignancy-associated stroke [13–19]:

- Anticoagulation in the setting of malignancy-related hypercoagulability may be recommendable.
    - However standard guidelines for the use of anticoagulants have not been established, and treatment approaches will vary depending on the primary.

- Unfractionated heparin (intravenous or subcutaneous) or enoxaparin is preferred over oral vitamin K antagonists or novel oral agents.

4.2.3   Treatment of Acute Stroke

- Use of thrombolytics within the therapeutic time window is not contraindicated by extracranial malignancy under current stroke guidelines [13–19].

- Multimodal MRI (DWI and perfusion-weighted imaging) can be used to identify patients for recanalization therapy:
  - However, the target mismatch profile is seldom seen in cancer-related stroke.

  - Normal perfusion-weighted imaging and angiographic findings are often observed in cancer patients with stroke, even when multiple infarcts and severe neurological deficits are present.

5   Fibromuscular Dysplasia

Fibromuscular dysplasia (FMD) was initially described in renal arteries as causative of secondary hypertension. FMD is characterized by hyperplasia of the different arterial tunicae, with medial hyperplasia comprising about 90% of FMD cases. This disease has been increasingly recognized in extrarenal vascular beds, including the carotid and vertebral arteries. One study found similar prevalences between these forms, with 79.7% renal and 74.3% extrarenal. Although prevalence estimates vary between 0.4% and 1.1% of the general population, FMD is often overlooked with a delay to diagnosis averaging 4–9 years. FMD may be responsible for nearly 20% of ischemic strokes in patients below 45 years of age [20–23].

5.1   Pathophysiology and Clinical Presentation

Manifestations vary by the vascular bed involved, and although secondary hypertension is classically described, it is not conditio sine qua non for this disease. Dissections are common in FMD, 75% of which are seen in carotid vessels and 17% in vertebrals. Inversely, 15–20% of spontaneous carotid dissections are associated with FMD [20–23].

Cerebrovascular events are reported at a high rate in FMD:

- 13% stroke.

- 10% TIA.

- 6% amaurosis fugax.

As for symptomatology, men are more likely to present with extracranial cerebrovascular disease, while women are more likely to have the more classically described renal manifestations of disease.

The US FMD registry study described symptoms of FMD as such [20–23]:

- 64% hypertension.
- 52% headaches.
- 28% pulsatile tinnitus.
- 26% dizziness.
- 16% flank or abdominal pain.
- 16% chest pain or dyspnea.
- 6% asymptomatic.

Other vascular systems involved can include the mesenterium and rarely, the coronaries. Family history of sudden death, stroke, aneurysmal disease, FMD, or combinations thereof may be present.

5.2 Management

5.2.1 Diagnostic Approach

- Screening is recommended for patients between the ages of 20 and 50 years with newly diagnosed or refractory hypertension, especially women or patients with suggestive family history.
- Cryptogenic stroke, notable family history, and new onset or refractory hypertension in a patient younger than 45 years should arouse suspicion for FMD.
- Full examination focusing on auscultation is warranted, as carotid bruits carry a positive predictive value of 95.4% and a negative predictive value of 37.4%.
  - Abdominal or flank bruits have similar values.
- While renal artery ultrasound has been deemed highly accurate in FMD diagnosis, this modality may be limited in extrarenal manifestations especially intracranial involvement.
- Symptoms suggestive of cervical artery involvement such as dizziness, pulsatile tinnitus, Horner's syndrome, TIA, or amaurosis fugax should prompt CTA or MRA evaluation of the said vasculature.
- Digital subtraction angiography (DSA) is considered the golden standard for diagnosis and can be employed if indirect vessel imaging is inconclusive (Fig. 3).
- Patients in whom initial presentation is stroke may have cerebrovascular FMD diagnosed during initial evaluation.
- Angiographically, FMD can be described as focal or multifocal (Fig. 3)
  - Multifocal disease is more prevalent at 80% and consists of the classically described "string-of-beads" appearance representing two or more areas of stenosis within a vessel segment.
  - Multifocal FMD most commonly affects the vessel media and is more likely to affect several vascular beds.

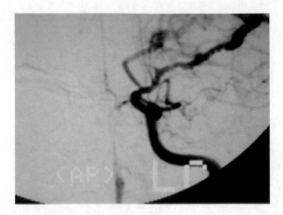

Fig. 3 Cerebral angiogram frontal projection showing classic appearance of FMD involving left middle cerebral artery. (Image courtesy of Dr Huey-Jen Lee, M.D. Professor of Radiology, Rutgers NJMS)

5.2.2 Treatment Approach

Rapid diagnosis and intervention in renal FMD is paramount as it may yield improved outcomes in terms of hypertension and stroke risk. Prompt recognition and treatment of FMD improves hypertension outcomes [20–23].

- ACE inhibitors are considered first line for FMD-related hypertension.
- Aspirin for primary stroke prevention is routinely recommended.
- Renal revascularization, via surgical or endovascular intervention, is recommended whenever feasible.

Treatment of stroke in FMD does not differ from routine.

- Both tPA and mechanical thrombectomy are feasible and have been successful in prior case reports.

6   Cervical Carotid Artery Dissection

Carotid artery dissection is a leading cause of ischemic stroke in young adults, possibly associated with up to 10–25% of cases. This entity is classified as traumatic or spontaneous when no trigger is identified. Carotid artery dissection is believed to be underestimated in incidence as it can present asymptomatically [24].

6.1 Pathophysiology and Clinical Presentation

Tearing of the intima may lead to blood penetrating between the tunicae intima and media, creating a false lumen or hematoma which impinges on the lumen causing stenosis or, in severe cases, occlusion. Dissection between media and adventitia, in turn, leads to pseudoaneurysms. The dissection may be covered by an intimal

Table 2
Clinical presentations of carotid dissection

Clinical presentations of carotid dissection
Unilateral headache or neck pain, which may have a sudden onset
Horner's syndrome
Carotid bruit
Involvement of cranial nerves IX, X, XI, and XII

Adapted from Caplan (2017)

flap and usually extends in an anterograde fashion, with extension reported as distally as into middle cerebral arteries. Within the false lumen, subendothelial proteins such as collagen are exposed, activating the coagulation cascade. Anterograde projection of thrombi or occlusion of the true lumen leads to stroke in 36–68% of cases [24].

The most common dissection is 2 cm distal to the carotid bulb in the ICA. It has been postulated that this region is susceptible to trauma as it is mobile versus the fixed point of entry to the cranium. Common carotid artery dissection is rare.

Cervical artery dissection manifestations range from asymptomatic to florid. A prior history of migraines may be present in up to 20% of cervical artery dissection patients, and thunderclap headaches have been reported.

Manifestations vary by affected artery (Table 2). Carotid dissections may cause ipsilateral anterior neck, facial, and periocular pain, whereas vertebral artery dissection tends to affect the posterior neck and occiput. Sudden onset of any of the above presentations can be quite suggestive. Cerebrovascular events occur in the dissected arteries' territory, most being embolic in nature. A majority of ischemic events occur soon after symptomatic debut of the dissection [24].

Clinical presentation may also include lower cranial neuropathies, including pairs IX, X, XI, and XII, with resulting dysphagia, hoarseness, or dysarthria.

6.2 Management

6.2.1 Diagnostic Approach

Differential diagnosis for cervical artery dissection includes migraine, cerebral venous thrombosis, or posterior reversible encephalopathy syndrome (PRES).

- If onset is sudden or thunderclap in nature, subarachnoid hemorrhage (SAH) and RCVS (reversible cerebral vasoconstriction syndrome) should be considered.

- Risk factors include:

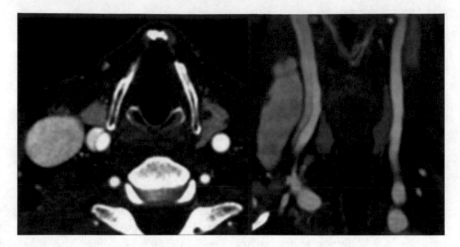

Fig. 4 CTA in axial and coronal vews showing long-segment dissecton of cervical right common carotid artery. (Images courtesy of Dr Huey-Jen Lee, M.D. Professor of Radiology, Rutgers NJMS)

- Trauma.
- Connective tissue disorders.
- Homocystinuria.
- Atherosclerosis.
- Aneurysmal disease.

- Imaging should always be pursued if cervical artery dissection is suspected. MRA and CTA are viable alternatives.
- A history of recent trauma should alert the physician to the possibility of cervical artery dissection.
  - Trauma is described as mild in 90% of cases and in some cases may be associated with chiropractic manipulation or strenuous exercise.
- Sudden-onset unilateral cervical pain or headache in the setting of known risk factors, especially trauma, or contralateral focal findings should prompt imaging.
- The cornerstone for cervical artery dissection diagnosis is vessel imaging.
  - CTA or MRA is appropriate, CTA having the advantage of wide clinical availability (Fig. 4).
- Angiography is highly sensitive and specific; however, it is reserved for diagnostic dilemmas or curative procedures.

6.2.2 Treatment Approach

- Standard therapy consists of anticoagulation or antithrombotics to prevent thrombotic events (Fig. 5).
  - The Cervical Artery Dissection in Stroke Study (CADISS) trial found no significant difference in stroke reduction between antithrombotics and anticoagulation [25].

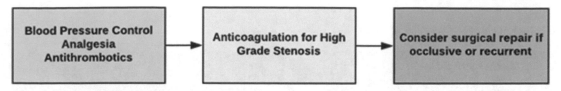

Fig. 5 Flowchart for cervical artery dissection

- Some advocate for 3–6 months of anticoagulation, while in our institution, we recommend anticoagulation for high-grade stenosis.

- High-grade stenosis or occlusion, as well as pseudoaneurysmal dilatation, may warrant surgical or interventional repair.

    Treatment of stroke in cervical artery stenosis:

- Standard stroke protocols apply in cervical dissection cases.
    - Both tPA and mechanical thrombectomy are deemed safe in cervical artery dissection.

    In cases where the initial manifestation is stroke, initial evaluation may itself yield a hitherto undiscovered cervical artery dissection. Prognosis is favorable with a low mortality rate of 4–5% and a resolution rate of 90%, with about 50% of patients experiencing no sequelae.

7   Reversible Cerebral Vasoconstriction Syndrome

RCVS is an entity with incompletely understood pathophysiology, characterized by arteriolar vasospasm. The most likely cause is transient cerebrovascular dysregulation with sympathetic overactivity, possibly progressing centripetally from distal arterioles. The average age of onset is 42 years, affecting women more often than men.

7.1 Pathophysiology and Clinical Presentation

The clinical hallmark of this entity is recurrent thunderclap headaches commonly associated with distinct triggers (Table 3). RCVS has been associated with cervical artery dissection as well as migraines.

Cohort studies have shown about 33–50% of patients will develop ischemic strokes, hemorrhage, or cerebral edema. The said lesions may be reflected in the physical examination, with signs and symptoms varying with location. These may include but are not limited to ataxia, paresis, aphasia, or decreased consciousness.

Table 3
Selected reported triggers of RCVS

Selected reported triggers	
• Valsalva	• PRES
• Physical activity	• Eclampsia
• Orgasm	• Sepsis
• Urination	• PRBC transfusion
• Bending down	• IVIG
• Showering/swimming/bathing	• Vasoactive drugs, notably serotonergics
• Puerperium	• Drugs of abuse

- The average patient will have four attacks lasting 1–3 h. Headaches tend to be severe and accompanied by nausea, emesis, sensitivity to light or sound, and agitation.
- Strokes are thought to result from vasospasm with up to 10% of patients exhibiting transient focal deficits, most commonly visual.
- Paresis, dysarthria, or sensory changes may also be observed, and if the said deficits are persistent, stroke should be suspected.
- Seizures may also complicate 1–17% of RCVS cases but tend not to recur after resolution. A pure cephalic form is also described with headaches as the lone symptom.

7.2 Management

7.2.1 Diagnostic Approach

The differential diagnosis for RCVS includes causes of thunderclap headache, including intracerebral or subacute hemorrhage. Other cerebrovascular events such as CVT or PRES, which at times may be concurrent with RCVS, should be considered. Cervical artery dissection, meningitis, or intracranial hypotension should also be considered. Recurrent thunderclap headaches are 98% specific for this disease [26–29].

- RCVS should be highly suspected in cryptogenic stroke accompanied by headache, recurrent thunderclap headaches, or non-aneurysmal convexity SAH (Tables 4 and 5).
  - If there is associated neck pain, cervical vasculature must be investigated as there is a link between RCVS and carotid dissection as well as unruptured aneurysms and migraines.
- Initial approach should be targeted at causes of thunderclap headaches, including brain and cervical vasculature imaging.
  - MRI/MRA or CT/CTA of the brain and cervical vessels is warranted (Fig. 6).

Table 4
Imaging findings in RCVS

Imaging findings in RCVS
Parenchymal hemorrhages are commonly single and lobar and more commonly seen in women and migraneurs. Hemorrhages may appear within days of onset
Infarctions are most commonly localized at watershed areas and tend to be symptomatic. They appear at an average of 9 days after debut. Cerebellar infarctions are also described
Patients may exhibit convexity cerebral hemorrhage, usually within 7 days
Edema may be observed promptly in the course and tends to reverse earlier than vasoconstriction

Table 5
Diagnostic criteria for RCVS

Diagnostic criteria for RCVS
• Thunderclap headache ± focal deficit
• Monophasic course
• Convexity SAH
• Segmental vasoconstriction evidenced on imaging
• CSF with <100 mg/dl proteins and < 15 WBC mg/L
• Resolution of vasoconstriction at 12 weeks

Modified from Ducros (2012)

- The sensitivity of CTA or MRA is about 70% that of angiography.
- Lumbar puncture should be pursued in order to rule out sentinel bleed.
  - In RCVS, it may be normal within the first week, and CSF parameters are <100 mg/dl proteins and < 15 WBC mg/L.
- Initial CT or MRI may be normal despite diffuse vasoconstriction on concomitant cerebral angiograms.
  - In one series, 55% had an initial scan without abnormalities, and of these, 81% exhibited findings at subsequent imaging.
- Cerebral angiography is considered the golden standard for diagnosis and will show areas of smooth segmental vasoconstriction or occlusions (Fig. 7).
  - Angiography may be normal during the first week, showing segmental narrowing and dilation in a string-of-beads appearance.
  - Vasospasm is observed usually in one or more arteries in a bilateral and diffuse distribution.

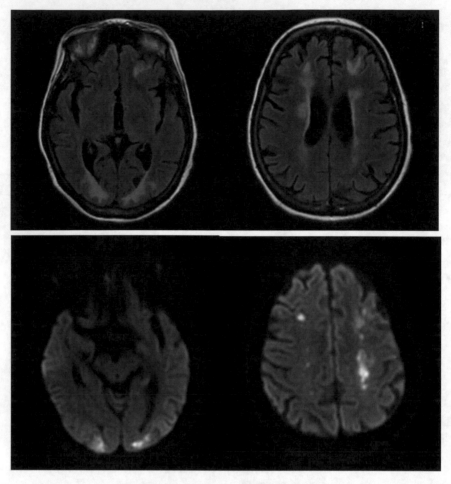

Fig. 6 MRI in a 62 y/o female with confirmed RCVS, with FLAIR sequences showing peri-ventricular white matter abnormal signal. DWI showing scattered focal areas of watershed infarctions

- Angiographic findings vary over time and have been described as dynamic, with maximum vasoconstriction of the MCA branches seen at 16 days on average, giving credence to those proposing that vasospasm progresses centripetally.

- Response to intra-arterial verapamil injection is diagnostic as it proves vasospasm (Figs. 7 and 8).

    Response to intra-arterial verapamil may help differentiate RCVS from primary angiitis of the central nervous system (PACNS).

- Up to 9% of patients may present with a TIA after angiography, and the procedure is not routinely recommended for patients presenting the pure cephalgic form of RCVS in which diagnostic certainty has already been achieved.

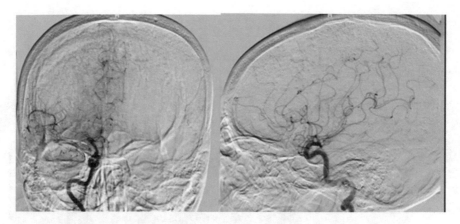

Fig. 7 Angiography for the patient described above, in frontal and lateral projections showing severe flow-limiting stenosis of intracranial right ICA, extending into both ACA and MCA

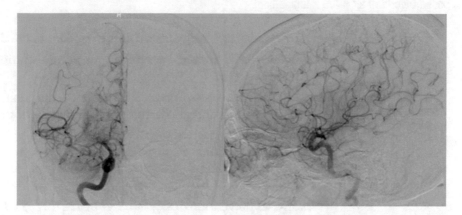

Fig. 8 Angiography for the patient described above, immediately after intra-arterial administration of Verapamil showing improvement of ICA, MCA and ACA stenosis with improved distal capillary opacification

- Laboratories obtained should include CBC, complete chem, and coagulation studies. In order to rule out inflammatory entities, workup should include Erythrocyte Sedimentation Rate (ESR), C-Reactive Protein (CRP), Antinuclear Antibody (ANA), Antineutrophil cytoplasm antibodies (ANCA), and Lyme.
- Urinalysis should include an illicit drug screen. Given the paroxysmal nature of RCVS, some authors recommend that pheochromocytoma be ruled out via urine S-hydroxyindoleacetic and vanyllic mandelic acid.

7.2.2 Management: Treatment Approach

Currently, no randomized control trials have been conducted regarding RCVS treatment. Current recommendations include symptomatic treatment, avoiding precipitating factors and exertion, and cessation of vasoactive drugs (Fig. 9).

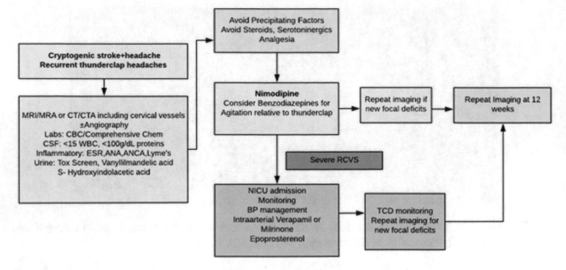

Fig. 9 Proposed algorithm for RCVS management

- Rest and avoidance of triggers is routinely recommended.
  - Glucocorticoids should be avoided, as administration may worsen outcomes.
- Nimodipine at vasospasm doses is considered the mainstay of treatment, noted to decrease frequency and severity of thunderclap headache attack.
  - Nimodipine is the most evidence-supported treatment, though verapamil and magnesium sulfate have also been employed.
  - Vasospasm prevention dosage is described as 60 mg q4 hours. Smaller doses at shorter intervals may be attempted if hypotension develops. Treatment is recommendable for 4–12 weeks.
- Benzodiazepines such as lorazepam have been utilized in the setting of agitation secondary to headache.
- Treatment of mild or purely cephalgic RCVS without imaging findings and without focal findings is controversial; observation versus further follow-up as an outpatient, as the disease is generally self-limiting.
  - Follow-up imaging should be obtained at 12 weeks to determine remission.
- RCVS course may be complicated by seizures, which should be treated routinely with anticonvulsants.

7.3 Severe RCVS

- Progressive cases with recurrent stroke or hemorrhages or diffuse bilateral large vessel involvement merit NICU admission and monitoring.

- BP control, goal systolic <220 mmHg, may be recommendable, as well as avoidance of hypotension.

- Serial Transcranial dopplers (TCDs) are recommendable for vasospasm monitoring. Peak velocities are observed around week 3 of the disease.

- Intra-arterial verapamil and milrinone are recommended for severe vasospasm rescue (Fig. 8). Epoprostenol and balloon angioplasty have also been utilized in the past.

Prognosis in RCVS is generally favorable, with most cases being self-limited. Residual deficits tend to improve over time, with most patients returning to baseline within weeks. Less than 5% of patients suffer a life-threatening course, with cerebral edema and several cerebrovascular events. Mortality in RCVS is tallied at less than 1%. Recurrent cases are described in the literature [26–29].

8 Sickle Cell Disease

Sickle cell disease (SCD) is one of the most common single-gene disorders worldwide, caused by a substitution in the hemoglobin beta chain. This generates HbS, a variant prone to polymerization when deoxygenated.

8.1 Pathophysiology and Clinical Presentation

SCD pathophysiology is complex, characterized by chronically elevated inflammatory cytokines, excessive endothelial adhesivity, dysregulated vascular tone, and chronic vasculopathy. These elements, along with hypercoagulability and RBC sickling, cause vascular events ranging from strokes to the often seen vaso-occlusive crises. Common cerebrovascular complications of SCD include silent or symptomatic cerebral infarcts.

Clinical hallmarks of SCD include painful vaso-occlusive crisis and acute chest syndrome. Hemolytic anemia is another mainstay of this disease, as well as a proclivity to infections with encapsulated bacteria. Neurological complications are well described in SCD [30–33]:

8.1.1 Silent Cerebral Infarcts (SCIs)

- Defined as infarcts visualized on MRI T2 sequence on two views, measuring at least 3x1 mm without attributable neurological deficits (Fig. 10).

- The most common cerebrovascular complication seen in children, and prevalence may be as high as 40% by age 18.

- Associated with low hemoglobin level, elevated systolic blood pressure, and male gender.

- SCIs are associated with neurocognitive deficits, recurrent silent infarcts, and overt strokes.

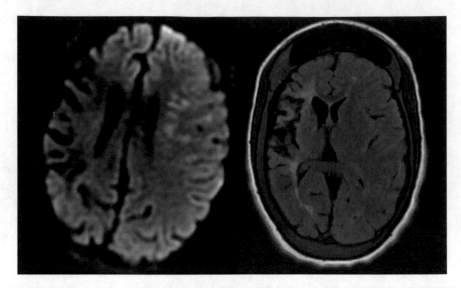

Fig. 10 MRI in a patient with SCD showing encephomalacia of the right hemisphere due to chronic ischemia. DWI and FLAIR images also showing cortical hyperintenity of the left hemisphere from leptomeningeal collateralization

8.1.2 Ischemic Stroke

- Increased incidence during the first decade of life and after the age of 30 years.
- Typically involves the large vessel distributions.
- Risk factors include HbSS genotype, increasing age, acute chest syndrome (recent or recurrent), elevated systolic blood pressure, low hemoglobin level, recent infection, nocturnal hypoxemia, prior TIAs, and elevated velocities on TCD screening [30–33].

8.1.3 Hemorrhagic Stroke

- Intracranial hemorrhage incidence peaks during the second decade of life, with some authors estimating that hemorrhage may supersede ischemia in this age group.
- Risk factors include increasing age, low hemoglobin, and leukocytosis.

8.2 Management

8.2.1 Diagnostic Approach

- Sickle cell disease is screened for in neonates in England and the United States.
- SCD should be suspected in patients presenting with vaso-occlusive crisis or stroke at a very early age.
- Evidence of silent infarcts should also raise suspicions.
- Diagnosis is readily made by identifying HbS hemoglobin via electrophoresis or other widely available methods.

8.2.2   Management: Treatment Approach

Stroke Prevention

Primary and secondary stroke prevention in SCD differs in the pediatric and adult populations.

In children and adolescents:

- Yearly TCDs are recommended for risk stratification.
- In the Stroke Prevention Trial in Sickle Cell Anemia (STOP), children with TCD velocities greater than 200 cm/s who received regular transfusion therapy with a goal of HbS <30% showed a relative risk reduction of 92% vs untreated [30–33].

There is currently a paucity of data for stroke prevention strategies in adults.

- Stroke prevention in adults relies on antiplatelet/anticoagulant agents and modification of classical risk factors.
- Ischemic strokes in adults with SCD tend to be cryptogenic in 25% of cases and cardioembolic in another 25%.
- Hydroxyurea prophylaxis has been investigated and is associated with fewer acute chest episodes and transfusions.
  - However, there was no change in stroke outcomes, mortality, or hepatic sequestration.

Treatment of Acute Stroke in SCD

Treatment of acute stroke also differs in pediatric and adult patients with SCD.

In children and adolescents, RBC transfusion is recommended.

- Simple transfusion can be performed initially.
- Once the patient is stable and adequate access obtained, manual exchange or erythrocytapheresis can be done, with goal of HbS <30% and total Hgb 10–12.5 g/dl.

In adults, management of both ischemic and hemorrhagic strokes in SCD patients follows that of the general population.

- One study performed by the GWTG Stroke program found that concurrent SCD with stroke had no bearing on tPA safety or outcomes, thus is routine tPA administration for eligible patients is generally recommended.
- Initial labs should include routine CBC, chemistry, and PT/PTT as well as INR. Laboratories pertaining to known SCD should include a reticulocyte count as well as type and cross-match in preparation for transfusion.
- Oxygen therapy should be started promptly with goal saturation > 95%.
- Some authors recommend RBC transfusion to goal Hb > 10 or HbS <30% in adults.

Incidence of ischemic stroke is increased in the first decade of life and after age 30, while hemorrhagic stroke is most likely to occur between the ages of 20 and 29 years. The mortality rate following hemorrhagic stroke is significantly higher than that of ischemic stroke. Recurrence of stroke can be very high without treatment [30–33].

9 CADASIL

Cerebral autosomal dominant arteriopathy with subcortical infarcts and leukoencephalopathy (CADASIL) is the most common hereditary cause of stroke and vascular dementia. Prevalence is estimated at nearly 2 per 100,000 in one cohort.

9.1 Pathophysiology and Clinical Presentation

The pathophysiology of the disease involves mutations in the NOTCH3 gene, which encodes an arterial smooth muscle cell transmembrane receptor. Mutations of *NOTCH3* change the number of cysteine residues in the extracellular domain of the receptor, causing accumulation of periodic acid-Schiff-positive material and damage to small arteries (Fig. 11). Manifestations of the disease beyond the CNS are rare for reasons that remain to be elucidated.

The natural history of CADASIL progresses from early headaches and mood disturbances or personality changes to early strokes which are usually subcortical. A family history of early strokes, dementia, or personality changes may be suggestive [34, 35].

- Migraine with aura is the most common initial manifestation, seen in 40–60% of CADASIL patients at an average age of 30 years.

  - Nearly half of patients may also present with prolonged auras, hemiplegic or basilar forms.

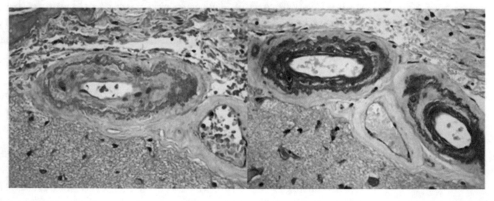

Fig. 11 H&E 400× Medium sized artery containing granular material (L) The granular material is highlighted by PAS tunica media (R) (Images courtesy of Dr Ada Baisre de Leon, Associate Professor of Pathology, Rutgers NJMS)

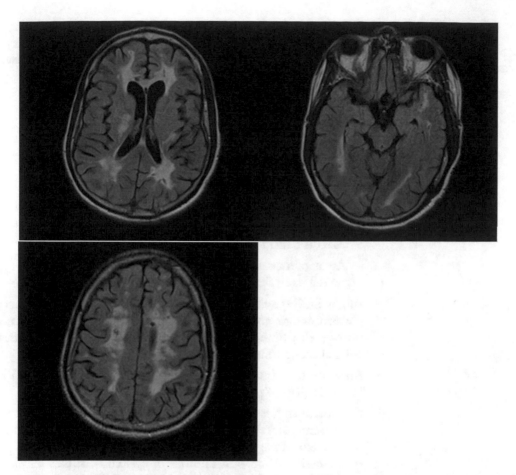

Fig. 12 MRI of a patient with CADASIL showing typical periventricular white matter T2/FLAIR hyperintensities with prevalaence in the anterior temporal pole on the left (Images courtesy of Dr Taha Nisar, Department of Neurology, Rutgers)

- Cognitive impairment is also commonly seen, initially as impaired executive function and processing speed, and culminating in dementia.
- Ischemic events are very common, reported in 60–85% of patients at an average age of 49 years.
  - Localization is usually subcortical, and imaging may show severe leukoaraiosis beyond expected for a patient's age (Fig. 12).
- Neuropsychiatric symptoms, presenting as severe depression or apathy, may be seen in 20 and 40% of cases, respectively.

CADASIL is a progressive disease. After the initial ischemic event, most patients have two to five recurrent strokes over several years. Life expectancy is reduced in patients with CADASIL, with an average age of death of 64 years in men and 70 years in women according to one large study.

9.2 Management

9.2.1 Diagnostic Approach

CADASIL should be considered in stroke accompanied by suggestive family history, early cognitive impairment, or a history of complicated migraines. Unexplained extensive white matter disease in young patients should also raise suspicion [34, 35].

- Differential diagnoses for CADASIL include:
  - Multiple sclerosis.
  - Hereditary or acquired leukoencephalopathies.
  - Fabry's disease.
  - CARASIL.
- MRI is the ideal imaging modality. Changes may precede symptoms by 10–15 years, appearing at an average age of 30 years.
  - MRI is considered 100% sensitive after the age of 35 years.
  - Areas of increased signal on the T2 or FLAIR sequences can be detected from the age of 20 years (Fig. 12).
- Initial findings may appear as punctate or nodular lesions in periventricular areas and centrum semiovale, becoming more diffuse with time. Other MRI findings include dilated perivascular spaces and microbleeds.
- Anterior temporal pole hyperintensities (O'Sullivan sign) are found in 90% of patients.
- Genetic testing is the gold standard for diagnosis. It is indicated for patients with a characteristic clinical presentation along with typical neuroimaging features or a positive family history.
  - Detection of an odd number of cysteine residues within an EGFR has 100% specificity and nearly 100% sensitivity.

9.2.2 Treatment Approach

To date, there is no efficacious treatment available for CADASIL, and treatment is supportive. Goals of care in CADASIL should be aimed at maintaining independence and functionality.

- In migraines, acute treatments with NSAIDs are recommended.
  - Acute treatment with serotonin agonists is not routinely utilized given the possible vasoconstrictive effects.
  - Prophylaxis with acetazolamide has proven effective in some cases.
- Dementia may be approached with donepezil.
  - Although it has not been shown to improve outcomes, there have been reports of improved cognition.
- For psychiatric disturbances such as depression, the usual agents may be used.

9.2.3 CADASIL and Stroke

- Aggressive management of classical risk factors is recommended.
- There is little data to support antithrombotic primary prevention, and anticoagulation is not routinely recommended due to risk of hemorrhage.
- tPA has been utilized successfully in the past.

10 CARASIL

Cerebral autosomal recessive arteriopathy with subcortical infarcts and leukoencephalopathy (CARASIL) is a very rare inherited cause of cerebral small vessel disease, with less than 100 reported cases. Prevalence is unknown, although there is a male predominance with an observed 3:1 ratio [36].

10.1 Pathophysiology and Clinical Presentation

CARASIL is a single-gene disorder caused by a mutation in the *HTRA1* gene encoding HtrA serine peptidase/protease 1 which results in a failure to repress TGF-β signaling. This generates severe arteriosclerosis in the small white matter and the basal ganglia penetrator vessels for poorly understood reasons.

CARASIL patients present similar to CADASIL, sans migraines and with added tegumental manifestations and intervertebral disk degeneration. Manifestations include:

- 90% of patients debut with very early alopecia, starting in the second decade.
- 80% of patients start suffering from back pain at ages 20–40 years.
- 50% of patients present with basal ganglia or brainstem lacunes, at ages 30–40.
- Progressive cognitive decline may also manifest between the ages of 30 and 40 years.
  - It progresses from initial memory disturbance to acalculia, personality changes, abulia, and akinetic mutism.
- Other symptoms include pseudobulbar palsy, gait disturbance, and pyramidal and extrapyramidal disorders.

CARASIL is a progressive disease and carries a poor prognosis, with nearly all patients becoming bedridden or deceased within 10 years.

10.2 Management

10.2.1 Diagnostic Approach

Differential diagnosis includes CADASIL, multiple sclerosis, Binswanger's disease, and PACNS based on similar clinical characteristics and MRI abnormalities. Notably, CARASIL manifests much earlier than CADASIL, at an average age of 32 versus 45 years.

Imaging and genetic testing are essential to the diagnosis of CARASIL.

- Although extremely rare, CARASIL should be suspected in new-onset or recurrent lacunes in a young patient without known risk factors, especially in the setting of premature baldness and a history of low back pain with disk disease.

- Hyperintense lesions initially appear in subcortical white matter by age 20 years, suggesting that white matter changes precede neurologic symptoms.

- MRI findings include leukoaraiosis of deep white matter and periventricular area and multiple lacunar infarcts in the basal ganglia and thalamus. The most characteristic finding for CARASIL is the "arc sign," a hyperintense lesion trailing the ponto-cerebellar tract from pons to middle cerebellar peduncles.

- On cerebral angiography, about half of patients show no significant abnormalities, while the remainder show findings compatible with arteriosclerosis.

- Definite diagnosis is made via detection of HTRA1 mutation.

10.2.2 Management: Treatment Approach

- No effective treatment exists.

- Secondary prevention of ischemic stroke, although the effects of antiplatelet and anticoagulation agents are unclear.

- Genetic counseling, supportive care, and treatment of dementia.

---

11   Fabry's Disease

Fabry's disease is an X-linked lysosomal storage disease caused by over 600 distinct mutations in α-galactosidase. The prevalence is 1 per 40,000–70,000 at birth and may be higher for late-onset disease. Being X-linked, this disorder affects men more commonly and severely, although phenotypical variability is reported. It has been estimated that Fabry's accounts for 1–4% of cryptogenic ischemic strokes in some cohorts [37–44].

11.1 Pathophysiology and Clinical Presentation

Decreased lysosomal α-galactosidase A activity leads to accumulation of glycosphingolipids in vascular endothelium, autonomic and dorsal root ganglia, and smooth muscle cells, leading to neurovascular and systemic manifestations.

Systemic manifestations such as angiokeratomas, corneal dystrophy, acroparesthesias, and hypohidrosis present earlier in the disease course. CNS complications occur later in life alongside cardiac conduction impairment, cardiomyopathy, and renal impairment. Vertigo, hearing loss, and neuropsychiatric symptoms are also reported.

- Painful neuropathies or acroparesthesias are common, seen in 60–75% of patients.
  - Hypohydrosis is also a diagnostic clue.
- The most common manifestation is early-onset ischemic stroke, occurring at an average age of 28 years in men and 43 years in women.
  - Strokes are more often seen in the posterior circulation.
  - In some studies, Fabry's accounts for 1–4% of cryptogenic stroke in young patients.

Fabry's disease lowers life expectancy to approximately 58 years in men and 75 years in women, the most common cause of death being cardiovascular disease. Cerebrovascular manifestations of Fabry's disease frequently recur and portend a poor prognosis [38–44].

11.2 Management

11.2.1 Diagnostic Approach

Data from the Fabry Registry showed nearly half of patients suffered an ischemic event prior to diagnosis.

- Fabry's should be considered in young patients presenting with cryptogenic ischemic stroke in the setting of early acroparesthesias or painful neuropathies, angiokeratomas, renal dysfunction, or a positive family history (Fig. 13).
- Dolichoectasia or posterior circulation stroke in young patients should also raise suspicions.

Patients suspected of Fabry's disease should undergo MRI.

- MRI findings in patients with Fabry's disease include leukoaraiosis, vertebrobasilar involvement with possible dilatations, or dolichoectasia.
- The pathognomonic finding is pulvinar hyperintensity on T1-weighted MRI sequences; however, it is not highly sensitive.
- Leukoaraiosis may be absent or minimal in young patients.

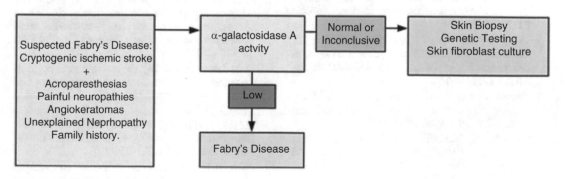

Fig. 13 Diagnostic workup flowchart for suspected Fabry's diseasec

- In a large study of young patients with stroke, MRI findings failed to detect the 1% of patients with Fabry's disease.

  Definitive diagnosis of Fabry's disease is via leukocyte α-galactosidase A activity testing with 100% sensitivity and specificity in males. However, this assay may only detect 50% of female carriers. Genetic testing, skin biopsy, or skin fibroblast culture may be obtained to finalize diagnosis in uncertain cases [38–44].

11.2.2 Treatment Approach

- Primary and secondary stroke prevention with antiplatelet agents and classical risk factor reduction.
- Enzyme replacement therapy may stabilize white matter lesions and decreases cardiac and renal accumulation.
  - However, it has not been found to reduce stroke risk or mortality or improve quality of life.
  - Its role may be limited in advanced disease, such as cardiac fibrosis and severe renal dysfunction.

11.2.3 Fabry's Disease and Stroke

- Diagnostic workup including standard workup and vessel imaging should be pursued.
- TEE is recommendable to assess cardiac involvement.
- Currently there are two reported cases of tPA administration in the literature: one patient improved, and one did not worsen or have hemorrhagic conversion.

References

1. Bogousslavsky J, Garazi S, Jeanrenaud X, Aebischer N, Van Melle G (1996) Stroke recurrence in patients with patent foramen ovale: the Lausanne study. Lausanne Stroke with Paradoxal Embolism Study Group. Neurology 46:1301–1305

2. De Castro S et al (2000) Morphological and functional characteristics of patent foramen ovale and their embolic implications. Stroke 31:2407–2413

3. Di Tullio M, Sacco RL, Gopal A, Mohr JP, Homma S (1992) Patent foramen ovale as a risk factor for cryptogenic stroke. Ann Intern Med 117:461–465

4. Furlan AJ et al (2012) Closure or medical therapy for cryptogenic stroke with patent foramen ovale. N Engl J Med 366:991–999

5. Handke M, Harloff A, Olschewski M, Hetzel A, Geibel A (2007) Patent foramen ovale and cryptogenic stroke in older patients. N Engl J Med 357:2262–2268

6. Kim BJ et al (2013) Imaging characteristics of ischemic strokes related to patent foramen ovale. Stroke 44:3350–3356

7. Mojadidi MK, Khalid Mojadidi M, Gevorgyan R, Tobis JM (2014) A comparison of methods to detect and quantitate PFO: TCD, TTE, ICE and TEE. Patent Foramen Ovale:55–65

8. Odunukan OW, Price MJ (2017) Current dataset for patent foramen ovale closure in cryptogenic stroke: randomized clinical trials and observational studies. Interv Cardiol Clin 6:525–538

9. Wu LA et al (2004) Patent foramen ovale in cryptogenic stroke: current understanding and management options. Arch Intern Med 164:950–956

10. Hankey GJ et al (2001) Inherited thrombophilia in ischemic stroke and its pathogenic subtypes. Stroke 32:1793–1799

11. Morris JG, Singh S, Fisher M (2010) Testing for inherited thrombophilias in arterial stroke:

can it cause more harm than good? Stroke 41:2985–2990

12. Pahus SH, Hansen AT, Hvas A-M (2016) Thrombophilia testing in young patients with ischemic stroke. Thromb Res 137:108–112

13. Bang OY et al (2011) Ischemic stroke and cancer: stroke severely impacts cancer patients, while cancer increases the number of strokes. J Clin Neurol 7:53–59

14. Kim SG et al (2010) Ischemic stroke in cancer patients with and without conventional mechanisms: a multicenter study in Korea. Stroke 41:798–801

15. Kwon H-M, Kang BS, Yoon B-W (2007) Stroke as the first manifestation of concealed cancer. J Neurol Sci 258:80–83

16. National Comprehensive Cancer Network (2017) NCCN clinical practice guidelines in oncology: occult primary (cancer of unknown primary [CUP])

17. Parikh NS, Burch JE, Kamel H, DeAngelis LM, Navi BB (2017) Recurrent thromboembolic events after ischemic stroke in patients with primary brain tumors. J Stroke Cerebrovasc Dis 26:2396–2403

18. Schwarzbach CJ et al (2012) Stroke and cancer: the importance of cancer-associated hypercoagulation as a possible stroke etiology. Stroke 43:3029–3034

19. Zhang Y-Y, Chan DKY, Cordato D, Shen Q, Sheng A-Z (2006) Stroke risk factor, pattern and outcome in patients with cancer. Acta Neurol Scand 114:378–383

20. Cohen JE, Itshayek E, Keigler G, Eichel R, Leker RR (2014) Endovascular thrombectomy and stenting in the management of carotid fibromuscular dysplasia presenting with major ischemic stroke. J Clin Neurosci 21:2021–2023

21. Peköz MT et al (2014) Fibromuscular dysplasia and intravenous thrombolytic treatment. Noro Psikiyatr Ars 51:175–177

22. Ruzicka M, Kucharski SE, Hiremath S (2017) Balancing overscreening and underdiagnosis in secondary hypertension: the case of fibromuscular dysplasia. Cardiol Clin 35:247–254

23. Sharma AM, Kline B (2014) The United States registry for fibromuscular dysplasia: new findings and breaking myths. Tech Vasc Interv Radiol 17:258–263

24. Stevic I, Chan HHW, Chan AKC (2011) Carotid artery dissections: thrombosis of the false lumen. Thromb Res 128:317–324

25. CADISS Trial Investigators et al (2015) Antiplatelet treatment compared with anticoagulation treatment for cervical artery dissection (CADISS): a randomised trial. Lancet Neurol 14, 361–367

26. Ducros A (2012) Reversible cerebral vasoconstriction syndrome. Lancet Neurol 11:906–917

27. Mawet J, Debette S, Bousser M-G, Ducros A (2016) The link between migraine, reversible cerebral vasoconstriction syndrome and cervical artery dissection. Headache 56:645–656

28. Salvarani C, Brown RD Jr, Hunder GG (2012) Adult primary central nervous system vasculitis. Lancet 380:767–777

29. Topcuoglu MA et al (2017) Cerebral vasomotor reactivity in reversible cerebral vasoconstriction syndrome. Cephalalgia 37:541–547

30. Adams RJ (2001) Stroke prevention and treatment in sickle cell disease. Arch Neurol 58

31. Adams RJ et al (1998) Prevention of a first stroke by transfusions in children with sickle cell anemia and abnormal results on transcranial Doppler ultrasonography. N Engl J Med 339:5–11

32. DeBaun MR, Kirkham FJ (2016) Central nervous system complications and management in sickle cell disease. Blood 127:829–838

33. Mack AK, Kyle Mack A, Thompson AA (2017) Primary and secondary stroke prevention in children with sickle cell disease. J Pediatr Health Care 31:145–154

34. Chabriat H, Joutel A, Dichgans M, Tournier-Lasserve E, Bousser M-G (2009) Cadasil. Lancet Neurol 8:643–653

35. Stojanov D et al (2015) Imaging characteristics of cerebral autosomal dominant arteriopathy with subcortical infarcts and leucoencephalopathy (CADASIL). Bosn J Basic Med Sci 15:1–8

36. Fukutake T (2011) Cerebral autosomal recessive arteriopathy with subcortical infarcts and leukoencephalopathy (CARASIL): from discovery to gene identification. J Stroke Cerebrovasc Dis 20:85–93

37. Arends M, Hollak CEM, Biegstraaten M (2015) Quality of life in patients with Fabry disease: a systematic review of the literature. Orphanet J Rare Dis 10:77

38. Biegstraaten M et al (2015) Recommendations for initiation and cessation of enzyme replacement therapy in patients with Fabry disease: the European Fabry Working Group consensus document. Orphanet J Rare Dis 10:36

39. Dubuc V et al (2013) Prevalence of Fabry disease in young patients with cryptogenic ischemic stroke. J Stroke Cerebrovasc Dis 22:1288–1292

40. Fazekas F et al (2015) Brain magnetic resonance imaging findings fail to suspect Fabry disease in young patients with an acute cerebrovascular event. Stroke 46:1548–1553

41. Fellgiebel A et al (2014) Enzyme replacement therapy stabilized white matter lesion progression in Fabry disease. Cerebrovasc Dis 38:448–456

42. Mehta A et al (2004) Fabry disease defined: baseline clinical manifestations of 366 patients in the Fabry Outcome Survey. Eur J Clin Investig 34:236–242

43. Sims K, Politei J, Banikazemi M, Lee P (2009) Stroke in Fabry disease frequently occurs before diagnosis and in the absence of other clinical events: natural history data from the Fabry Registry. Stroke 40:788–794

44. Waldek S, Patel MR, Banikazemi M, Lemay R, Lee P (2009) Life expectancy and cause of death in males and females with Fabry disease: findings from the Fabry Registry. Genet Med 11:790–796

Chapter 19

Endovascular Techniques for Emergent Large Vessel Occlusion

Krishna Amuluru and Fawaz Al-Mufti

Abstract

Care for acute ischemic stroke is one of the most rapidly evolving fields due to the robust outcomes achieved by mechanical thrombectomy. Stroke care was revolutionized in 2015 with the publication of the first randomized controlled trials (RCTs) showing that endovascular treatment is more effective than intravenous thrombolysis alone for patients with large vessel occlusion (LVO) stroke and later trials showing benefit up to 24 h in select patients. While endovascular therapy is highly effective, several devices and techniques exist, and each possesses unique benefits and risks. In recent years, techniques have become more standardized with the uses of thrombus aspiration, stent retrievers, and/or combinations thereof, and revascularization and patient outcomes have accordingly shown improvement in a meaningful manner. This review chapter will present a brief overview of LVO management, selection criteria with focus on symptomatology and imaging, and techniques and devices used.

Keywords Stroke, Ischemic stroke, Stent retriever, Aspiration catheter, Large vessel occlusion, Mechanical thrombectomy

1 Introduction

The standard of care for management of anterior circulation emergent large vessel occlusion (ELVO) ischemic stroke has been revolutionized over the past several years, promulgated by technological improvements in intravascular thrombectomy. For several years after the National Institutes of Neurological Disease and Stroke (NINDS) Intravenous Recombinant Tissue Plasminogen Activator (r-tPA) Study published in the *New England Journal of Medicine* in December 1995, intravenous tPA had been the only FDA-approved medical therapy for acute ischemic stroke and was the dominant treatment paradigm [1]. Given the time-sensitive efficacy of IV

The original version of this chapter was revised. The correction to this chapter is available at https://doi.org/10.1007/978-1-0716-1530-0_23

Fawaz Al-Mufti and Krishna Amuluru (eds.), *Cerebrovascular Disorders*, Neuromethods, vol. 170, https://doi.org/10.1007/978-1-0716-1530-0_19, © Springer Science+Business Media, LLC, part of Springer Nature 2021, Corrected Publication 2021

tPA, the effects rapidly diminish with increasing symptoms, and longer onset to treatment time leads to worse outcomes. Thus, endovascular treatments have been explored to manage ELVO.

In recent years, several trials have emerged in which endovascular techniques have been tested against tPA and subsequently shown to have defined clinical benefits. Initial efforts to manage ELVO with endovascular technique, similar to acute myocardial infarction models, suffered from a lack of defined inclusion and exclusion criteria, non-standardized techniques, and a lack of consensus on the concurrent usage of systemic or catheter-based pharmacologic intervention [2]. In recent years, techniques have become more standardized with the uses of thrombus aspiration, stent retrievers, and/or combinations thereof, and revascularization and patient outcomes have accordingly shown improvement in a meaningful manner.

2   Pathophysiology and Clinical Presentation

2.1   Pathophysiology

Large vessel occlusions can develop through four mechanisms: occlusion at the primary arterial site secondary to the development of atherosclerosis of an intracranial artery, extracranial artery atherosclerotic embolism or plaque rupture resulting in the occlusion of an intracranial vessel, cardioembolic events resulting in intracranial vessel occlusion, and cryptogenic causes [3]. Large vessel occlusions often result in insufficient blood flow to brain parenchyma, causing cellular failure and inflammatory cascades leading to death of neurons, glia, and endothelial cells. Although ischemic changes occur within minutes, the ultimate volume of infarcted tissue is multifactorial and predominantly determined by the chronicity of hypoperfusion and the degree of collateral flow to an ischemic area [4].

The ischemic penumbra, an area of hypoperfused cerebral tissue outside of an irreversibly damaged infarct core, represents clinically important "at-risk" tissue that can either evolve with time or potentially be salvaged with revascularization. In the initial RCTs establishing the efficacy of thrombectomy for ELVOs up to 6 h, patients with large infarct cores were excluded in order to maximize the volume of salvageable penumbra to increase the chances of interventional benefit [4]. Subsequent RCTs have utilized CT perfusion or MR perfusion imaging to differentiate the ischemic penumbra from the core infarct and demonstrated the efficacy of thrombectomy for select LVO patients with favorable penumbral/core ratios up to 24 h from symptom onset [5].

2.2   Clinical Presentation

While not pathognomonic, the clinical presentation of patients with LVOs is often stereotyped and anatomically matched to the site of occlusion and downstream affected cerebrum. Specifically, internal carotid artery (ICA) or proximal middle cerebral artery

(MCA) occlusions often present with contralateral hemibody and face weakness and/or numbness, contralateral homonymous hemianopsia, and ipsilateral gaze deviation, as well as aphasia for dominant hemispheric lesions and neglect for lesions of the nondominant hemisphere [4]. One notable distal occlusion site is the M3 branch to the angular gyrus of the dominant hemisphere. Due to the involvement of this vascular territory in speech processing and complex cognition, focal occlusions in this location may be more aggressively treated in an attempt to preserve speech and cognition, when compared to similarly distant occlusions in arteries supplying less eloquent cortex.

Posterior circulation ELVOs are often more challenging to diagnose than those of the anterior circulation due to the nonspecificity of symptoms, which can lead to significant delays in management of this potentially devastating pathology. Nonetheless, common symptoms encountered with occlusion of the vertebral and basilar arteries include hemibody weakness or numbness, dizziness, nausea, vomiting, gait and balance issues, or alterations in consciousness [4]. Occlusions of the posterior cerebral artery typically result in contralateral vision changes such as a homonymous hemianopsia or quadrantanopia.

3 Management Strategies

3.1 Diagnostic Approach

Defining the criteria to evaluate and select patients with ELVO for endovascular treatment is critically important, since between 3% and 22% of patients with acute ischemic stroke (AIS) are potentially eligible for mechanical thrombectomy, depending on the specific selection criteria used [6]. Time of symptom onset (or last known well), magnitude of early ischemic change on imaging, clinical severity of stroke symptoms, pre-stroke level of functioning, and anatomic location of the ELVO are the most important determinants of candidacy for mechanical thrombectomy, according to the Society of Neurointerventional Surgery Standards and Guidelines Committee [6].

3.1.1 Time from Symptom Onset

For anterior circulation acute ischemic stroke, thrombectomy is indicated in select patients up to 16 h from symptom onset or time last known well for unwitnessed strokes, including wake-up strokes [class I, level A], and is indicated in select patients up to 24 h from last known normal [6].

3.1.2 Imaging

In patients with anterior circulation AIS within the first 6 h of symptom onset and *either* CT ASPECTS ≥ 6, MRI-DWI Alberta Stroke Program Early CT Score (ASPECTS) ≥ 6, moderate-to-good collateral status on multiphase CTA (mCTA) (>50% MCA territory), small (<50–70 mL)-core infarct volumes, or significant penumbral to core mismatch on advanced perfusion imaging (CTP or MRI-DWI-PWI), thrombectomy is indicated [class I, level A] [6].

In patients with anterior circulation AIS due to intracranial ICA and/or M1 occlusion within 6–24 h of symptom onset who meet the advanced MRI-DWI-PWI or CT perfusion (CTP) imaging criteria for DAWN or DEFUSE 3, thrombectomy is indicated.

3.1.3   Location of ELVO

Thrombectomy is indicated in patients with occlusions of the ICA (including intracranial, cervical segments or tandem occlusion) and M1/M2 MCA [class I, level A] [6].

The benefit of thrombectomy in more distal segments such as the M3 segment of the MCA or the A2-A3 segments of the anterior cerebral artery is unclear. Thrombectomy of such patients may be reasonable in some cases and should be considered on a case-by-case basis.

3.1.4   Stroke Severity

Thrombectomy is indicated in patients with anterior circulation ELVO with National Institute of Health Stroke Score (NIHSS) score ≥ 6 [class I, level A] [6].

Thrombectomy may be considered in patients with anterior circulation AIS and National Institute of Health Stroke Score (NIHSS) score < 6 when associated with disabling symptoms. However, care should be taken when treating these patients to keep complication and hemorrhagic rates below those reported in RCTs.

3.1.5   Age and Baseline Level of Functioning

Age > 80 years should not be used as a contraindication for thrombectomy [class IIa, level A]. The benefit of thrombectomy in patients with baseline modified Rankin score (mRS) score > 1 is unknown.

3.2   Treatment Approach

While there are no trials comparing specific thrombectomy techniques, multiple theories on the most efficacious tools and techniques now exist. Stent retrievers remain the recommended first-line thrombectomy approach, as they were the predominant devices used in the original major RCTs [7]. As thrombectomy experience continues to evolve, techniques employing aspiration, both alone and in combination with stent retrievers, have evolved.

3.2.1   Aspiration Techniques

One strategy to manage the thrombus is using aspiration-based techniques via an intermediate or aspiration catheter. This technique has been termed ADAPT (A Direct Aspiration first-Pass Technique).

The triaxial system includes a guide catheter (such as Neuron MAX 088, Infinity 088, and/or Cook Shuttle), through which a large-bore inner diameter aspiration catheter (such as Penumbra ACE, Medtronic React, and Sofia Plus) is inserted. The third part of the triaxial system is a standard microcatheter and microwire.

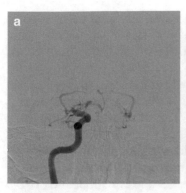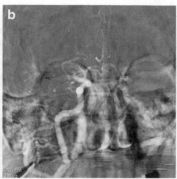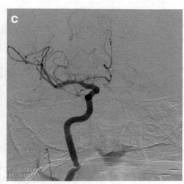

Fig. 1 (a) AP view of right ICA terminus occlusion without anterograde flow into either ACA or MCA. (b) Roadmap image showing Penumbra Ace 068 aspiration catheter tracked over a Marathon 027 microcatheter, with aspiration catheter embedded within the intraluminal thrombus. (c) Post aspiration angiogram showing TICI 3 recanalization with ADAPT technique

Larger microcatheters are preferred for increased stability and ability to transverse the thrombus such as a Marksman 027 or Velocity 025. The guide catheter is advanced into the distal common carotid artery (CCA) or cervical or proximal petrous segment of the ICA. In the posterior circulation, the catheter may be navigated as distally as possible in the vertebral artery while avoiding arterial dissection. The intermediate catheter is advanced coaxially over a microwire and a microcatheter to the proximal face of the thrombus. Aspiration is then applied either manually with a 60 mL syringe or through an aspiration pump, depending on the operator experience. As soon as the absence of flow is noted within the aspiration system, the catheter is slowly advanced to ensure solid engagement with the thrombus. Static aspiration is then allowed to proceed for approximately 1 min. When no flow through the system is found, the catheter is withdrawn. This technique has been shown to achieve recanalization in nearly 75% of cases and, in certain reports, can be accomplished in half the time of the use of a stent retriever (Fig. 1) [8].

The ADAPT is a relatively simple technique, but until recently, such a technique was difficult to perform. Cerebrovascular catheters previously lacked the navigability to gain distal intracranial access while also simultaneously being able to withstand the forces needed to maintain aspiration. However, with increasing advances in catheter technology, catheters are now able to generate an aspiration rate that allows for removal of thrombus without the use of a separator or retriever. As such, ADAPT has now become the first-line therapy for many operators [8].

3.2.2 Stent Retriever Techniques

In 2008, Wakhloo published a series of in vitro and in vivo studies in pigs, demonstrating that a retrievable intracerebral stent could ensnare mis-deployed embolization coils and in situ thrombus

[9]. This work suggested that such devices, delivered to the site of a thrombus via a microcatheter, could be a useful means of retrieving clots from the cerebral circulation. Two such commonly used devices are TREVO and Solitaire.

The TREVO device was specifically engineered to engage with clot in the cerebral circulation. It is a hydrophilic device that is integrally attached to a core wire. The device is deployed via a microcatheter and has a platinum distal tip, so it can be visualized with angiography. It is a helical device with a mesh-like configuration that allows thrombus incorporation into the basket. The distal end of the device is 1 cm past the basket and is radiopaque [10].

The Solitaire revascularization device features a parametric design with overlapping stent retriever-based technology. The overlapping design allows the device to expand in larger vessels and compress in smaller vessels during deployment and retrieval while maintaining consistent cell size and structure, limiting device elongation and foreshortening, providing multiple planes of clot integration contact [11].

The system used to deploy stent retrievers is similar to the aspiration technique, in that a triaxial system is usually employed. A large guide catheter is advanced into the ICA as distally as possible to reach the cervical or proximal petrous segment of the ICA. After access is obtained to the ICA circulation, an intermediate catheter can be used for increased support. Wire access distal to the clot is obtained with a microwire and microcatheter. When used during thrombectomy, a microcatheter is advanced distally to the thrombus, under angiographic guidance. The stent retriever is then positioned within the microcatheter in a location either distal to or across the suspected location of the thrombus. The stent is then deployed by retracting the microcatheter, allowing the stent to assume its full shape and associate with the thrombus. Operators may allow approximately 3–5 min for the thrombus to adhere to the stent retriever. Once ensnared, the thrombus and the device are withdrawn into the proximal catheter and removed.

The Thrombectomy Revascularization of Large Vessel Occlusions in Ischemic Stroke (TREVO and TREVO 2) trials were the first trials examining the second generation of stent retrieval techniques. Although randomization was against the older Merci device, the TREVO 2 trial showed over 90% rates of recanalization and 40–55% rates of favorable outcomes (mRS < 2) [10].

Retrievable stents represented a major advancement in endovascular therapy for ischemic large vessel stroke. In light of the previous setbacks, several published trials in the 2010s were devised using modern thrombectomy technology. These trials included ESCAPE, EXTEND-IA, SWIFT-PRIME, and REVASCAT. ESCAPE was a randomization between thrombectomy with any available retrievable stent and tPA. EXTEND-IA, SWIFT-PRIME, and REVASCAT were all randomizations between the Solitaire device and tPA. With the exception of REVASCAT, these studies all showed at least an odds ratio of 2.5 or more favoring endovascular therapy – REVASCAT did not reach statistical significance [2]. These studies all showed a clear benefit when examining final mRS at 90 days, as well as revascularization rates.

3.2.3 Combination Techniques

Current theories regarding thrombectomy in the management of ELVO involve a combination of techniques: stent retriever devices, mechanical thrombus aspiration, and/or a combination thereof. A series of recent trials have shown faster, safer, and better rates of recanalization compared to prior studies. In comparing the two techniques, although direct aspiration may be faster, cheaper, and intuitively safer, consensus does not exist as to whether it achieves comparable outcomes compared to stent retrieval. Further, many authors postulate that a combination approach is the best means of achieving recanalization.

In 2016, Massari and colleagues at the University of Massachusetts published an interesting retrospective study of 42 patients managed with an Aspiration-Retriever Technique for Stroke (ARTS) [12]. ARTS involves deploying a stent retriever device across the clot and then utilizing a Penumbra device to provide continuous aspiration of the clot and gradual concurrent withdrawal of both devices. ARTS is based on a combined large lumen aspiration catheter and a partially resheathed stent retriever; when resistance is felt while retracting the stent retriever, the entire assembly is locked and removed in toto under continuous aspiration with additional flow arrest. Though retrospective, the authors showed a tremendous improvement in their ability to achieve recanalization. TICI 2b–3 flow was established in 97.6% of cases and TICI 3 flow in 54.7%. Almost half of patients had recanalization in one pass of the device, and the average number of passes dropped to 1.4 per patient in the final 15 patients in their series – overall the rate was 2.2 passes per patient. In addition, the average time to TICI 2b flow or better was 65 min, with a range of 17–182 min [12].

Finally, investigators at Brown have described a technique called continuous aspiration prior to intracranial vascular embolectomy (CAPTIVE) [13]. This technique also involves engaging the thrombus with an aspiration catheter and then ensnaring it with a stent retrieval device. The authors report a series of 95 patients in a retrospective and non-randomized manner. The CAPTIVE group achieved 80% TICI 2c/3 flow in an average of 14 min and with 59% of patients needing only one pass, as compared to 31 min and 40% rates of TICI 2c/3 flow [13].

The CAPTIVE technique is also a combination of an aspiration and stent retrieval technique. After access is obtained, the aspiration

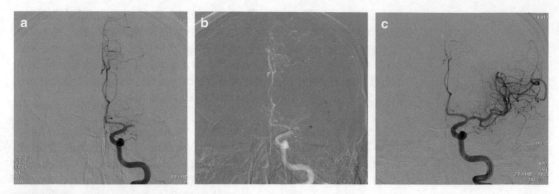

Fig. 2 (**a**) DSA in AP view showing left MCA M1 occlusion. (**b**) Roadmap image showing Solitaire stent retriever deployed through thrombus with concurrent placement of Penumbra 068 aspiration catheter at the proximal face of thrombus. (**c**) Post intervention DSA showing TICI 3 recanalization

catheter is placed proximal to the thrombus, as in ADAPT. Next aspiration is initiated, but the clot is not engaged until a stent retrieval device is deployed. Next, the microcatheter is withdrawn from the set in order to increase the surface area for aspiration force, and simultaneously, the aspiration catheter is advanced through the thrombus. Once the thrombus is engaged in both the stent retriever device and aspiration force, both the retriever device and the aspiration catheter are withdrawn into the sheath in the proximal carotid artery. This maneuver must occur in concert, in order to avoid the complication of propagating distal emboli (Fig. 2).

4   Conclusion

Endovascular therapy for stroke has become a first-line management technique in the management of intracranial large vessel occlusion. Similar to the early days of percutaneous intervention for myocardial infarction, this field is witnessing a remarkable growth in evidence-based implementation. Yet further research is still needed to improve techniques and patient outcome. Firstly, high-quality randomized trials comparing the technique of using a stent retriever plus aspiration to either technique alone will be needed to examine the most efficacious technique. Secondly, terminology will need to be standardized, as many operators are likely performing similar or identical procedures with differing terminology. Finally, the specific inclusion and exclusion criteria for these techniques will need further definition.

References

1. Hacke W et al (2008) Thrombolysis with alteplase 3 to 4.5 hours after acute ischemic stroke. N Engl J Med 359(13):1317–1329

2. Badhiwala JH et al (2015) Endovascular thrombectomy for acute ischemic stroke: a meta-analysis. JAMA 314(17):1832–1843

3. Brouns R, De Deyn PP (2009) The complexity of neurobiological processes in acute ischemic stroke. Clin Neurol Neurosurg 111 (6):483–495

4. Rennert RC et al (2019) Epidemiology, natural history, and clinical presentation of large vessel ischemic stroke. Neurosurgery 85(Suppl 1): S4–S8

5. Nogueira RG et al (2018) Thrombectomy 6 to 24 hours after stroke with a mismatch between deficit and infarct. N Engl J Med 378 (1):11–21

6. Mokin M et al (2019) Indications for thrombectomy in acute ischemic stroke from emergent large vessel occlusion (ELVO): report of the SNIS standards and guidelines committee. J Neurointerv Surg 11(3):215–220

7. Maingard J et al (2019) Endovascular treatment of acute ischemic stroke. Curr Treat Options Cardiovasc Med 21(12):89

8. Vargas J et al (2017) Long term experience using the ADAPT technique for the treatment of acute ischemic stroke. J Neurointerv Surg 9 (5):437–441

9. Wakhloo AK, Gounis MJ (2008) Retrievable closed cell intracranial stent for foreign body and clot removal. Neurosurgery 62(5 Suppl 2): ONS390–ONS393. discussion ONS393–4

10. Nogueira RG et al (2012) Trevo versus merci retrievers for thrombectomy revascularisation of large vessel occlusions in acute ischaemic stroke (TREVO 2): a randomised trial. Lancet 380(9849):1231–1240

11. Saver JL et al (2015) Solitaire with the intention for thrombectomy as primary endovascular treatment for acute ischemic stroke (SWIFT PRIME) trial: protocol for a randomized, controlled, multicenter study comparing the solitaire revascularization device with IV tPA with IV tPA alone in acute ischemic stroke. Int J Stroke 10(3):439–448

12. Massari F et al (2016) ARTS (aspiration-retriever technique for stroke): initial clinical experience. Interv Neuroradiol 22(3):325–332

13. McTaggart RA et al (2017) Continuous aspiration prior to intracranial vascular embolectomy (CAPTIVE): a technique which improves outcomes. J Neurointerv Surg 9(12):1154–1159

Chapter 20

Spontaneous Intracerebral Hemorrhage

James Lee and Igor Rybinnik

Abstract

Spontaneous, nontraumatic intracerebral hemorrhage (ICH) accounts for approximately 10–25% of annual stroke cases and often portends poor neurological outcomes. Despite advances in education and medical care, the overall incidence has remained essentially unchanged over many years, and patients are often left with long-term functional deficits. Overall morbidity and mortality further increase when associated with intraventricular hemorrhage (IVH). Overall ICH case fatality rates have remained constant over the past few decades, although in-hospital rates have substantially decreased, likely due to improvements in neuro-critical care. Herein, we discuss the epidemiology, classification, differential diagnoses, and management options for ICH.

Key words Intracerebral, Intraparenchymal, Spontaneous, Hemorrhage, Nontraumatic, Intraventricular

1 Introduction

Spontaneous, nontraumatic intracerebral hemorrhage (ICH) is a potentially devastating disease that accounts for approximately 10–25% of annual stroke cases and often portends poor neurological outcomes [1–5]. Despite advances in education and medical care, the overall incidence has remained essentially unchanged over many years, with about 15 to 40 cases per 100,000 patient-years [1, 6]. Case fatality ratios for ICH range from 13% to 61% (median 40%) in the first month and >50% in the first year from disease onset and >70% within 5 years [6–8]. Less than 40% of patients achieve long-term functional independence. The mortality rate when associated with intraventricular hemorrhage (IVH) increases to 50% within the first month and 80% at 6 months from onset [9, 10]. Overall ICH case fatality rates have remained constant over the past few decades, although in-hospital rates have substantially decreased, likely due to improvements in neurocritical care [11].

Various epidemiological analyses have demonstrated that ICH carries a predilection for particular subgroups, including advanced

Fawaz Al-Mufti and Krishna Amuluru (eds.), *Cerebrovascular Disorders*, Neuromethods, vol. 170,
https://doi.org/10.1007/978-1-0716-1530-0_20, © Springer Science+Business Media, LLC, part of Springer Nature 2021

age (over 75 years), male gender, Asian and African ethnicities, and patients of low socioeconomic status [1, 2, 6, 12–15].

2 Pathophysiology and Clinical Presentation

ICH is primarily classified based upon the location of the bleed—deep (60–70%) or lobar (30–40%). Deep bleeds are the most common and are frequently associated with chronic arterial hypertension, which causes degenerative changes in the cerebral vessels, resulting in fibrinoid necrosis and reactive changes (i.e., lipohyalinosis), which subsequently lead to increased vessel wall fragility and rupture. The most common locations for these hemorrhages include the basal ganglia, internal capsule, and thalamus, which account for over two thirds of cases, with less than 10% occurring in the brainstem and cerebellum [16, 17]. Lobar bleeds occur in the subcortical white matter of the cerebral lobes and are most commonly associated with cerebral amyloid angiopathy (CAA), which is due to deposition of amyloid protein in the walls of the leptomeningeal vasculature, leading to friable vessels that have a higher propensity to spontaneously bleed [18]. Important differential diagnoses to consider as potential sources for spontaneous ICH are summarized in Table 1.

3 Management Strategies

3.1 Diagnostic Approach

- While the diagnosis of hemorrhagic stroke may be suspected by a history of abrupt onset focal neurological deficits especially in the setting of rapid progression with early signs of increased intracranial pressure, such as headache, lethargy, confusion, and decreased level of consciousness, neuroimaging confirmation is ultimately essential.

- Management of intracerebral hemorrhage begins in the pre-hospital setting (Fig. 1). Pre-hospital stroke care is common across all stroke types with an emphasis on rapid and accurate identification of stroke symptoms and timely transport to a nearby stroke center, albeit with one important difference—over one third of intracerebral hemorrhage patients deteriorate in the first several hours after symptom onset, especially during transport, emphasizing the need for early aggressive care [30, 31]. Pre-hospital interventions should not delay rapid transport to designated stroke centers, and pre-hospital notification by EMS should be implemented whenever possible, since it accelerates diagnosis and increases the intensity and speed of hospital stroke treatment [32–34].

Table 1

Common causes of intracerebral hemorrhage

	Presumed mechanism of bleeding	Annual risk of *re-bleeding*	Treatment
Primary causes			
Arterial hypertension	Rupture of arterioles secondary to chronic degenerative changes Microaneurysms of perforating arteries (i.e., Charcot-Bouchard aneurysms)	2% [19]	Blood pressure control [20, 21]
Cerebral amyloid angiopathy (CAA)	Amyloid beta (Aβ) protein deposition in small-to-medium leptomeningeal and cortical vessels Sporadic (elderly): Variations in apolipoprotein E e2/e4 on chromosome 19 Familial (Flemish, Dutch): Duplication of amyloid precursor protein (APP) on chromosome 21; presenilin (PS) 1 and 2 mutations	10.5% [22]	Avoidance of antithrombotics Preliminary studies indicate possible benefit from perindopril [23]
Secondary causes			
Sympathomimetic drugs	Usually due to sudden, transient increase in BP [24, 25] Usually not associated with an underlying vascular abnormality		Discontinuation of inciting agent
Coagulopathy/bleeding diatheses (inherited or acquired)	Usually due to use of anticoagulants and antithrombotics	1–2%	Rapid reversal/correction of coagulopathy
Vascular malformations	Rupture of saccular aneurysms of medium-sized artery, arteriovenous malformation (including dural arteriovenous fistula), and cavernous angioma	Up to 16%, depending on the particular vascular malformation and its features	Addressing underlying cause usually by endovascular or surgical means
Vasculitis	Primary CNS angiitis: Inflammation and subsequent degeneration of cerebral vessels resulting in rupture Secondary: Large vessel (giant cell arteritis, Takayasu arteritis), medium vessel (polyarteritis nodosa, Kawasaki disease), and small vessel (Churg-Strauss, Wegener's granulomatosis, microscopic polyangiitis, cryoglobulin-associated vasculitis, and Behcet's)		Immunosuppression

(continued)

Table 1
(continued)

	Presumed mechanism of bleeding	Annual risk of *re-bleeding*	Treatment
Moyamoya vasculopathy	Primary: Thought to be due to inflammatory processes that result in weakness and abnormal formation of lenticulostriate and thalamostriate collaterals [26] Secondary: Diseases causing narrowing of carotid terminus (e.g., atherosclerosis)		Addressing underlying cause Extracranial to intracranial direct and indirect bypass may be considered in certain cases
Intracranial neoplasms (primary or metastatic)	Due to necrosis and bleeding from hypervascular lesions		Treatment tailored to the underlying lesion
Hemorrhagic transformation of ischemic strokes (arterial or venous)	Due to disruption of the blood-brain barrier (BBB) from interruption of the sodium/potassium (Na/K) pump from depletion of adenosine triphosphate (ATP); this results in a rise in intracellular potassium, lactic acidosis, and release of extracellular glutamate [27]		Blood pressure control Reversal of coagulopathy if present
Cerebral venous sinus thrombosis	Hemorrhagic venous infarction	10% within the first 12 months; 1% thereafter [28]	Anticoagulation, despite hemorrhage
Hyperperfusion syndrome	Occurs when cerebral blood flow 100% or greater increases compared to baseline [29] Usually encountered post-procedure (i.e., carotid endarterectomy, stenting), but can also occur after thrombolytic reperfusion therapy in acute ischemic stroke		Strict blood pressure control

Acute focal neurological deficits identified
Hemorrhage suspected when early headache, lethargy, confusion are present

Stroke Code Activation
Called where appropriate

EARLY RESUSCITATION

Airway: Consider intubating if GCS ≤8 and pCO2 >45 torr

Breathing: maintain SaO2 > 90% and pCO2 <40 torr

Circulation: Maintain MAP > 70mm Hg

Access, Labs: Place two large bore proximal intravenous lines, draw labs for CBC, CMP, PT, aPTT, INR

Check *thrombin time* if history of Dabigatran use, and *anti-Xa* if Rivaroxaban, Apixaban, or Edoxaban use

Check glucose level

Rapid neurological assessment
(Recommend peforming the standardized NIH Stroke Scale)

EARLY ICP CRISIS MANAGEMENT

If signs of elevated intracranial pressure:
Raise head of bed to >30 degrees, and straighten neck
Urgent hyperventilation to 30 breaths per minute for at least 30 minutes (PaCO2 26-30 mmHg)
Bolus intravenous hyperosmolar therapy with Mannitol 1.0-1.5gm/kg, 250cc 3% NaCl *or* 30mL 23.4% NaCl
May repeat bolus osmotherapy until signs of herniation improve

DIAGNOSTIC IMAGING

Non-contrast head CT
CT Angiogram of the head with contrast if ICH is identified

Evaluate for vascular malformation, spot sign
Estimate hematoma volume by AxBxC/2 method

Spontaneous ICH confirmed

PREVENTION OF HEMATOMA EXPANSION

URGENT MANAGEMENT OF MASS EFFECT

NEUROSURGICAL EVALUATION

ICP monitor:

Consider ICP monitor placement when acute signs of mass effect, IVH ≥ 20cc +/- hydrocephalus with GCS ≤ 8

Evacuation via craniotomy:

Consider where appropriate if age ≤ 65, 50-100cc lobar hemorrhage within 1 cm of cortical surface or >3 cm (≥ 30cc) supratentorial hemorrhage associated with neuro deterioration, brainstem compression, hydrocephalus.

Hypertension:
Aggressive blood pressure control to goal systolic blood pressure 120-140 with Nicardipine IVPB 25mg/250cc NS @ 5-15mg/hr or Labetalol 10-20mg IVP q10 min to max 300mg, followed by infusion rate 2-8mg/min

Coagulopathy:
Reversal coagulopathy rapidly wherever present. Consider 4-factor PCC 1500 Units to start, and additional dosing as needed
Confirm reversal by repeating coags within 30 minutes of dose

PREVENTION OF SECONDARY INJURY

Close monitoring for neurological deterioration in a neuro intensive care unit
Maintain homeostasis (normoglycemia glucose goal 120-180 mg/dL, normothermia, normocarbia)
Low threshold for EEG monitoring when depressed mental status out of proportion to brain injury

Follow up stability non-**contrast head CT at 6 hours and 24 hours** after index event
Consider performing MRI **Brain w/ and w/o Gad** during the first 72 hours for etiological workup

Consider off label minimally invasive evacuation techniques for functional patients (baseline mRS 0-1), aged 18-80, diagnosed within 24h of symptom onset and accompanied by neuro deterioration (GCS ≤14 or NIHSS ≥ 6) with secondary causes of ICH and/or IVH carefully excluded and at least 6hr stability documented with repeat CT.

Additional considerations:
Minimally Invasive Hematoma Evacuation
ICH volume ≥ 30mL, absent active herniation.
Intraventricular Thrombolysis
Supratentorial ICH component of ≤ 30 mL, IVH volume ≥ 20 mL, and absent traditional rtPA contraindications

Fig. 1 ICH management algorithm

- The best initial diagnostic test is a non-contrast brain CT (NCCT), commonly used emergently due to high sensitivity for acute intracranial hemorrhage in non-anemic patients, wide availability, and rapid acquisition. NCCT also allows for estimation of hematoma volume by ABC/2 method (estimated volume of an ellipsoid clot) to communicate hemorrhage severity and prognosticate outcomes. In the ABC/2 formula, A is the greatest hemorrhage diameter, B is the largest diameter perpendicular to A, and C is the approximate number of slices multiplied by slice thickness [35].

- Contrast-enhanced CT angiogram (CTA) can also be performed at the time of the initial NCCT and is useful in identifying spot signs and vascular malformations. Spot sign—a small enhancing focus within the acute hematoma—occurs in one third of patients within 3 h of symptom onset, indicates active contrast extravasation, and is a potent imaging biomarker of hematoma expansion with a positive predictive value of over 70% and negative predictive value of over 80% [36, 37]. Spot sign is also helpful in identifying patients who may potentially benefit from acute hemostatic treatment, such as recombinant activated factor VII, although evidence of benefit is lacking [38]. Early CTA is also very valuable in the diagnosis of underlying vascular malformations, with sensitivity and specificity of over 95%, and a macrovascular cause identified in up to one third of the patients with ICH [39, 40]. The disadvantages of CTA include radiation, contrast-induced nephropathy (CIN), and allergic reactions, although serious and permanent side effects are rare [41]. For this reason, we advocate performing CTA with NCCT as standard imaging for patients with acute stroke symptoms, even when intracerebral hemorrhage is suspected.

- Multimodal MRI is equivalent to NCCT for the diagnosis of acute ICH and significantly more accurate for the identification of chronic ICH, with sensitivity and specificity approaching 100% [42]. MRI's sensitivity in the acute period is largely achieved by the addition of gradient-recalled echo (GRE) sequence, which detects by-products of hemoglobin metabolism [43]. Additionally, diffusion-weighted imaging (DWI) sequence best correlates with CT hematoma size and helps distinguish primary hemorrhage from hemorrhagic transformation of acute ischemic infarction. Contrast-enhanced MRI with MR angiography (MRA) and MR venography (MRV) is also an excellent tool for diagnosing underlying etiologies of secondary hemorrhages, including vascular malformations, tumors, cerebral venous thrombosis, and cerebral amyloid angiopathy [44].

- Despite the abovementioned advantages, MRI in the hyperacute setting remains somewhat impractical due to lack of widespread timely availability and various patient-related factors, including clinical instability, presence of metal (i.e., cardiac pacemaker), diminished consciousness, and agitation [45]. When utilized in the acute period to exclude neoplastic pathology, MRI should be performed within the first several days from symptom onset, before the hematoma becomes hyperintense on T1 (hemoglobin is converted to methemoglobin) hindering identification of abnormal contrast enhancement. We recommend that the study includes the following sequences: T1 with and without gadolinium, GRE, DWI, and FLAIR (fluid-attenuated inversion recovery). Non-enhanced MRA of the head and enhanced MRV of the head may be added in the appropriate clinical setting.

- Catheter angiogram is the gold standard for the diagnosis of vascular malformations and should be considered for spontaneous ICH patients aged less than 45 years and those without a history of hypertension [46]. This diagnostic modality is discussed in-depth elsewhere in this book.

- Urgent NCCT and CTA on all patients with suspected acute stroke. Multimodal imaging can be useful in diagnosing underlying structural lesion but may be performed on a non-emergent basis when there is sufficient clinical or radiological suspicion.

3.2 Treatment Approach

Management of intracerebral hemorrhage begins in the pre-hospital setting. Pre-hospital stroke care is common across all stroke types with an emphasis on rapid and accurate identification of stroke symptoms and timely transport to a nearby stroke center, albeit with one important difference—over one third of intracerebral hemorrhage patients deteriorate in the first several hours after symptom onset, especially during transport, emphasizing the need for early aggressive care [30, 31]. Pre-hospital interventions should not delay rapid transport to designated stroke centers, and pre-hospital notification by EMS should be implemented whenever possible, since it accelerates diagnosis and increases the intensity and speed of hospital stroke treatment [32–34].

Assessment and management of airway, breathing, and circulation is paramount in accordance with general principles of resuscitation, especially considering early deterioration in patients with acute ICH. Intravenous access should be established, and advanced airway placement should be considered whenever airway protection is suspect (e.g., when Glasgow Coma Scale ≤8, or significant respiratory distress is present) or there is failure of oxygenation or anticipated cardiopulmonary deterioration [47]. Maintenance of oxygen saturation >94% is reasonable for patients with suspected stroke. Blood pressure modifications, short of addressing symptomatic hypotension, should be delayed until the diagnosis of intracerebral hemorrhage is made [48]. Hypoglycemia should be rapidly corrected to a goal of >60 mg/dL. After immediate stabilization, validated screening tools, such as BE-FAST (Balance, Eyes, Face, Arm, Speech, Time), may be used in the emergency department setting to identify neurological deficits, prompting activation of a standardized stroke care pathway with a "stroke code," which alerts a multidisciplinary stroke team and imaging department to a potential stroke.

Once the diagnosis of ICH has been established, prevention of hematoma expansion and management of associated complications become the immediate goals of care in the acute period. Hematoma enlargement occurs in approximately one third of the patients with ICH over the first 24 h after onset and is independently associated with neurological deterioration and worsening outcome. Over half of expansions occur within the first 3–6 h, further underscoring the need of early aggressive care [30, 49]. There are essentially three strategies for the prevention of hematoma expansion: blood pressure control, reversal of coagulopathy, and use of hemostatic agents such as activated factor VII.

- Blood pressure control: There is a U-shaped relationship between blood pressure in the acute setting and outcomes post ischemic and hemorrhagic stroke, where both high and low blood pressures portend higher mortality [50]. Furthermore, high systolic blood pressure (SBP) is associated with increased risk of hematoma expansion, which in turn is a surrogate marker for poor outcomes [51]. This relationship was further underscored by recent landmark trials of blood pressure management in acute spontaneous ICH. Rapid and intensive blood pressure lowering to a goal of <140/90 mmHg was found to be safe and feasible and had modest effect on improvement of functional outcomes without reducing death or severe disability [20]. However, lowering systolic blood pressure (SBP) to a more aggressive target of <120 mmHg in the acute setting abolished the abovementioned modest benefits and was associated with higher rate of early serious adverse events, such as renal and cardiac insufficiency [20].

  A 2017 meta-analysis reaffirmed the safety of aggressive blood pressure lowering without improvement in mortality or functional outcomes while observing an association between intensive blood pressure reduction and lower risk for significant ICH expansion [52]. Although a single intervention, such as blood pressure lowering, may not improve outcomes, it is still worth pursuing for the possibility of synergistic effect with other interventions in the treatment of acute spontaneous ICH. Therefore, it is reasonable to recommend early aggressive SBP lowering to a target of 120–140 mmHg within 3 h of symptom onset in patients without a contraindication to acute blood pressure treatment. We prefer treating with nicardipine intravenous infusion (25 mg per 250 cc of normal saline) at a rate of 5–15 mg/h due to its efficacy, titratability, and favorable side effect profile, although labetalol infusion at a rate of 2–8 mg/min may be used as a substitute.

3.2.1 Reversal of Coagulopathy and use of Hemostatic Agents

It is important to establish history of antithrombotic use, especially direct oral anticoagulants, in a patient with acute stroke symptoms. Patients often do not volunteer that information and require prompting with a list of generic and brand names of high-risk medications. Oral anticoagulant use increases the risk of index hemorrhage, large hematoma size at presentation, early neurological deterioration, and both early and late hematoma enlargements in patients with ICH [53–55]. Therefore, efforts should be made to aggressively and rapidly reverse coagulopathy within 3 h from symptom onset.

Four-factor inactivated prothrombin complex concentrate (PCC) is favored over fresh frozen plasma (FFP) due to more rapid onset of effect, relatively small risk of clotting complications, and small volume of infusion reducing the risk of volume overload and heart failure [56]. For a patient with acute ICH while on vitamin K antagonists (VKAs) with INR elevated above 1.4, we recommend holding further anticoagulant dosing and administering vitamin K 10 mg intravenously, in addition to four-factor PCC at a dose of 25 U/kg for INR greater than 1.4 but less than 4, 35 U/kg for INR greater than or equal to 4 but less than 6, and 50 U/kg for INR at or above 6. PT and INR must be repeated 30 min after PCC administration, and patient may be re-dosed with PCC up to a maximum dose of 5000 U if INR is not at goal <1.4.

For direct factor Xa inhibitors, including apixaban, rivaroxaban, and edoxaban, we recommend four-factor PCC at a dose of 50 U/kg up to a maximum dose of 5000 U. While rivaroxaban may be qualitatively assessed with prothrombin time (PT), apixaban and edoxaban may not affect traditional coagulation profile and need confirmatory anti-factor Xa testing to ensure appropriate reversal. A specific antidote andexanet alfa showed promise in earlier trials and will likely replace PCC for reversal of direct factor Xa inhibitors in the near future [57].

Acute ICH patients on dabigatran should receive the specific antidote idarucizumab, administered as two separate doses of 2.5 mg intravenously no more than 15 min apart. Activated partial thromboplastin time (aPTT) offers a qualitative measure of dabigatran activity and should be assessed before and after reversal. We also recommend obtaining thrombin time due to its superior sensitivity [58].

It is our practice to reverse heparinoids (both unfractionated and low-molecular-weight heparins) with variable dosing of protamine. For intravenous unfractionated heparin (UFH) with baseline elevated aPTT and low-molecular-weight heparin (LMWH) dose within 8 h, any further dosing should be held, and immediately after the diagnosis of ICH, the patient should receive 1 mg of protamine per 100 U of UFH or per 1 mg LMWH up to a maximum dose of 50 mg. Protamine dose should be halved if the time from the last UFH dose is more than 30 min but less than 2 h or 8–12 h from the last LMWH dose. Patients who received UFH >2 h from symptom onset should be reversed with one quarter of the protamine dose. Protamine is not likely to be helpful for patients who received LMWH >12 h from symptom onset. aPTT and anti-factor Xa should be repeated to confirm reversal of UFH and LMWH, respectively [59].

Regardless of specific anticoagulant, confirmation of reversal is imperative and should not be neglected. In cases of incomplete reversal, treatment with the appropriate reversal agent may be repeated up to the maximal dose of that agent.

Antiplatelet use may increase the frequency of initial hemorrhage and may be independently associated with increased mortality in patients with ICH [60–62]. However, while anticoagulation is clearly detrimental in this patient population, the effect of antiplatelet agents on acute hematoma is more controversial, with retrospective studies suggesting no increase in hematoma volume, degree of expansion, or short-term clinical outcome [63]. The largest trial of platelet transfusion after acute spontaneous primary ICH in patients taking antiplatelet therapy concluded that platelet transfusion is at best ineffective and potentially harmful, although patients with smaller hemorrhages were overrepresented, and platelets may have been administered beyond the critical 6-h window. Therefore, platelet transfusion should be used with caution in this patient population, and routine transfusions cannot be recommended [64].

Hemostatic agents have been considered for the treatment of spontaneous non-coagulopathic ICH, with recombinant activated factor VII (rFVIIa) being one of the most promising and best studied agents. Although rFVIIa reduced hematoma expansion in early trials, that improvement did not translate into clinical benefit in phase 3 trials, even with careful selection of patients at high risk for early hematoma expansion, and rFVIIa was associated with increased frequency of thromboembolic events [38, 65, 66]. Therefore, routine use of hemostatic agents in the absence of a proven coagulopathy in patients with acute spontaneous ICH cannot be recommended.

While the pressure in and around the hematoma may be increased, global intracranial pressure (ICP) is generally normal, especially in patients with smaller ICHs, and a vast majority do not require monitoring or treatment. Intracranial hypertension is more common in younger patients (aged less than 60 years) with large

(>30 cc) supratentorial hematomas and those with intraventricular hemorrhage [67]. In unmonitored patients with acute neurological deterioration due to suspected ICP crisis, aggressive approaches include: head of bed elevation above 30° while keeping head in mid-position, therapeutic hyperventilation to $PaCO2$ target of 26–30 mmHg, and hyperosmolar therapy (bolus of mannitol 1.0–1.5 g/kg, or 250 cc of 3% NaCl or 30 cc of 23.4% NaCl) to a target sodium of 145–160 mmol/L and serum osmolarity of 300–320 mOsm/L; the maneuvers rapidly reduce ICP and reverse herniation [68]. Of note, glucocorticoid dexamethasone is not recommended as it was shown to be harmful, increasing infection rates without demonstrable beneficial effect on outcome [69]. However, medical therapy alone is ultimately inadequate at preventing recurrent intracranial hypertension and improving outcomes, thus promoting continuing interest in more definitive surgical treatments [70].

Intracranial monitoring indications and management goals are ill-defined in the spontaneous ICH population and are largely adapted from the traumatic brain injury guidelines. Placement of an ICP monitor is recommended in comatose patients (Glasgow Coma Scales 3–8) with suspected mass effect, and externally placed intraventricular catheter (external ventricular drain or EVD) is preferred for its therapeutic capability of lowering ICP by increasing cerebrospinal fluid drainage, despite the modest complications of infection and catheter occlusion [71]. ICP should be maintained at a goal of <20 mmHg while targeting cerebral perfusion pressure (CPP = MAP-ICP) at 70–110 mmHg, depending on the status of cerebral autoregulation [72]. Notably, intracranial hypertension does not necessarily portend poor outcome, and a substantial number of ICH patients with elevated ICP recover well with aggressive management [67].

3.2.2 Surgical
Hematoma Evacuation

In addition to primary injury from mechanical compression of the surrounding brain tissue by the hematoma, perihematomal edema, mass effect, and primary metabolic depression due to perihematomal mitochondrial dysfunction may cause secondary brain injury, contributing to high early mortality and morbidity in patients with this disease [73, 74]. Surgical intervention for spontaneous ICH aims to mitigate secondary brain injury by evacuating the hematoma by three main approaches—conventional craniotomy, catheter drainage, and endoscopic treatment. Modern large-scale attempts at early supratentorial clot evacuation via craniotomy have been inconclusive or failed to significantly improve outcomes, although these were criticized for possible selection bias, significant crossover rates to surgical intervention post randomization, and focus on early surgery [75, 76]. Meta-analysis of surgical trials for supratentorial hematoma evacuation eventually showed significant advantage to surgery, although robustness of this result was questioned due to heterogeneity of data [77]. A small patient population with a median age of 65 years (interquartile range 55–74), with neurological deterioration due to superficial lobar hematomas 10–100 cc in volume within 1 cm of the cortex, and without associated intraventricular hemorrhage may benefit from clot evacuation within 48 hrs. of ictus after secondary causes of hemorrhage (such as vascular malformation or tumor) have been excluded. The idea that hematoma evacuation would be beneficial is pathologically sound, although benefit has not been substantiated by robust data presumably due to the negative effects of obligatory surgical trauma with traditional craniotomy, late timing of surgery beyond a yet to be defined optimal therapeutic window, incomplete evacuation, inadequate sample sizes, and presence of intraventricular hemorrhage, which is an independent risk factor for poor outcome [78].

Modern minimally invasive approaches, thrombolytic-assisted evacuation and endoscopic techniques chief among them, have emerged to standardize surgical treatment of ICH, minimizing surgical brain injury and addressing the above-mentioned concerns with early surgical trials. The MISTIE (Minimally Invasive Surgery plus recombinant Tissue plasminogen activator for Intracerebral Hemorrhage Evacuation) technique adapts a combined mechanical and pharmacological approach involving image-guided single-pass cannulation of the intracerebral hematoma via a standard burr hole and aspiration at the time of catheter placement, followed by serial doses of recombinant tissue plasminogen activator (rtPA) and continuous passive drainage. Feasibility and safety of this technique have been confirmed in early trials, and recent data hints at potential efficacy with robust surgical performance and reduction of perihematomal edema [79, 80]. Phase 3 trial has completed enrollment and entered follow-up phase, and final results are eagerly anticipated. In the meantime, we generally avoid traditional craniotomy for evacuation of intracerebral hemorrhage at our institution and currently offer the MISTIE procedure off-label to those that meet the phase 3 trial enrollment criteria: functional patients (baseline mRS 0–1) aged 18–80 years, with intracerebral hemorrhage ≥30 mL diagnosed within 24 hrs. of symptom onset and accompanied by neurological deterioration (GCS ≤14 or NIHSS ≥6), and absent active herniation. Early aggressive medical management is required, and great care is taken to exclude secondary causes and document radiographic and clinical stability with requisite serial CT scans, vascular imaging (typically CTA, but also conventional angiogram where appropriate), and MRI with and without contrast before the MISTIE procedure.

ICES (Intraoperative Stereotactic Computer Tomography-Guided Endoscopic Surgery) is a purely mechanical approach to hematoma evacuation, where hematoma is accessed in a manner similar to the MISTIE procedure, but instead of a gradual volume reduction over days with thrombolysis and passive drainage, the hematoma is rapidly removed over hours with the aspiration and irrigation technique followed by endoscope-guided hemostasis. The safety and efficacy of the ICES procedure have been established in early study, but MISTIE is further along in the process of providing meaningful outcome data [81].

Despite the progress in surgical evacuation of supratentorial hematomas, the surgical management of infratentorial hematomas where outcomes are poor, and early deterioration is common due prominent local mass effect and obstructive hydrocephalus in the narrow confines of the posterior fossa, remains uncertain [82]. Based on limited available data and expert opinion, we recommend evacuating cerebellar hematomas via early suboccipital craniotomy in patients with ICH volume >30 cc and neurological deterioration (GCS ≤14) due to severe brainstem compression and/or hydrocephalus from ventricular obstruction. Ventriculostomy may need to be performed in conjunction with craniotomy where indicated but is not recommended as an alternative to surgical decompression as it may cause upward herniation [83].

Intraventricular hemorrhage (IVH) occurs in up one third of ICH patients, aggravated by antithrombotic use, and is independently associated with impaired consciousness at presentation (by damaging reticular activating system) and worse long-term outcomes [82, 84–86]. IVH frequently causes acute obstructive hydrocephalus, especially with blood in third and fourth ventricles barring CSF outflow through cerebral aqueduct and various ventricular foramina. Obstructive (non-communicating) hydrocephalus can cause life-threatening mass effect with increasing risk of herniation, and cerebral perfusion deficit, and usually requires urgent CSF flow diversion by an EVD [75, 87]. Moreover, the natural resolution of intraventricular blood follows first-order kinetics and may last for weeks especially with larger IVHs due to insufficient CSF fibrinolytic activity, long enough for blood degradation products to occlude and damage arachnoid granulations, inhibiting CSF absorption, and causing delayed non-obstructive (communicating) hydrocephalus. Incidentally, the degree and incidence of communicating hydrocephalus is not improved by EVD, which does not affect the rate of blood clot resolution [88, 89].

With these considerations, intraventricular thrombolysis with rtPA was developed to accelerate IVH clearance. A large, prospective, multicenter, randomized, placebo-controlled phase 3 trial (CLEAR III) showed increased survival albeit with severe disability and functional dependence. However, probability of favorable outcome paralleled the efficiency of clot removal especially in patients with larger IVH burden (≥ 20 mL), and safety parameters favored alteplase treatment [90–92]. While intraventricular thrombolysis cannot yet be recommended in all patients with IVH, it may be carefully considered for off-label use in highly functional patients (historical mRS of 0–1) aged 18–80 years diagnosed within 24 h of symptom onset, with supratentorial intraparenchymal hemorrhage component of no more than 30 mL, IVH volume of equal to or greater than 20 mL, documented radiographic clot stability at least 6 h after EVD placement, careful exclusion of secondary causes of IVH, and absent traditional rtPA contraindications, such as ongoing bleeding and coagulopathy. Up to 12 doses of 1 mg alteplase every 8 h may be administered until at least 80% of the clot is removed.

3.2.3 Management of Complications

ICH patients are critically ill and are at risk for hemodynamic and respiratory compromise requiring intubation and mechanical ventilation, hematoma expansion with possible intraventricular extension, worsening mass effect resulting in intracranial pressure crises, increased seizure frequency, and venous thromboembolism, among other medical issues and comorbidities. These patients should therefore be transferred from the emergency department without delay and managed in dedicated neurointensive care units, which have been shown to reduce mortality and improve functional outcomes [93–95].

- *Hyperglycemia.*
  - Hyperglycemia on admission is associated with poor outcome and increased mortality in patients with acute stroke and intracerebral hemorrhage, although optimal management of glucose in this patient population is still unclear, and aggressive glucose control increases the risk of detrimental hypoglycemia [96]. Targeting normoglycemia (glucose goal 120–180 mg/dL) while avoiding hypoglycemia is a reasonable approach [97, 98].

- *Fever.*
  - Fever (defined as temperature over 38.3 °C) is common with ICH likely secondary to damaged temperature homeostasis pathways and occurs in up to one third of the patients especially with larger hematomas and IVH. Hyperthermia is an independent predictor of worsened neurological outcome, and maintenance of normothermia is reasonable especially in the first 72 h, although unproven to alter outcome [99, 100]. Optimal temperature-lowering approach is unfortunately unknown, and round-the-clock antipyretics and surface cooling are both reasonable.
  - Targeted temperature management (therapeutic hypothermia), which has become commonplace in the neurocritical care setting, although not routinely recommended in ICH

patients, reduces perihematomal edema and may be considered in the cases of refractory intracranial hypertension [101]. Finally, up to one third of the patients with fever may be a documented infection, and care should be taken to exclude infection before making the diagnosis hyperthermia of central origin, especially in patients with intracranial drains [102].

- *Seizures:* Early seizures are reported in up to one third of the patients with acute spontaneous ICH. Seizures are often non-convulsive, more frequently related to lobar hemorrhages, and are independently associated with worsening midline shift, hematoma expansion, and neurological deterioration [103–106]. Electrographic seizures are an important cause of depressed mental status disproportionate to the degree of brain injury in ICH patients, and clinicians should have a low threshold for initiating continuous EEG monitoring. Clinical seizures or electrographic seizures in patients with altered sensorium should be treated with antiepileptics.

- There is paucity of evidence to recommend routine seizure prophylaxis, especially since antiepileptics may worsen outcomes by presumably inhibiting neuroplasticity and interfering with neurological recovery [107, 108]. However, a 7-day course of seizure prophylaxis with the lowest dose of an antiepileptic with relatively favorable side effect profile (e.g., levetiracetam 500 mg twice daily) may be reasonable in patients with large lobar hematomas where seizure-induced worsening of cerebral edema may result in an intracranial pressure crisis [109, 110]. Unfortunately, the habitual use of prophylactic antiepileptics, specifically levetiracetam, in ICH patients is widespread even in academic centers despite contrary recommendations in the American Heart Association/American Stroke Association guidelines and may be a difficult habit to break [111].

- *Venous thromboembolism prophylaxis and treatment:* ICH victims are at higher risk for venous thromboembolic disease, which afflicts up to 15% of immobilized patients in the first month post hemorrhage onset. Intermittent pneumatic compression early in the hospital course was found to be a safe and inexpensive method of reducing thromboembolism risk and improving survival in this patient population and is recommended beginning the day of hospital admission [112].

In the absence of coagulopathy, enlargement of hematoma in patients with spontaneous ICH beyond 24 h from onset is extremely rare, while the rate of venous thromboembolism rises quickly even in the first 10 days of hospitalization [49, 113]. In a meta-analysis, early use of heparinoids in the first week after admission was safe and reduced the rate of pulmonary embolism by nearly 40%, with a nonsignificant reduction in mortality [114]. In immobilized patients without coagulopathy, Low-dose low-molecular-weight or unfractionated heparin prophylaxis should be started after 24 hrs. from onset of spontaneous hemorrhage, provided confirmation of hemorrhage stability. Unfortunately, the choice and timing of full-dose anticoagulation treatment in ICH patients remains challenging, as there are no randomized trials evaluating anticoagulant use after ICH. The largest meta-analysis of available literature to date showed that early anticoagulation (median time of initiation 10–39 days) most commonly prescribed for patients with atrial fibrillation, prosthetic heart valves, and acute venous thromboembolism significantly reduced the risks of thromboembolic events by a third, without increasing ICH recurrence, although the possibility of significant selection bias should be carefully considered [115]. Our approach to anticoagulation treatment in spontaneous ICH patients typically involves an individualized risk-benefit analysis and transparent communication with the patient (where possible) and decision-making parties. The risks of further brain injury include index hematoma size greater than 30 cc, lobar location, early hematoma expansion (or presence of "spot sign" on CTA signifying increased risk of expansion), presence of significant mass effect or IVH, uncontrolled hypertension, coagulopathic state, and presence of microbleeds on GRE MRI sequence [116]. In the absence of these factors, there is a moderate risk of hemorrhage recurrence in ICH survivors of up to 2% per year, and anticoagulation use roughly doubles that risk [117]. Anticoagulation may be cautiously restarted after 10–14 days post index ICH in this patient population with documented radiographic and clinical stability and aggressive blood pressure control, if the perceived morbidity or mortality from thromboembolic events is higher than 4%. Direct oral anticoagulants (DOACs) are the preferred anticoagulation method, since they demonstrate significantly lower ICH risk compared to warfarin. Furthermore, DOAC-associated ICHs tend to be smaller with more favorable functional outcomes [118–120]. Dabigatran may be considered due to ready availability of a specific reversal agent, although reversibility of direct factor Xa inhibitors is forthcoming [58, 121].

4   Outcome

As previously mentioned, ICH survivors are at risk for ICH recurrence, especially if the underlying cause is not treated or remains unknown. A meta-analysis of 122 studies surveying long-term outcome after ICH demonstrated an annual ICH recurrence risk of 1.3–7.4%, with risk being higher in lobar vs non-lobar ICH [122].

Modifiable risk factors for ICH recurrence include blood pressure, heavy alcohol consumption, smoking, diet, illicit drug use, obesity, physical activity, socioeconomic status, and use of antithrombotic agents. Non-modifiable factors include age, sex, ethnicity, family history, and genetics. Hypertension is by far the strongest risk factor, accounting for approximately 70% of the population-associated risk of ICH recurrence [13]. Overall ICH incidence has decreased in younger patients (age < 60 years) due to earlier detection and treatment of hypertension, along with improved risk factor control. Conversely however, older patients (age > 75 years) suffer increased incidence of ICH due to prevalent use of antithrombotic medications for the management of other pre-existing comorbidities [123, 124].

Although less common than ischemic stroke, ICH continues to bear higher rates of morbidity and mortality, despite advances in critical care management [125]. Outcomes are substantially variable and depend on several crucial factors, including location of bleed, hematoma volume (i.e., clot burden), and involvement of the ventricles, among others [10]. Although several scales exist for outcome stratification, the ICH score is the most commonly used validated grading scale and is highly predictive of 30-day mortality [10]. Components of the ICH score include age, initial Glasgow Coma Scale (GCS) score, ICH volume, ICH location (supratentorial versus infratentorial), and presence of IVH. ICH scores range from 0 to 6, with each successive increase in score correlating with increased risk of mortality and poor functional outcome. When utilizing the ICH score, it is important to remember that it is helpful in predicting outcome for only spontaneous ICH cases and only after 48 h of aggressive care.

5 Conclusion

Although spontaneous ICH is a diverse disease with significant mortality and morbidity, clinicians armed with advancing understanding of timely targeted blood pressure management, rapid effective coagulopathy reversal for prevention of hematoma expansion, uncompromising ICP management, and the importance of comprehensive stroke centers with specialized neurological intensive care units to combat ICH complications should advocate for aggressive time-sensitive treatment analogous to ischemic stroke, contradicting entrenched nihilistic attitudes. Despite disappointing performance in earlier trials, hemostatic agents remain an interesting area of study. Although traditional surgical techniques failed to improve outcomes, minimally invasive approaches aim to limit requisite surgical trauma and forecast a hopeful future for early

ICH evacuation and prevention of secondary trauma. Finally, while no single intervention modifies outcomes in ICH, the possibility of synergistic effect of multiple interventions is too promising to ignore. There is a heightened expectation that we will see marked strides in ICH care in the near future.

References

1. Feigin VL, Lawes CM, Bennett DA, Barker-Collo SL, Parag V (2009) Worldwide stroke incidence and early case fatality reported in 56 population-based studies: a systematic review. Lancet Neurol 8:355–369. Available from: http://www.ncbi.nlm.nih.gov/pubmed/19233729. Accessed 6 Oct 2017

2. Flaherty ML, Woo D, Haverbusch M, Sekar P, Khoury J, Sauerbeck L et al (2005) Racial variations in location and risk of intracerebral hemorrhage. Stroke 36:934–937. Available from: http://www.ncbi.nlm.nih.gov/pubmed/15790947. Accessed 6 Oct 2017

3. Hemphill JC, Greenberg SM, Anderson CS, Becker K, Bendok BR, Cushman M et al (2015) Guidelines for the management of spontaneous intracerebral hemorrhage. Stroke 46:2032–2060. Available from: http://www.ncbi.nlm.nih.gov/pubmed/26022637. Accessed 6 Oct 2017

4. Hong K-S, Bang OY, Kang D-W, Yu K-H, Bae H-J, Lee JS et al (2013) Stroke statistics in Korea: Part I. Epidemiology and risk factors: a report from the korean stroke society and clinical research center for stroke. J Stroke 15:2. Available from: http://www.ncbi.nlm.nih.gov/pubmed/24324935. Accessed 6 Oct 2017

5. Sacco S, Marini C, Toni D, Olivieri L, Carolei A (2009) Incidence and 10-year survival of intracerebral hemorrhage in a population-based registry. Stroke 40:394–399. Available from: http://www.ncbi.nlm.nih.gov/pubmed/19038914. Accessed 6 Oct 2017

6. van Asch CJ, Luitse MJ, Rinkel GJ, van der Tweel I, Algra A, Klijn CJ (2010) Incidence, case fatality, and functional outcome of intracerebral haemorrhage over time, according to age, sex, and ethnic origin: a systematic review and meta-analysis. Lancet Neurol 9:167–176. Available from: http://www.ncbi.nlm.nih.gov/pubmed/20056489. Accessed 6 Oct 2017

7. Gonzalez-Perez A, Gaist D, Wallander M-A, McFeat G, Garcia-Rodriguez LA (2013) Mortality after hemorrhagic stroke: Data from general practice (The Health Improvement Network). Neurology 81:559–565. Available from: http://www.ncbi.nlm.nih.gov/pubmed/23843467. Accessed 6 Oct 2017

8. Liotta EM, Prabhakaran S (2013) Warfarin-associated intracerebral hemorrhage is increasing in prevalence in the United States. J Stroke Cerebrovasc Dis 22:1151–1155. Available from: http://www.ncbi.nlm.nih.gov/pubmed/23287421. Accessed 6 Oct 2017

9. Bhattathiri PS, Gregson B, Prasad KSM, Mendelow AD, STICH Investigators (2006) Intraventricular hemorrhage and hydrocephalus after spontaneous intracerebral hemorrhage: results from the STICH trial. Acta Neurochir Suppl 96:65–68. Available from: http://www.ncbi.nlm.nih.gov/pubmed/16671427. Accessed 6 Oct 2017

10. Hemphill JC, Bonovich DC, Besmertis L, Manley GT, Johnston SC (2001) The ICH score: a simple, reliable grading scale for intracerebral hemorrhage. Stroke 32:891–897. Available from: http://www.ncbi.nlm.nih.gov/pubmed/11283388. Accessed 6 Oct 2017

11. Chan S, Hemphill JC (2014) Critical care management of intracerebral hemorrhage. Crit Care Clin 30(4):699–717

12. Broderick JP, Brott T, Tomsick T, Miller R, Huster G (1993) Intracerebral hemorrhage more than twice as common as subarachnoid hemorrhage. J Neurosurg 78:188–191. Available from: http://www.ncbi.nlm.nih.gov/pubmed/8421201. Accessed 6 Oct 2017

13. Jolink WMT, Klijn CJM, Brouwers PJAM, Kappelle LJ, Vaartjes I (2015) Time trends in incidence, case fatality, and mortality of intracerebral hemorrhage. Neurology 85:1318–1324. Available from: http://www.ncbi.nlm.nih.gov/pubmed/26377254. Accessed 6 Oct 2017

14. Krishnamurthi RV, Feigin VL, Forouzanfar MH, Mensah GA, Connor M, Bennett DA et al (2013) Global and regional burden of first-ever ischaemic and haemorrhagic stroke during 1990–2010: findings from the Global Burden of Disease Study 2010. Lancet Glob Health 1:e259–e281. Available from: http://

www.ncbi.nlm.nih.gov/pubmed/25104492. Accessed 6 Oct 2017

15. Stein M, Misselwitz B, Hamann GF, Scharbrodt W, Schummer DI, Oertel MF (2012) Intracerebral hemorrhage in the very old. Stroke 43:1126–1128. Available from: http://www.ncbi.nlm.nih.gov/pubmed/22282880. Accessed 6 Oct 2017

16. Grysiewicz RA, Thomas K, Pandey DK (2008) Epidemiology of ischemic and hemorrhagic stroke: incidence, prevalence, mortality, and risk factors. Neurol Clin 26:871–895. Available from: http://www.ncbi.nlm.nih.gov/pubmed/19026895. Accessed 22 December 2017

17. Labovitz DL, Halim A, Boden-Albala B, Hauser WA, Sacco RL (2005) The incidence of deep and lobar intracerebral hemorrhage in whites, blacks, and Hispanics. Neurology 65:518–522. Available from: http://www.ncbi.nlm.nih.gov/pubmed/16116109. Accessed 22 Dec 2017

18. Massaro AR, Sacco RL, Mohr JP, Foulkes MA, Tatemichi TK, Price TR et al (1991) Clinical discriminators of lobar and deep hemorrhages: the Stroke Data Bank. Neurology 41:1881–1885. Available from: http://www.ncbi.nlm.nih.gov/pubmed/1745342. Accessed 22 Dec 2017

19. Arakawa S, Saku Y, Ibayashi S, Nagao T, Fujishima M (1998) Blood pressure control and recurrence of hypertensive brain hemorrhage. Stroke 29:1806–1809. Available from: http://www.ncbi.nlm.nih.gov/pubmed/9731599. Accessed 3 Nov 2017

20. Anderson CS, Heeley E, Huang Y, Wang J, Stapf C, Delcourt C et al (2013) Rapid blood-pressure lowering in patients with acute intracerebral hemorrhage. N Engl J Med 368:2355–2365. Available from: http://www.ncbi.nlm.nih.gov/pubmed/23713578. Accessed 2 Jan 2018

21. Anderson CS, Huang Y, Wang JG, Arima H, Neal B, Peng B et al (2008) Intensive blood pressure reduction in acute cerebral haemorrhage trial (INTERACT): a randomised pilot trial. Lancet Neurol 7:391–399. Available from: http://www.ncbi.nlm.nih.gov/pubmed/18396107. Accessed 3 Nov 2017

22. O'Donnell HC, Rosand J, Knudsen KA, Furie KL, Segal AZ, Chiu RI et al (2000) Apolipoprotein E genotype and the risk of recurrent lobar intracerebral hemorrhage. N Engl J Med 342:240–245. Available from: http://www.ncbi.nlm.nih.gov/pubmed/10648765. Accessed 3 Nov 2017

23. Arima H, Anderson C, Omae T, Woodward M, MacMahon S, Mancia G et al (2012) Effects of blood pressure lowering on intracranial and extracranial bleeding in patients on antithrombotic therapy: the PROGRESS trial. Stroke 43:1675–1677. Available from: http://www.ncbi.nlm.nih.gov/pubmed/22535269. Accessed 3 Nov 2017

24. Aggarwal SK, Williams V, Levine SR, Cassin BJ, Garcia JH (1996) Cocaine-associated intracranial hemorrhage: absence of vasculitis in 14 cases. Neurology 46:1741–1743. Available from: http://www.ncbi.nlm.nih.gov/pubmed/8649582. Accessed 3 Nov 2017

25. Martin-Schild S, Albright KC, Hallevi H, Barreto AD, Philip M, Misra V et al (2010) Intracerebral hemorrhage in cocaine users. Stroke 41:680–684. Available from: http://www.ncbi.nlm.nih.gov/pubmed/20185779. Accessed 3 Nov 2017

26. Huang S, Guo Z-N, Shi M, Yang Y, Rao M (2017) Etiology and pathogenesis of Moyamoya disease: an update on disease prevalence. Int J Stroke 12:246–253. Available from: http://journals.sagepub.com/doi/10.1177/1747493017694393. Accessed 4 Nov 2017

27. Kulik T, Kusano Y, Aronhime S, Sandler AL, Winn HR (2008) Regulation of cerebral vasculature in normal and ischemic brain. Neuropharmacology 55:281–288. Available from: http://www.sciencedirect.com/science/article/pii/S0028390808001135?via%3Dihub. Accessed 4 Nov 2017

28. Preter M, Tzourio C, Ameri A, Bousser MG (1996) Long-term prognosis in cerebral venous thrombosis. Follow-up of 77 patients. Stroke 27:243–246. Available from: http://www.ncbi.nlm.nih.gov/pubmed/8571417. Accessed 4 Nov 2017

29. Sundt TM, Sharbrough FW, Piepgras DG, Kearns TP, Messick JM, O'Fallon WM (1981) Correlation of cerebral blood flow and electroencephalographic changes during carotid endarterectomy: with results of surgery and hemodynamics of cerebral ischemia. Mayo Clin Proc 56:533–543. Available from: http://www.ncbi.nlm.nih.gov/pubmed/7266064. Accessed 22 Dec 2017

30. Brott T, Broderick J, Kothari R, Barsan W, Tomsick T, Sauerbeck L et al (1997) Early hemorrhage growth in patients with intracerebral hemorrhage. Stroke 28:1–5. Available from: http://www.ncbi.nlm.nih.gov/pubmed/8996478. Accessed 2 Jan 2018

31. Moon J-S, Janjua N, Ahmed S, Kirmani JF, Harris-Lane P, Jacob M et al (2008) Prehospital neurologic deterioration in patients with intracerebral hemorrhage. Crit Care Med

36:172–175. Available from: http://content. wkhealth.com/linkback/openurl? sid=WKPTLP:landingpage&an=00003246-200801000-00023. Accessed 2 Jan 2018

32. Abdullah AR, Smith EE, Biddinger PD, Kalenderian D, Schwamm LH (2008) Advance hospital notification by EMS in acute stroke is associated with shorter door-to-computed tomography time and increased likelihood of administration of tissue-plasminogen activator. Prehosp Emerg Care 12:426–431. Available from: http://www. tandfonline.com/doi/full/10.1080/ 10903120802290828. Accessed 2 Jan 2018

33. Kim SK, Lee SY, Bae HJ, Lee YS, Kim SY, Kang MJ et al (2009) Pre-hospital notification reduced the door-to-needle time for iv t-PA in acute ischaemic stroke. Eur J Neurol 16:1331–1335. Available from: http://doi. wiley.com/10.1111/j.1468-1331.2009. 02762.x. Accessed 2 Jan 2018

34. McKinney JS, Mylavarapu K, Lane J, Roberts V, Ohman-Strickland P, Merlin MA (2013) Hospital prenotification of stroke patients by emergency medical services improves stroke time targets. J Stroke Cerebrovasc Dis 22:113–118. Available from: http://linkinghub.elsevier.com/retrieve/pii/S1052305711001777. Accessed 2 Jan 2018

35. Kothari RU, Brott T, Broderick JP, Barsan WG, Sauerbeck LR, Zuccarello M et al (1996) The ABCs of measuring intracerebral hemorrhage volumes. Stroke 27:1304–1305. Available from: http://www.ncbi.nlm.nih. gov/pubmed/8711791. Accessed 2 Jan 2018

36. Demchuk AM, Dowlatshahi D, Rodriguez-Luna D, Molina CA, Blas YS, Dzialowski I et al (2012) Prediction of haematoma growth and outcome in patients with intracerebral haemorrhage using the CT-angiography spot sign (PREDICT): a prospective observational study. Lancet Neurol 11:307–314. Available from: http://www.ncbi.nlm.nih.gov/ pubmed/22405630. Accessed 2 Jan 2018

37. Murai Y, Ikeda Y, Teramoto A, Goldstein JN, Greenberg SM, Smith EE et al (2007) Contrast extravasation on ct angiography predicts hematoma expansion in intracerebral hemorrhage. Neurology 69:617–617. Available from: http://www.ncbi.nlm.nih.gov/ pubmed/17679688. Accessed 2 Jan 2018

38. Gladstone DJ, Aviv RI, Demchuk AM, Hill MD, Thorpe KE KJ et al (2017) Randomized trial of hemostatic therapy for "spot sign" positive intracerebral hemorrhage: primary results from the SPOTLIGHT/STOP-IT study collaboration. In International stroke

conference 2017, Houston, TX, Available from: http://www.abstractsonline.com/ pp8/#!/4172/presentation/13152. Accessed 2 Jan 2018

39. Wong GKC, Siu DYW, Abrigo JM, Poon WS, Tsang FCP, Zhu XL et al (2011) Computed tomographic angiography and venography for young or nonhypertensive patients with acute spontaneous intracerebral hemorrhage. Stroke 42:211–213. Available from: http:// www.ncbi.nlm.nih.gov/pubmed/21088241. Accessed 2 Jan 2018

40. Yoon DY, Chang SK, Choi CS, Kim W-K, Lee J-H (2009) Multidetector row CT angiography in spontaneous lobar intracerebral hemorrhage: a prospective comparison with conventional angiography. AJNR Am J Neuroradiol 30:962–967. Available from: http:// www.ajnr.org/cgi/doi/10.3174/ajnr. A1471. Accessed 2 Jan 2018

41. Wilhelm-Leen E, Montez-Rath ME, Chertow G (2017) Estimating the risk of radiocontrast-associated nephropathy. J Am Soc Nephrol 28:653–659. Available from: http://www. jasn.org/lookup/doi/10.1681/ASN. 2016010021. Accessed 2 Jan 2018

42. Kidwell CS, Chalela JA, Saver JL, Starkman S, Hill MD, Demchuk AM et al (2004) Comparison of MRI and CT for detection of acute intracerebral hemorrhage. JAMA 292:1823. Available from: http://www.ncbi.nlm.nih. gov/pubmed/15494579. Accessed 2 Jan 2018

43. Patel MR, Edelman RR, Warach S (1996) Detection of hyperacute primary intraparenchymal hemorrhage by magnetic resonance imaging. Stroke 27:2321–2324. Available from: http://www.ncbi.nlm.nih.gov/ pubmed/8969800. Accessed 2 Jan 2018

44. Kamel H, Navi BB, Hemphill JC (2013) A rule to identify patients who require magnetic resonance imaging after intracerebral hemorrhage. Neurocrit Care 18:59–63. Available from: http://www.ncbi.nlm.nih.gov/ pubmed/21761271. Accessed 2 Jan 2018

45. Singer OC, Sitzer M, du Mesnil de Rochemont R, Neumann-Haefelin T (2004) Practical limitations of acute stroke MRI due to patient-related problems. Neurology 62:1848–1849. Available from: http://www. ncbi.nlm.nih.gov/pubmed/15159492. Accessed 2 Jan 2018

46. Zhu XL, Chan MS, Poon WS (1997) Spontaneous intracranial hemorrhage: which patients need diagnostic cerebral angiography? A prospective study of 206 cases and review of the literature. Stroke 28:1406–1409. Available from: http://www.

ncbi.nlm.nih.gov/pubmed/9227692. Accessed 2 Jan 2018

47. Brown CA, Sakles JC, Mick NW (2017) The walls manual of emergency airway management, 5th edn. Lippincott Williams & Wilkins, Philadelphia

48. Jauch EC, Cucchiara B, Adeoye O, Meurer W, Brice J, Chan YY-F et al (2010) Part 11: adult stroke: 2010 American Heart Association guidelines for cardiopulmonary resuscitation and emergency cardiovascular care. Circulation 122:S818–S828. Available from: http://circ.ahajournals.org/cgi/doi/10.1161/CIRCULATIONAHA.110.971044. Accessed 2 Jan 2018

49. Kazui S, Naritomi H, Yamamoto H, Sawada T, Yamaguchi T (1996) Enlargement of spontaneous intracerebral hemorrhage. Incidence and time course. Stroke 27:1783–1787. Available from: http://www.ncbi.nlm.nih.gov/pubmed/8841330. Accessed 2 Jan 2018

50. Vemmos KN, Tsivgoulis G, Spengos K, Zakopoulos N, Synetos A, Manios E et al (2004) U-shaped relationship between mortality and admission blood pressure in patients with acute stroke. J Intern Med 255:257–265. Available from: http://www.ncbi.nlm.nih.gov/pubmed/14746563. Accessed 2 Jan 2018

51. Rodriguez-Luna D, Piñeiro S, Rubiera M, Ribo M, Coscojuela P, Pagola J et al (2013) Impact of blood pressure changes and course on hematoma growth in acute intracerebral hemorrhage. Eur J Neurol 20:1277–1283. Available from: http://www.ncbi.nlm.nih.gov/pubmed/23647568. Accessed 2 Jan 2018

52. Boulouis G, Morotti A, Goldstein JN, Charidimou A (2017) Intensive blood pressure lowering in patients with acute intracerebral haemorrhage: clinical outcomes and haemorrhage expansion. Systematic review and meta-analysis of randomised trials. J Neurol Neurosurg Psychiatry 88:339–345. Available from: http://www.ncbi.nlm.nih.gov/pubmed/28214798. Accessed 2 Jan 2018

53. Flaherty ML, Tao H, Haverbusch M, Sekar P, Kleindorfer D, Kissela B et al (2008) Warfarin use leads to larger intracerebral hematomas. Neurology 71:1084–1089. Available from: http://www.ncbi.nlm.nih.gov/pubmed/18824672. Accessed 2 Jan 2018

54. Flibotte JJ, Hagan N, O'Donnell J, Greenberg SM, Rosand J (2004) Warfarin, hematoma expansion, and outcome of intracerebral hemorrhage. Neurology 63:1059–1064. Available from: http://www.ncbi.nlm.nih.

gov/pubmed/15452298. Accessed 2 Jan 2018

55. Kuramatsu JB, Gerner ST, Schellinger PD, Glahn J, Endres M, Sobesky J et al (2015) Anticoagulant reversal, blood pressure levels, and anticoagulant resumption in patients with anticoagulation-related intracerebral hemorrhage. JAMA 313:824. Available from: http://www.ncbi.nlm.nih.gov/pubmed/25710659. Accessed 2 Jan 2018

56. Hickey M, Gatien M, Taljaard M, Aujnarain A, Giulivi A, Perry JJ (2013) Outcomes of urgent warfarin reversal with frozen plasma versus prothrombin complex concentrate in the emergency department. Circulation 128:360–364. Available from: http://circ.ahajournals.org/cgi/doi/10.1161/CIRCULATIONAHA.113.001875. Accessed 2 Jan 2018

57. Connolly SJ, Milling TJ, Eikelboom JW, Gibson CM, Curnutte JT, Gold A et al (2016) Andexanet alfa for acute major bleeding associated with factor Xa inhibitors. N Engl J Med 375:1131–1141. Available from: http://www.nejm.org/doi/10.1056/NEJMoa1607887. Accessed 2 Jan 2018

58. Pollack CV, Reilly PA, Eikelboom J, Glund S, Verhamme P, Bernstein RA et al (2015) Idarucizumab for dabigatran reversal. N Engl J Med 373:511–520. Available from: http://www.ncbi.nlm.nih.gov/pubmed/26095746. Accessed 2 Jan 2018

59. Frontera JA, Lewin JJ III, Rabinstein AA, Aisiku IP, Alexandrov AW, Cook AM et al (2016) Guideline for reversal of antithrombotics in intracranial hemorrhage. Neurocrit Care 24:6–46. Available from: http://www.ncbi.nlm.nih.gov/pubmed/26714677. Accessed 2 Jan 2018

60. Investigators ACTIVE, Connolly SJ, Pogue J, Hart RG, Hohnloser SH, Pfeffer M et al (2009) Effect of clopidogrel added to aspirin in patients with atrial fibrillation. N Engl J Med 360:2066–2078. Available from: http://www.ncbi.nlm.nih.gov/pubmed/19336502. Accessed 2 Jan 2018

61. Roquer J, Vivanco Hidalgo RM, Ois A, Rodríguez Campello A, Cuadrado Godia E, Giralt Steinhauer E et al (2017) Antithrombotic pretreatment increases very-early mortality in primary intracerebral hemorrhage. Neurology 88:885–891. Available from: http://www.ncbi.nlm.nih.gov/pubmed/28148636. Accessed 2 Jan 2018

62. Thompson BB, Béjot Y, Caso V, Castillo J, Christensen H, Flaherty ML et al (2010) Prior antiplatelet therapy and outcome following intracerebral hemorrhage: a systematic

review. Neurology 75:1333–1342. Available from: http://www.neurology.org/cgi/doi/10.1212/WNL.0b013e3181f735e5. Accessed 2 Jan 2018

63. Sansing LH, Messe SR, Cucchiara BL, Cohen SN, Lyden PD, Kasner SE et al (2009) Prior antiplatelet use does not affect hemorrhage growth or outcome after ICH. Neurology 72:1397–1402. Available from: http://www.ncbi.nlm.nih.gov/pubmed/19129506. Accessed 2 Jan 2018

64. Baharoglu MI, Cordonnier C, Salman RA-S, de Gans K, Koopman MM, Brand A et al (2016) Platelet transfusion versus standard care after acute stroke due to spontaneous cerebral haemorrhage associated with antiplatelet therapy (PATCH): a randomised, open-label, phase 3 trial. Lancet 387:2605–2613. Available from: http://www.ncbi.nlm.nih.gov/pubmed/27178479. Accessed 2 Jan 2018

65. Mayer SA, Brun NC, Begtrup K, Broderick J, Davis S, Diringer MN et al (2008) Efficacy and safety of recombinant activated factor VII for acute intracerebral hemorrhage. N Engl J Med 358:2127–2137. Available from: http://www.ncbi.nlm.nih.gov/pubmed/18480205. Accessed 2 Jan 2018

66. Mayer SA, Brun NC, Begtrup K, Broderick J, Davis S, Diringer MN et al (2005) Recombinant activated factor VII for acute intracerebral hemorrhage. N Engl J Med 352:777–785. Available from: http://www.ncbi.nlm.nih.gov/pubmed/15728810. Accessed 2 Jan 2018

67. Kamel H, Hemphill JC (2012) Characteristics and sequelae of intracranial hypertension after intracerebral hemorrhage. Neurocrit Care 17:172–176. Available from: http://www.ncbi.nlm.nih.gov/pubmed/22833445. Accessed 2 Jan 2018

68. Qureshi AI, Geocadin RG, Suarez JI, Ulatowski JA (2000) Long-term outcome after medical reversal of transtentorial herniation in patients with supratentorial mass lesions. Crit Care Med 28:1556–1564. Available from: http://www.ncbi.nlm.nih.gov/pubmed/10834711. Accessed 2 Jan 2018

69. Poungvarin N, Bhoopat W, Viriyavejakul A, Rodprasert P, Buranasiri P, Sukondhabhant S et al (1987) Effects of dexamethasone in primary supratentorial intracerebral hemorrhage. N Engl J Med 316:1229–1233. Available from: http://www.ncbi.nlm.nih.gov/pubmed/3574383. Accessed 2 Jan 2018

70. Mayer SA, Chong JY (2002) Critical care management of increased intracranial pressure. J Intensive Care Med 17:55–67. Available from: http://journals.sagepub.com/doi/10.1177/088506660201700201. Accessed 2 Jan 2018

71. Holloway KL, Barnes T, Choi S, Bullock R, Marshall LF, Eisenberg HM et al (1996) Ventriculostomy infections: the effect of monitoring duration and catheter exchange in 584 patients. J Neurosurg 85:419–424. Available from: http://thejns.org/doi/10.3171/jns.1996.85.3.0419. Accessed 2 Jan 2018

72. Hemphill JC, Greenberg SM, Anderson CS, Becker K, Bendok BR, Cushman M et al (2015) Guidelines for the management of spontaneous intracerebral hemorrhage: a guideline for healthcare professionals from the American Heart Association/American Stroke Association. Stroke 46(7):2032–2060

73. Kim-Han JS, Kopp SJ, Dugan LL, Diringer MN (2006) Perihematomal mitochondrial dysfunction after intracerebral hemorrhage. Stroke 37:2457–2462. Available from: http://stroke.ahajournals.org/cgi/doi/10.1161/01.STR.0000240674.99945.4e. Accessed 2 Jan 2018

74. Zazulia AR, Diringer MN, Videen TO, Adams RE, Yundt K, Aiyagari V et al (2001) Hypoperfusion without ischemia surrounding acute intracerebral hemorrhage. J Cereb Blood Flow Metab 21:804–810. Available from: http://www.ncbi.nlm.nih.gov/pubmed/11435792. Accessed 2 Jan 2018

75. Mendelow AD, Gregson BA, Fernandes HM, Murray GD, Teasdale GM, Hope DT et al (2005) Early surgery versus initial conservative treatment in patients with spontaneous supratentorial intracerebral haematomas in the International Surgical Trial in Intracerebral Haemorrhage (STICH): a randomised trial. Lancet 365:387–397. Available from: http://www.ncbi.nlm.nih.gov/pubmed/15680453. Accessed 2 Jan 2018

76. Mendelow AD, Gregson BA, Rowan EN, Murray GD, Gholkar A, Mitchell PM et al (2013) Early surgery versus initial conservative treatment in patients with spontaneous supratentorial lobar intracerebral haematomas (STICH II): a randomised trial. Lancet 382:397–408. Available from: http://www.ncbi.nlm.nih.gov/pubmed/23726393. Accessed 2 Jan 2018

77. Prasad K, Mendelow AD, Gregson B (2008) Surgery for primary supratentorial intracerebral haemorrhage. In: Prasad K (ed) Cochrane database of systematic reviews. John Wiley & Sons, Ltd, Chichester, UK, p CD000200. Available from: http://www.ncbi.nlm.nih.gov/pubmed/18843607. Accessed 2 Jan 2018

78. Vespa PM, Martin N, Zuccarello M, Awad I, Hanley DF (2013) Surgical trials in intracerebral hemorrhage. Stroke 44:S79–S82. Available from: http://www.ncbi.nlm.nih.gov/pubmed/23709739. Accessed 2 Jan 2018

79. Fam MD, Hanley D, Stadnik A, Zeineddine HA, Girard R, Jesselson M et al (2017) Surgical performance in minimally invasive surgery plus recombinant tissue plasminogen activator for intracerebral hemorrhage evacuation phase III clinical trial. Neurosurgery 81:860–866. Available from: http://www.ncbi.nlm.nih.gov/pubmed/28402516. Accessed 2 Jan 2018

80. Hanley DF, Thompson RE, Muschelli J, Rosenblum M, McBee N, Lane K et al (2016) Safety and efficacy of minimally invasive surgery plus alteplase in intracerebral haemorrhage evacuation (MISTIE): a randomised, controlled, open-label, phase 2 trial. Lancet Neurol 15:1228–1237. Available from: http://linkinghub.elsevier.com/retrieve/pii/S1474442216302344. Accessed 2 Jan 2018

81. Vespa P, Hanley D, Betz J, Hoffer A, Engh J, Carter R et al (2016) ICES (intraoperative stereotactic computed tomography-guided endoscopic surgery) for brain hemorrhage: a multicenter randomized controlled trial. Stroke 47:2749–2755. Available from: http://stroke.ahajournals.org/lookup/doi/10.1161/STROKEAHA.116.013837. Accessed 2 Jan 2018

82. Delcourt C, Sato S, Zhang S, Sandset EC, Zheng D, Chen X et al (2017) Intracerebral hemorrhage location and outcome among INTERACT2 participants. Neurology 88:1408–1414. Available from: http://www.ncbi.nlm.nih.gov/pubmed/28235817. Accessed 2 Jan 2018

83. Kobayashi S, Sato A, Kageyama Y, Nakamura H, Watanabe Y, Yamaura A (1994) Treatment of hypertensive cerebellar hemorrhage—surgical or conservative management? Neurosurgery 34:246–250. Available from: http://www.ncbi.nlm.nih.gov/pubmed/8177384. Accessed 2 Jan 2018

84. Hanley DF (2009) Intraventricular hemorrhage: severity factor and treatment target in spontaneous intracerebral hemorrhage. Stroke 40:1533–1538. Available from: http://www.ncbi.nlm.nih.gov/pubmed/19246695. Accessed 2 Jan 2018

85. Maas MB, Nemeth AJ, Rosenberg NF, Kosteva AR, Prabhakaran S, Naidech AM (2013) Delayed intraventricular hemorrhage is common and worsens outcomes in intracerebral hemorrhage. Neurology 80:1295–1299. Available from: http://www.ncbi.nlm.nih.gov/pubmed/23516315. Accessed 2 Jan 2018

86. Tuhrim S, Horowitz DR, Sacher M, Godbold JH (1999) Volume of ventricular blood is an important determinant of outcome in supratentorial intracerebral hemorrhage. Crit Care Med 27:617–621. Available from: http://www.ncbi.nlm.nih.gov/pubmed/10199544. Accessed 2 Jan 2018

87. Passero S, Ulivelli M, Reale F (2002) Primary intraventricular haemorrhage in adults. Acta Neurol Scand 105:115–119. Available from: http://www.ncbi.nlm.nih.gov/pubmed/11903121. Accessed 2 Jan 2018

88. Naff NJ, Williams MA, Rigamonti D, Keyl PM, Hanley DF (2001) Blood clot resolution in human cerebrospinal fluid: evidence of first-order kinetics. Neurosurgery 49:614–619. Available from: http://www.ncbi.nlm.nih.gov/pubmed/11523671. Accessed 2 Jan 2018

89. Whitelaw A (1993) Endogenous fibrinolysis in neonatal cerebrospinal fluid. Eur J Pediatr 152:928–930. Available from: http://www.ncbi.nlm.nih.gov/pubmed/8276026. Accessed 2 Jan 2018

90. Awad IA, Hanley DF (2016) CLEAR III: efficiency of IVH removal determines mRS. In: International stroke conference. Available from: https://professional.heart.org/idc/groups/ahamah-public/@wcm/@sop/@scon/documents/downloadable/ucm_481659.pdf. Accessed 2 Jan 2018

91. Hanley DF, Lane K, McBee N, Ziai W, Tuhrim S, Lees KR et al (2017) Thrombolytic removal of intraventricular haemorrhage in treatment of severe stroke: results of the randomised, multicentre, multiregion, placebo-controlled CLEAR III trial. Lancet 389:603–611. Available from: http://www.ncbi.nlm.nih.gov/pubmed/28081952. Accessed 2 Jan 2018

92. Ziai WC, Tuhrim S, Lane K, McBee N, Lees K, Dawson J et al (2014) A multicenter, randomized, double-blinded, placebo-controlled phase III study of clot lysis evaluation of accelerated resolution of intraventricular hemorrhage (CLEAR III). Int J Stroke 9:536–542. Available from: http://www.ncbi.nlm.nih.gov/pubmed/24033910. Accessed 2 Jan 2018

93. Diringer MN, Edwards DF (2001) Admission to a neurologic/neurosurgical intensive care unit is associated with reduced mortality rate after intracerebral hemorrhage. Crit Care Med 29:635–640. Available from: http://www.ncbi.nlm.nih.gov/pubmed/11373434. Accessed 2 Jan 2018

356   James Lee and Igor Rybinnik

94. Jeong J-H, Bang J, Jeong W, Yum K, Chang J, Hong J-H et al (2017) A dedicated neurological intensive care unit offers improved outcomes for patients with brain and spine injuries. J Intensive Care Med:885066617706675. Available from: http://journals.sagepub.com/doi/10.1177/0885066617706675. Accessed 2 Jan 2018

95. Rincon F, Mayer SA, Rivolta J, Stillman J, Boden-Albala B, Elkind MSV et al (2010) Impact of delayed transfer of critically ill stroke patients from the emergency department to the neuro-ICU. Neurocrit Care 13:75–81. Available from: http://link.springer.com/10.1007/s12028-010-9347-0. Accessed 2 Jan 2018

96. Passero S, Ciacci G, Ulivelli M (2003) The influence of diabetes and hyperglycemia on clinical course after intracerebral hemorrhage. Neurology 61:1351–1356. Available from: http://www.ncbi.nlm.nih.gov/pubmed/14638954. Accessed 2 Jan 2018

97. Kimura K, Iguchi Y, Inoue T, Shibazaki K, Matsumoto N, Kobayashi K et al (2007) Hyperglycemia independently increases the risk of early death in acute spontaneous intracerebral hemorrhage. J Neurol Sci 255:90–94. Available from: http://www.ncbi.nlm.nih.gov/pubmed/17350046. Accessed 2 Jan 2018

98. Stead LG, Gilmore RM, Bellolio MF, Mishra S, Bhagra A, Vaidyanathan L et al (2009) Hyperglycemia as an independent predictor of worse outcome in non-diabetic patients presenting with acute ischemic stroke. Neurocrit Care 10:181–186. Available from: http://www.ncbi.nlm.nih.gov/pubmed/18357419. Accessed 2 Jan 2018

99. Honig A, Michael S, Eliahou R, Leker RR (2015) Central fever in patients with spontaneous intracerebral hemorrhage: predicting factors and impact on outcome. BMC Neurol 15:6. Available from: http://www.ncbi.nlm.nih.gov/pubmed/25648165. Accessed 2 Jan 2018

100. Schwarz S, Häfner K, Aschoff A, Schwab S (2000) Incidence and prognostic significance of fever following intracerebral hemorrhage. Neurology 54:354–361. Available from: http://www.ncbi.nlm.nih.gov/pubmed/10668696. Accessed 2 Jan 2018

101. Kollmar R, Staykov D, Dorfler A, Schellinger PD, Schwab S, Bardutzky J (2010) Hypothermia reduces perihemorrhagic edema after intracerebral hemorrhage. Stroke 41:1684–1689. Available from: http://www.ncbi.nlm.nih.gov/pubmed/20616317. Accessed 2 Jan 2018

102. Georgilis K, Plomaritoglou A, Dafni U, Bassiakos Y, Vemmos K (1999) Aetiology of fever in patients with acute stroke. J Intern Med 246:203–209. Available from: http://www.ncbi.nlm.nih.gov/pubmed/10447789. Accessed 2 Jan 2018

103. Claassen J, Jetté N, Chum F, Green R, Schmidt M, Choi H et al (2007) Electrographic seizures and periodic discharges after intracerebral hemorrhage. Neurology 69:1356–1365. Available from: http://www.neurology.org/cgi/doi/10.1212/01.wnl.0000281664.02615.6c. Accessed 2 Jan 2018

104. De Herdt V, Dumont F, Hénon H, Derambure P, Vonck K, Leys D et al (2011) Early seizures in intracerebral hemorrhage: incidence, associated factors, and outcome. Neurology 77:1794–1800. Available from: http://www.neurology.org/cgi/doi/10.1212/WNL.0b013e31823648a6. Accessed 2 Jan 2018

105. Passero S, Rocchi R, Rossi S, Ulivelli M, Vatti G (2002) Seizures after spontaneous supratentorial intracerebral hemorrhage. Epilepsia 43:1175–1180. Available from: http://www.ncbi.nlm.nih.gov/pubmed/12366733. Accessed 3 Jan 2018

106. Vespa PM, O'Phelan K, Shah M, Mirabelli J, Starkman S, Kidwell C et al (2003) Acute seizures after intracerebral hemorrhage: a factor in progressive midline shift and outcome. Neurology 60:1441–1446. Available from: http://www.ncbi.nlm.nih.gov/pubmed/12743228. Accessed 2 Jan 2018

107. Goldstein LB (1995) Common drugs may influence motor recovery after stroke. The Sygen In Acute Stroke Study Investigators. Neurology 45:865–871. Available from: http://www.ncbi.nlm.nih.gov/pubmed/7746398. Accessed 2 Jan 2018

108. Messé SR, Sansing LH, Cucchiara BL, Herman ST, Lyden PD, Kasner SE et al (2009) Prophylactic antiepileptic drug use is associated with poor outcome following ICH. Neurocrit Care 11:38–44. Available from: http://www.ncbi.nlm.nih.gov/pubmed/19319701. Accessed 2 Jan 2018

109. Gilad R, Boaz M, Dabby R, Sadeh M, Lampl Y (2011) Are post intracerebral hemorrhage seizures prevented by anti-epileptic treatment? Epilepsy Res 95:227–231. Available from: http://www.ncbi.nlm.nih.gov/pubmed/21632213. Accessed 2 Jan 2018

110. Hu X, Fang Y, Li H, Liu W, Lin S, Fu M et al (2014) Protocol for seizure prophylaxis following intracerebral hemorrhage study (SPICH): a randomized, double-blind, placebo-controlled trial of short-term sodium

valproate prophylaxis in patients with acute spontaneous supratentorial intracerebral hemorrhage. Int J Stroke 9:814–817. Available from: http://journals.sagepub.com/doi/10.1111/ijs.12187. Accessed 2 Jan 2018

111. Naidech AM, Beaumont J, Jahromi B, Prabhakaran S, Kho A, Holl JL (2017) Evolving use of seizure medications after intracerebral hemorrhage: a multicenter study. Neurology 88:52–56. Available from: http://www.neurology.org/lookup/doi/10.1212/WNL.0000000000003461. Accessed 2 Jan 2018

112. CLOTS (Clots in Legs Or sTockings after Stroke) Trials Collaboration, Dennis M, Sandercock P, Reid J, Graham C, Forbes J, et al (2013) Effectiveness of intermittent pneumatic compression in reduction of risk of deep vein thrombosis in patients who have had a stroke (CLOTS 3): a multicentre randomised controlled trial. Lancet 382:516–524, Available from: http://www.ncbi.nlm.nih.gov/pubmed/23727163. Accessed 2 Jan 2018

113. Boeer A, Voth E, Henze T, Prange HW (1991) Early heparin therapy in patients with spontaneous intracerebral haemorrhage. J Neurol Neurosurg Psychiatry 54:466–467. Available from: http://www.ncbi.nlm.nih.gov/pubmed/1865215. Accessed 2 Jan 2018

114. Paciaroni M, Agnelli G, Venti M, Alberti A, Acciarresi M, Caso V (2011) Efficacy and safety of anticoagulants in the prevention of venous thromboembolism in patients with acute cerebral hemorrhage: a meta-analysis of controlled studies. J Thromb Haemost 9:893–898. Available from: http://www.ncbi.nlm.nih.gov/pubmed/21324058. Accessed 2 Jan 2018

115. Murthy SB, Gupta A, Merkler AE, Navi BB, Mandava P, Iadecola C et al (2017) Restarting anticoagulant therapy after intracranial hemorrhage: a systematic review and meta-analysis. Stroke 48:1594–1600. Available from: http://stroke.ahajournals.org/lookup/doi/10.1161/STROKEAHA.116.016327. Accessed 2 Jan 2018

116. Charidimou A, Karayiannis C, Song T-J, Orken DN, Thijs V, Lemmens R et al (2017) Brain microbleeds, anticoagulation, and hemorrhage risk: meta-analysis in stroke patients with AF. Neurology 89:2317–2326. Available from: http://www.neurology.org/lookup/doi/10.1212/WNL.0000000000004704. Accessed 2 Jan 2018

117. Hill MD, Silver FL, Austin PC, Tu JV (2000) Rate of stroke recurrence in patients with primary intracerebral hemorrhage. Stroke 31:123–127. Available from: http://www.ncbi.nlm.nih.gov/pubmed/10625726. Accessed 2 Jan 2018

118. Chatterjee S, Sardar P, Biondi-Zoccai G, Kumbhani DJ (2013) New oral anticoagulants and the risk of intracranial hemorrhage. JAMA Neurol 70:1486–1490. Available from: http://www.ncbi.nlm.nih.gov/pubmed/24166666. Accessed 2 Jan 2018

119. Hagii J, Tomita H, Metoki N, Saito S, Shiroto H, Hitomi H et al (2014) Characteristics of intracerebral hemorrhage during rivaroxaban treatment. Stroke 45:2805–2807. Available from: http://www.ncbi.nlm.nih.gov/pubmed/25082810. Accessed 2 Jan 2018

120. Wilson D, Charidimou A, Shakeshaft C, Ambler G, White M, Cohen H et al (2016) Volume and functional outcome of intracerebral hemorrhage according to oral anticoagulant type. Neurology 86:360–366. Available from: http://www.ncbi.nlm.nih.gov/pubmed/26718576. Accessed 2 Jan 2018

121. Siegal DM, Curnutte JT, Connolly SJ, Lu G, Conley PB, Wiens BL et al (2015) Andexanet alfa for the reversal of factor Xa inhibitor activity. N Engl J Med 373:2413–2424. Available from: http://www.ncbi.nlm.nih.gov/pubmed/26559317. Accessed 2 Jan 2018

122. Poon MTC, Bell SM, Al-Shahi Salman R (2015) Epidemiology of Intracerebral Haemorrhage. Front Neurol Neurosci 37:1–12. Available from: http://www.ncbi.nlm.nih.gov/pubmed/26588164. Accessed 6 Oct 2017

123. Bejot Y, Cordonnier C, Durier J, Aboa-Eboule C, Rouaud O, Giroud M (2013) Intracerebral haemorrhage profiles are changing: results from the Dijon population-based study. Brain 136:658–664. Available from: http://www.ncbi.nlm.nih.gov/pubmed/23378220. Accessed 6 Oct 2017

124. Lovelock C, Molyneux A, Rothwell P, Oxford Vascular Study (2007) Change in incidence and aetiology of intracerebral haemorrhage in Oxfordshire, UK, between 1981 and 2006: a population-based study. Lancet Neurol 6:487–493. Available from: http://www.ncbi.nlm.nih.gov/pubmed/17509483. Accessed 6 Oct 2017

125. Keep RF, Hua Y, Xi G (2012) Intracerebral haemorrhage: mechanisms of injury and therapeutic targets. Lancet Neurol 11:720–731. Available from: http://www.ncbi.nlm.nih.gov/pubmed/22698888. Accessed 6 Oct 2017

Chapter 21

Central Nervous System Vasculitis: Primary Angiitis of the Central Nervous System

Aaron M. Gusdon, Mackenzie P. Lerario, Ehud Lavi, and Gary L. Bernardini

Abstract

Primary angiitis of the central nervous system (CNS) is a rare disorder involving idiopathic inflammation of small- and medium-sized leptomeningeal and parenchymal arteries within the CNS. The disease may cause headaches, multifocal strokes of either ischemic or hemorrhagic subtypes, cognitive impairment, seizures, and visual changes; however, there is a wide variation in symptomatology and clinical course of patients presenting with primary angiitis of the CNS. This chapter reviews the standard diagnostic evaluation for primary angiitis of the CNS, which includes serological testing for signs of inflammation and potential alternative diagnoses, vascular neuroimaging, lumbar puncture, diagnostic cerebral angiogram, and possible brain biopsy. Although the disease carries high mortality if left untreated, it is relatively responsive to corticosteroids as part of an acute treatment course. This chapter also reviews several immunosuppressive agents used for maintenance therapy.

Key words Stroke, Ischemic stroke, Hemorrhagic stroke, Vasculitis, Angiitis, Inflammation, Central nervous system, Cerebral angiography, Brain biopsy, Immunosuppressive medications

Abbreviations

ABRA	Aβ-related angiitis
ACE	Angiotensin-converting enzyme
ANCA	Antineutrophil cytoplasmic antibodies
CNS	Central nervous system
CRP	C-reactive protein
CSF	Cerebrospinal fluid
EBV	Epstein-Barr virus
ESR	Erythrocyte sedimentation rate
HHV-6	Human herpesvirus 6
HIV	Human immunodeficiency virus
HSV	Herpes simplex virus
MRA	Magnetic resonance angiography
MRI	Magnetic resonance imaging
mRS	Modified Rankin Scale
PACNS	Primary angiitis of the CNS

Fawaz Al-Mutti and Krishna Amuluru (eds.), *Cerebrovascular Disorders*, Neuromethods, vol. 170,
https://doi.org/10.1007/978-1-0716-1530-0_21, © Springer Science+Business Media, LLC, part of Springer Nature 2021

PRES   Posterior reversible encephalopathy syndrome
RCVS   Reversible cerebral vasoconstriction syndrome
SSA/B   Sjögren's syndrome-related antigen A/B
VDRL   Venereal disease research laboratory test
VZV   Varicella zoster virus

1   Introduction

Angiitis isolated to the central nervous system (CNS) is a rare and incompletely understood disease. Angiitis affecting the CNS may be primary, with no identifiable underlying infectious or inflammatory etiology, or it may be secondary to a number of other processes, several of which will be discussed in this chapter. Both primary and secondary CNS angiitis are characterized by inflammation of arteries in the brain, spinal cord, or both. This chapter will refer to idiopathic, arterial inflammation isolated to the CNS as primary angiitis of the CNS (PACNS) and will highlight the typical presentation, differential diagnosis, diagnostic challenges, and treatment regimens for the disease.

Due to rarity of the disease, limited data exist regarding the epidemiology of PACNS. Data from the Mayo Clinic suggest that incidence is 2.4 cases per 1000,000 person-years [1, 2]. The median age for diagnosis is 50 years, and gender distribution is similar between men and women [1, 2].

While vascular inflammation in secondary CNS angiitis can be due to a variety of different processes, inciting factor(s) driving PACNS remains poorly understood. Given known links with viral infections, an infectious etiology has been postulated for PACNS. Mycoplasma has been shown to be a cause of PACNS in animal models [3] with studies demonstrating features consistent with mycoplasma detected by electron microscopy after biopsy [3, 4], while others with biopsy material from PACNS patients have suggested an undiagnosed viral etiology [5].

2   Pathophysiology and Clinical Presentation

The clinical manifestations of PACNS are broad and variable in terms of acuity of onset, location of involvement, clinical symptomatology, and imaging characteristics. Furthermore, most patients develop multiple manifestations of disease either concurrently or over the course of their illness. The most common clinical findings are headache and cognitive dysfunction, with constitutional symptoms representing relatively rare findings in PACNS (Table 1) [1, 2]. Of note, headache as the initial manifestation of disease is very rarely described as thunderclap in nature [1]. Signs and

Table 1
Frequency of clinical manifestations in PACNS

Frequent
Headache (60%)
Cognitive impairment (encephalopathy, dementia, and personality changes) (55%)
Persistent neurological deficits or stroke syndromes (weakness or aphasia) (40%)
Transient ischemic attack (25%)
Common
Seizures (20%)
Ataxia (20%)
Visual field defect (20%)
Diplopia (15%)
Intracranial hemorrhage (10%)
Uncommon
Spinal involvement (paraparesis or quadriplegia) (5%)
Papilledema (5%)
Constitutional symptoms and fever (<10%)

symptoms of systemic vasculitis, including peripheral neuropathy or rash, are not typically present in PACNS. During the course of illness, any part of the central nervous system may be involved including the brain (both cortical and subcortical structures), spine, and optic nerves. Recurrent multiple strokes and transient ischemic attacks are common and may occur in up to half of patients [6]. Spinal involvement is relatively rare, with paraparesis or quadraparesis only affecting 5–7% of PACNS patients [1, 2]. As would be expected from variable location of involvement, clinical symptoms can be protean and nonspecific, and further confirmatory testing is typically needed to make the diagnosis.

The clinical onset can be acute, subacute, chronic, or recurrent with relapses and remissions. Most cases are characterized by insidious, subacute onset and long prodromal period, with relatively few patients presenting acutely [6]. At the worst end of clinical spectrum, PACNS can be rapidly progressive with often fatal outcomes [1, 7]. Biopsy specimens in patients with rapidly progressive disease usually demonstrate granulomatous or necrotizing histopathological patterns of angiitis. About one-quarter of patients with biopsy-positive PACNS have evidence of cerebral β-amyloid vascular deposition, a condition known as Aβ-related angiitis (ABRA) [1]. ABRA is more commonly associated with certain histopathological subtypes of PACNS as described below. The mean age of presentation for ABRA is 67 years, older than that seen with PACNS [8]. Patients with ABRA often have cognitive dysfunction, headaches, seizures, or focal neurological deficits at presentation. Additionally, hallucinations are present in over 10% of cases [8]. While patients with predominantly granulomatous or necrotizing histological subtypes often have rapidly progressive or fatal clinical courses, those with lymphocytic angiitis or ABRA tend to respond better to corticosteroids resulting in a more benign disease course [1].

The natural history of PACNS can vary, depending on several clinical, radiographic, and pathological characteristics. Without treatment, outcomes are generally poor. However, certain histological subtypes of disease, including ABRA in particular, can be highly responsive to immunosuppressive therapy. In the largest cohort ($n = 163$) of PACNS patients, the Mayo Clinic reported that mortality during treatment was 15% at median follow-up of 12 months [1]. The cause of death was due to stroke in nearly half of patients [1, 2]. For those patients who survive their initial disease presentation, severe disability (defined as a modified Rankin Scale [mRS] score of 4–5) on follow-up is rare [2, 9] with milder disability (mRS score of 1–3) being more common. However, many survivors experience persistent neurological deficits, cognitive impairments, seizures, speech disorders, reduced autonomy, and chronic headaches [9]. Characteristics associated with a high rate of disability on follow-up are advanced age, large vessel involvement, and presence of focal neurological deficits or cerebral infarction at presentation [1, 2]. Patients with poorer outcomes tended to have angiographic evidence of bilateral, multiple, large vessel lesions and imaging with multiple cerebral infarctions [1]. Alternatively, patients with negative angiography and involvement of small cortical and leptomeningeal vessels or prominent leptomeningeal enhancement on magnetic resonance imaging (MRI) tend to have more benign disease courses, which respond more favorably to treatment [1].

3   Diagnostic Approach

3.1   Differential Diagnosis

Given the rarity of PACNS and often variable and nonspecific presentation, broad differential diagnoses must be kept in mind (Table 2). Diseases to consider in the differential of PACNS include inflammatory vasculopathies caused by various neurotropic infections, autoimmune conditions, or malignancies, in which cases CNS angiitis would be considered secondary and not primary. Non-inflammatory vasculopathies caused by intracranial atherosclerosis, reversible cerebral vasoconstriction syndrome (RCVS), atypical cerebral embolism, or Moyamoya-like states may also be confused for diagnosis of PACNS.

Table 2
Differential diagnoses for PACNS

Infectious	Inflammatory/connective tissue disorders	Malignancy	RCVS
Varicella zoster virus (VZV)	Behçet's disease	Hodgkin's lymphoma	Intracranial atherosclerosis
Neurosyphilis	Granulomatosis with polyangiitis	Non-Hodgkin's lymphoma	Intravascular lymphoma
HIV	Microscopic polyangiitis		Embolic strokes
Hepatitis C	Eosinophilic granulomatosis with polyangiitis (Churg-Strauss)		Moyamoya disease
Cytomegalovirus	Henoch-Schönlein purpura		
Parvovirus B19	Kawasaki disease		
Lyme	Giant cell arteritis		
Tuberculosis	Takayasu's arteritis		
Mycoplasma	Sjögren's syndrome		
Bartonella	Rheumatoid arthritis		
Rickettsia	Neuropsychiatric lupus		
Aspergillosis			
Mucormycosis			
Coccidioidomycosis			
Candidiasis			
Cysticercosis			
Bacterial meningitis			

The most common non-inflammatory vasculopathy included in the differential of PACNS is RCVS. RCVS is defined by severe headaches and often other focal neurological deficits in combination with diffuse segmental constriction of cerebral arteries seen on cerebral angiography [10]. While both RCVS and PACNS share clinical presentation with headaches, the headaches in RCVS are commonly thunderclap in nature and short-lived (1–3 h) [10]. RCVS can be drug-induced, secondary to migrainous vasospasm, or the result of postpartum angiopathy; however, in up to one-half of cases, there will not be an identifiable trigger. While MRI of the brain can be normal in RCVS, subarachnoid hemorrhage may be present, and cerebral angiography classically shows constriction of cerebral arteries identical to that observed in PACNS. For a comparison between RCVS and PACNS, *see* Table 3. Other non-inflammatory vasculopathies such as

Table 3
Comparison of RCVS and PACNS

	PACNS	RCVS
Epidemiology		
Gender	No gender preference	More common in females
Age (median)	50 years	42 years
Pathophysiology	Granulomatous, lymphocytic, or necrotizing vessel wall inflammation	Abnormal regulation of cerebral arterial tone
Clinical features		
Headache	Insidious and progressive	Thunderclap
Course	Progressive, relapsing	Monophasic and self-limited
Diagnostic testing		
CSF	Leukocytosis and/or elevated protein	Normal or nearly normal
Angiography	Multifocal arterial stenoses, more often irregular; may be normal	Abnormal by definition
MRI	Abnormal in almost all cases	Normal in up to 70% of cases, can show ischemia, edema, SAH, or ICH
Treatment	Steroids, immunosuppressive, or cytotoxic agents	Nimodipine, supportive care

intracranial atherosclerosis, intravascular lymphoma, and radiation vasculopathy often enter into the differential diagnosis of PACNS. While many imaging findings are similar among these conditions, some important differences can be seen on vessel wall imaging (discussed separately in "*Imaging*"). Laboratory studies can also be helpful in narrowing this differential. For example, cerebrospinal fluid (CSF) with pleocytosis is rare in RCVS but present with PACNS, while elevated serum lactate dehydrogenase and CSF β2-microglobulin are often seen in intravascular lymphoma.

Secondary causes of CNS angiitis, including various infectious and inflammatory etiologies, should be thoroughly evaluated for before diagnosing PACNS. The most common infectious etiology of secondary CNS angiitis is varicella zoster virus (VZV), which can affect large to small cerebral arteries [11]. The CNS distribution of VZV vasculopathy is more commonly multifocal in immunocompromised patients. In syphilitic angiitis, spirochetes are thought to directly invade vascular endothelial cells, unlike in other infectious causes of PACNS, where inflammation secondary to infection is thought to cause angiitis [12]. Human immunodeficiency virus (HIV) has been associated with CNS angiitis either in the presence or absence of coexisting infection or lymphoproliferative disorders [13]. A variety of other infectious etiologies should be considered prior to diagnosing PACNS, including hepatitis C, cytomegalovirus, parvovirus B19, Lyme, tuberculosis, mycoplasma, bartonella, rickettsia, aspergillosis, mucormycosis, coccidioidomycosis, candidiasis, and cysticercosis. Bacterial meningitis has been documented to contribute to secondary CNS angiitis as well [14].

A variety of systemic inflammatory vasculitides and connective tissue disorders can also result in secondary CNS angiitis. Behçet's disease is characterized by systemic manifestations including oral and genital ulcers, erythema nodosum, uveitis, and arthritis [15]. About 5–14% of patients with Behçet's disease develop neurological involvement, which can involve either the parenchyma or meninges [16]. In secondary CNS angiitis due to Behçet's disease, inflammatory changes to cerebral vessels are secondary to perivascular infiltration of inflammatory cells rather than direct infiltration of vessel wall [17]. Behçet's can have a number of neurological manifestations including stroke, seizure, optic neuropathy, movement disorders, cerebral venous thromboses, as well as neuropsychiatric symptoms [15]. However, patients with PACNS typically do not display systemic signs seen in Behçet's.

Antineutrophil cytoplasmic antibodies (ANCA) are associated with granulomatosis with polyangiitis, microscopic polyangiitis, and eosinophilic granulomatosis with polyangiitis (i.e., Churg-Strauss associated with severe asthma). Collectively these conditions result in necrotizing inflammation of medium to small vessel walls systemically, which can also affect arteries in the CNS [18]. In addition to CNS vasculitis, peripheral and cranial nerves can be affected [19, 20]. Secondary vasculitides can occur less commonly from a number of other conditions including Henoch-Schönlein purpura, Kawasaki disease, and Takayasu's arteritis [21]. Although more commonly affecting the extracranial circulation, giant cell arteritis has also been shown to affect intracranial vessels with possible resultant posterior circulation infarctions [22, 23]. Recently, it has been suggested that VZV may play a role in development of giant cell arteritis [24]. However, specificity of antibodies used to detect VZV in these studies might have been relatively low raising concerns about the exact role of VZV in cases of giant cell arteritis [25].

Connective tissue disorders are rarely associated with the development of CNS angiitis. Angiitis is seen in fewer than 7% of patients with neuropsychiatric lupus [26, 27] and occurs even less commonly in patients with Sjögren's syndrome [28] or long-standing rheumatoid arthritis [29]. Similar to Behçet's disease, vasculitis in Sjögren's syndrome is thought to be related to perivascular inflammation [30].

Another important consideration in the differential diagnosis of PACNS is lymphoma. Secondary angiitis due to Hodgkin's or non-Hodgkin's lymphoma can be the initial manifestation of disease [31]. Rarely, an aggressive form of non-Hodgkin's B cell lymphoma can selectively proliferate in the lumen of blood vessels (i.e., intravascular lymphoma), and its angiographic appearance can mimic PACNS [32, 33]. The diagnosis can be particularly challenging, as even appearance on vessel wall imaging can be similar to CNS angiitis [34]. However, the pathological appearance is distinct with presence of neoplastic lymphocytes and is often characterized by diffuse CD20 positivity [32].

Embolic strokes secondary to bacterial endocarditis can mimic PACNS on imaging making it important to perform echocardiography when there is high suspicion for endocarditis [35].

3.2 Management Strategies

3.2.1 Diagnostic Approach

Diagnostic criteria have been proposed for PACNS [2, 21, 36]. The diagnosis of PACNS is generally accepted when a patient has neurological deficit of unknown origin, cerebral angiogram demonstrates characteristic findings, and brain biopsy is consistent with PACNS in combination with no evidence of systemic vasculitis after complete evaluation. Definitive diagnosis is typically reserved for those with positive biopsy, while a probable diagnosis can be made in the context of characteristic imaging and CSF findings (discussed below) [37]. The diagnostic tests useful for PACNS are discussed below and summarized in Table 4.

3.2.2 Lumbar Puncture and Laboratory Analysis

CSF analysis is usually one of the initial tests obtained when pursuing diagnosis of PACNS and can yield important diagnostic information. The CSF typically shows lymphocyte-predominant pleocytosis along with elevated protein but normal glucose [6, 21]. One study showed that out of 75 patients with PACNS, 66 patients (88%) had at least one abnormal CSF finding [2]. CSF cultures and viral serologies are also helpful in assessing for infectious etiologies of secondary CNS vasculitis. As well, CSF cytology and flow cytometry may be performed to evaluate for presence of lymphoma.

Specialized serum laboratory studies are recommended to assess for systemic vasculitic processes leading to secondary CNS angiitis. All patients presenting with signs and symptoms consistent with potential diagnosis of PACNS should undergo testing for acute phase reactants, including erythrocyte sedimentation rate (ESR) and C-reactive protein (CRP). ESR can be particularly helpful in discriminating secondary from primary CNS angiitis because ESR is often elevated with systemic vasculitis but can be normal in over 90% of cases of PACNS [2]. However, elevated ESR with PACNS may be useful adjunct to follow responsiveness to therapy. Other helpful studies include serum viral polymerase chain reaction testing for VZV, herpes simplex viruses 1 and 2, Epstein-Barr virus, human herpesvirus 6, hepatitis serologies, Lyme antibodies, and HIV. Systemic presence of rheumatological disorders should be assessed by sending serum studies for antinuclear antibodies, antiphospholipid antibodies, rheumatoid factor, cryoglobulins, anti-

Table 4
Diagnostic workup for PACNS

Serum	CSF	Imaging	Pathology
ESR/CRP	Cell count	CT angiography	Brain biopsy including dura, leptomeninges, cortex, and white matter
VZV PCR	Protein	MR angiography (MRA)	Staining with amyloid beta, CD3, CD20, CD68
HSV-1/ 2 PCR	Glucose	MRA with vessel wall imaging protocol	
EBV PCR	Cytology	Cerebral angiography	
HHV-6 PCR	Flow cytometry		
RPR	Bacterial cultures		
Hepatitis serologies	Viral serologies/ PCR		
Lyme antibodies	VDRL		
HIV	ACE		
ANA antibodies			
Antiphospholipid antibodies			
Rheumatoid factor			
SSA/B			
ANCA			
ACE			

Sjögren's syndrome-related antigen A/B (SSA/B) antibodies, and ANCA.

3.2.3 Imaging

Several imaging techniques are useful in the diagnosis of CNS angiitis. Cerebral angiography is the gold standard imaging technique for diagnosis of CNS angiitis [35, 38]. Typical findings include segmental stenosis with alternating normal segments of the artery (Fig. 1a). Involvement is typically bilateral and can involve large arteries as well as medium and smaller branches. Blood flow tends to be slower in narrowed vessels; however, non-affected vessels may demonstrate compensatory dilation suggesting hyper-vascularity and rapid arteriovenous shunting [39]. In case series from Yale and the Mayo Clinic, angiography was only

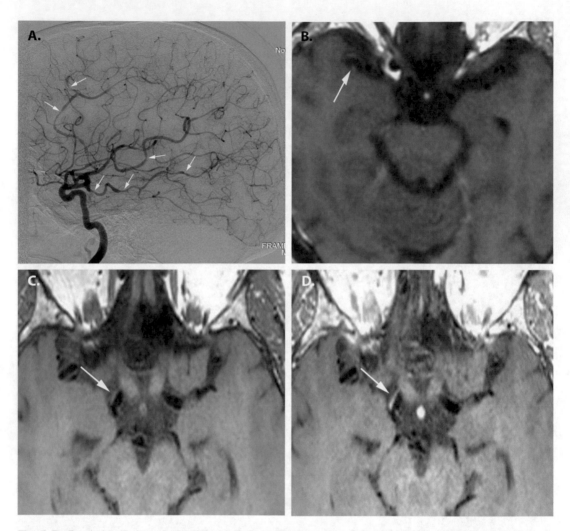

Fig. 1 Radiographic findings in PACNS. (**a**) Catheter angiography of the right internal carotid artery demonstrating multiple sites of arterial stenoses (arrows). (**b**) Post-gadolinium MRI vessel wall imaging of PACNS demonstrating thickening and concentric enhancement (arrow). Vessel wall both pre- (**c**) and post-gadolinium (**d**) showing eccentric vessel wall thickening and diffuse enhancement due to intracranial atherosclerosis (arrows)

positive in 27–43% of patients with histologically confirmed PACNS [1, 2, 40]. This may be due to inflammation and constriction in small arteries below the detection threshold for angiography [41]. An additional pitfall of angiography may be its low specificity, with similar findings being seen in RCVS, diffuse atherosclerosis, infections, and small embolic vascular occlusions [21]. Indeed, in one case series, with both angiography and biopsy performed, 49% had positive angiography findings without positive biopsy with only 17% having positive angiography and biopsy [1]. However, this may also have been due to biopsy being non-diagnostic.

Table 5
Spectrum of MRI findings in PACNS

Normal (rare)
Progressive confluent white matter lesions
Cortical and subcortical MRI T2-weighted lesions
Multiple diffusion-positive lesions
Single or multiple intraparenchymal hematomas
Multiple microhemorrhages
Single or multiple small enhancing lesions
Single and multiple large enhancing mass lesions
Enhancing vessel walls or perivascular spaces
Leptomeningeal enhancement

MRI and magnetic resonance angiography (MRA) can be diagnostically useful and are obtained as part of the workup for PACNS. Similar to conventional angiography, a pattern of focal arterial stenoses of large- and medium-sized arteries may be demonstrated with time-of-flight MRA.

Newer diagnostic technique with MRI using multicontrast high-resolution vessel wall imaging (VWI) may offer additional diagnostic accuracy over cerebral angiography in being able to differentiate PACNS from RCVS and intracranial atherosclerosis [42]. For example, in PACNS, circumferential vessel wall thickening is seen with smooth, homogenous isointense signal on MRI T2-weighted image, with post-contrast T1-weighted image demonstrating homogenous, circumferential vessel wall enhancement. However, with RCVS, there may be diffuse and uniform wall thickening, while post-contrast T1-weighted images show minimal circumferential wall thickening with minimal to no vessel wall enhancement (Fig. 1b) [43]. Other conditions such as arterial dissection, which demonstrates a dissection flap, and atherosclerosis, with more eccentric wall thickening, can also be distinguished from PACNS (Fig. 1c, d) [44]. Similar to PACNS, Moyamoya disease can demonstrate concentric vessel wall enhancement. However, in Moyamoya disease, there is usually minimal or no wall thickening, and it is usually characterized by more focal stenosis of distal internal carotid and middle cerebral arteries, collapsed vessel lumens, and well-developed collaterals [44, 45].

Brain MRI is abnormal in most cases of PACNS. Impaired blood flow secondary to vascular stenosis can result in infarcts, the most common finding, which are typically multiple and bilateral involving both cortical and subcortical territories [46, 47]. Other findings that can be seen include intracranial and subarachnoid hemorrhage, increased FLAIR signal, confluent white matter disease [48], and leptomeningeal enhancement (Table 5) [49, 50]. However, only 8% of patients present with intracerebral hemorrhage and 3% with subarachnoid hemorrhage [1]. Interestingly, leptomeningeal enhancement can be associated with improved prognosis and is more likely found in patients with normal vessel imaging [50]. The constellation of abnormal findings on brain MRI seen with PACNS can have high sensitivity but may be nonspecific.

3.2.4 Brain Biopsy

The treatment of intracranial angiitis often requires the use of powerful immunosuppressant therapy; therefore, diagnostic brain biopsy is often recommended to attain more clinical certainty. Given the overall low specificity of findings of CSF and vessel imaging [41], brain biopsy is often a critical component of diagnostic criteria.

For biopsy to be diagnostic, it should include dura, leptomeninges, cortex, and white matter. Ideally, biopsy should be targeted to radiographically abnormal area. Miller and colleagues showed that out of 46 patients with PACNS, 17 (35%) were non-diagnostic; furthermore, all non-targeted biopsies were non-diagnostic [51].

In addition to confirming diagnosis of PACNS, biopsy can yield valuable information regarding alternative diagnoses. One recent study found that out of 79 patients with suspected PACNS, only 9 (11%) had biopsy evidence of angiitis [52]. Importantly, biopsy results for 24 patients (30%) in this series demonstrated evidence of alternative diagnoses, such as cerebral amyloid angiopathy, viral meningoencephalitis, CNS lymphoma, posterior reversible encephalopathy syndrome (PRES), and progressive multifocal leukoencephalopathy [52].

3.2.5 Histopathology

There are three main histological categories of PACNS: granulomatous, lymphocytic, and necrotizing angiitis [51, 53]. The granulomatous category is most common, accounting for close to 60% of all cases of PACNS. Histologically, granulomatous angiitis appears as primarily transmural mononuclear infiltrate, with evidence of granulomas and multinucleated cells (Fig. 2a, b). Lymphocytic angiitis displays vessel wall thickening and destruction of vessel lumen but with predominantly lymphocytic infiltrate and without granulomatous changes (Fig. 2c–e). Necrotizing PACNS is least common and is characterized by transmural fibrinoid necrosis and is more commonly associated with underlying hemorrhage in brain parenchyma. Cerebral amyloid angiopathy, in condition known as ABRA, is most commonly associated with granulomatous PACNS [54], with up to 50% of cases showing β4-amyloid deposition [21]. In ABRA, deposition of βA4 amyloid in vessel can trigger

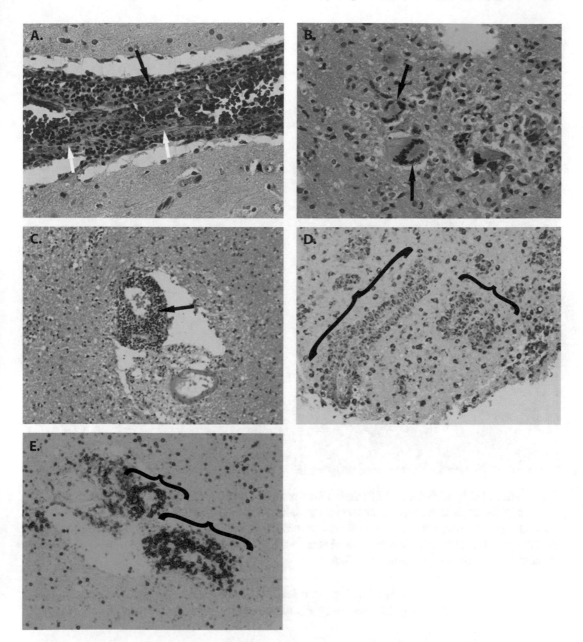

Fig. 2 Histopathology of PACNS. (**a**, **b**) H&E staining from brain biopsy of patient with granulomatous PACNS. In (**a**), mononuclear cells can be seen infiltrating the vessel wall (black arrow) along with areas of granulomatous inflammation (open arrows). In (**b**), prominent multinucleated giant cells (arrows) can be seen amidst granulomatous inflammation. (**c**) H&E staining from the brain biopsy of patient with lymphocytic PACNS. Vessel wall thickening along with a lymphocytic infiltrate can be seen (arrow). (**d**) CD68 staining from biopsy of patient with PACNS demonstrating infiltration by macrophages (brackets). (**e**) CD3 staining from biopsy of patient with PACNS demonstrating infiltration by T lymphocytes (brackets)

varying amounts of inflammation, including perivascular and vasculitic inflammation. As can be seen in Fig. 3, vascular accumulation

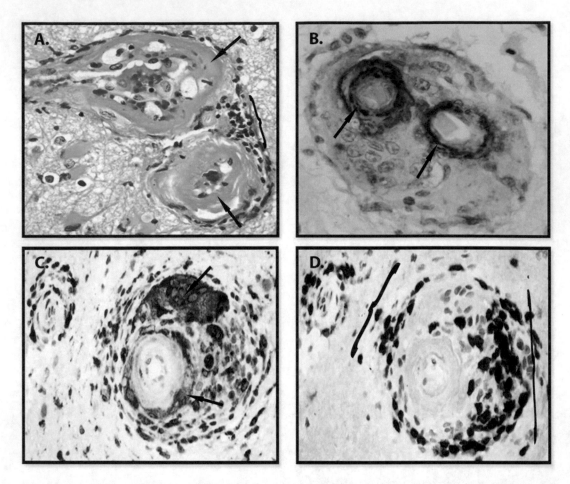

Fig. 3 Histopathology of ABRA. (**a**) H&E staining from brain biopsy of patient with ABRA. Vessel wall thickening by eosinophilic material (arrows) can be seen along with granulomatous infiltration (bracket). (**b**) Beta-amyloid staining demonstrating amyloid in vessel wall (arrows) of patient with ABRA. (**c**) CD68 staining showing macrophagic infiltration in vessel wall (arrows) of patient with ABRA. (**d**) CD3 staining showing T cell infiltration in vessel wall (brackets) of patient with ABRA

of Aβ (Fig. 3b) can lead to an inflammatory infiltrate characterized by the presence of both macrophages (Fig. 3c) and T cells (Fig. 3d).

4 Treatment Approach and Outcomes

PACNS is a heterogeneous condition with considerable variation in clinical presentation, outcomes, and response to treatment. Although PACNS may result in high mortality if diagnosed late, patients can be relatively responsive to immunosuppressive therapy. The majority of patients in modern case series receive treatment with corticosteroids alone or a combination of corticosteroids and other immunosuppressives such as cyclophosphamide, azathioprine, mycophenolate mofetil, or rituximab [1, 9]. Of these corticosteroid-sparing agents, cyclophosphamide has been used most frequently. Patients more likely to be treated with immuno-suppressive agents, in addition to corticosteroids, often display higher disability and stroke burden at presentation, consistent with more aggressive disease course. Unfortunately, given the rarity of PACNS, there are no randomized clinical trials comparing effectiveness of these treatment options. Therefore, treatment is guided by expert consensus and data from relatively small case series [1, 9]. Patients with secondary CNS angiitis caused by infectious and autoimmune conditions, or malignancy, should have specific treatment aimed at the underlying source of angiitis, which may include antimicrobials or chemotherapeutic agents.

Based on the available evidence, we propose the following guidelines for treatment of PACNS. Figure 4 also shows a summary of the recommended treatment algorithm.

- During the acute phase of illness, intravenous methylprednisolone (1 g daily) is administered over 3–5 days.

- Following the steroid pulse above, oral prednisone (1 mg/kg) is administered for maintenance therapy. There is some variability in the duration and total dose of steroids, with patients in the Mayo Clinic cohort receiving an average of 60 mg prednisone per day for a median of 9 months [1].

- A second agent can be added in addition to prednisone. In the Mayo Clinic cohort, approximately half of patients received corticosteroid treatment alone, while the other half received an additional agent [1]. Additional agents include the following:
  - Cyclophosphamide: used in 87% of patients in the Mayo Clinic cohort and can be given either orally or via intermittent pulses [1]. The average starting dose of intravenous cyclophosphamide was a monthly pulse dose of 1000 mg per month, while the average starting dose of oral cyclophosphamide was 150 mg per day for a median duration of 7 months of treatment [1].
  - Azathioprine: can be used at a starting dose of 100 mg per day as an alternative to cyclophosphamide [1].

Notable side effects for immunosuppressive agents used to combat PACNS are listed in Table 6. It is important to monitor patients intermittently for side effects. If present, risk and benefit discussion should be entertained to decide whether to continue or change therapy based on expected severity of potential adverse events. Clinicians should routinely prescribe prophylaxis against *Pneumocystis jirovecii* pneumonia in patients on high-dose or prolonged immunosuppressive therapy. Careful assessment for and preventative efforts against glucocorticoid-induced bone loss

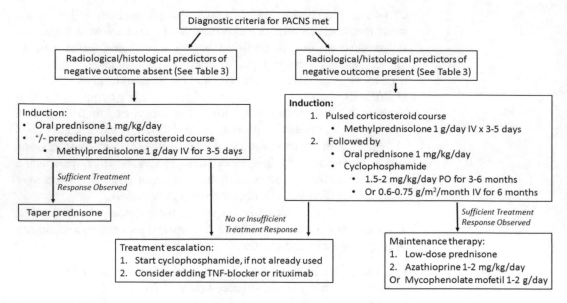

Fig. 4 Recommended treatment algorithm for PACNS

should be instituted in patients on maintenance therapy with oral prednisone.

Treatment response can be followed with periodic evaluation of patient's symptoms, neurological examination, and neuroimaging with dedicated MRI. Clinical decline, as demonstrated by worsening or new symptoms, development of focal neurological deficit, and progression of brain lesions are concerning for relapse. Depending on the nature and severity of abnormality, the decision to change or escalate treatment is reasonable in such a scenario. Treatment with corticosteroids followed by maintenance immunosuppressive medications has an 80–85% rate of achieving remission, although 25% of patients will eventually have relapse which will result in change or increase in therapy [1, 9]. Over 30% of relapsed patients will experience two or more relapses [1]. Nevertheless, more than 50% of patients should expect sustained therapeutic responses without relapse [1, 9].

Survival rates in recent series of patients on treatment are 85% at 1 year and 65% at 10 years [1]. No differences in survival have been reported based on whether the patient was treated with prednisone alone or in combination with other immunosuppressive agents. However, in a French cohort of patients who were treated with corticosteroids for a longer duration (median 23 months), there appeared to be lower mortality rates (6% at 3 years of follow-up) when compared to the Mayo Clinic cohort. Given the small sample sizes, indirect comparisons, and baseline differences in these patient cohorts, it is unclear if PACNS patients would benefit from extended treatment courses once remission has been obtained. As mentioned previously, certain histological subtypes

Table 6
Notable side effects of immunosuppressive agents used in the treatment of PACNS

	Corticosteroids	Cyclophosphamide	Azathioprine	Mycophenolate mofetil	Rituximab
Serious	• Adrenal suppression, crisis • Secondary infections • Psychosis • Myopathy • Cataracts, glaucoma • Osteonecrosis, tendon rupture • Gastrointestinal ulcers, perforation • Pseudotumor cerebri	• Myelosuppression • Secondary infections • Amenorrhea, infertility • Bladder cancer, hemorrhagic cystitis, bladder fibrosis • Pulmonary fibrosis, pneumonitis • Heart failure, arrhythmias	• Myelosuppression • Secondary infections • Secondary malignancies • Hepatotoxicity • Pancreatitis • Pulmonary edema	• Myelosuppression • Secondary infections • Secondary malignancies • Gastrointestinal ulcers, perforation, and colitis • Acute kidney injury • Interstitial lung disease	• Myelosuppression • Secondary infections • Stevens-Johnson syndrome, toxic epidermal necrolysis • Hypogammaglobulinemia • Hepatitis, other hepatotoxicity • Cardiopulmonary toxicity • Gastrointestinal obstruction, perforation
Common	• Mood change, including depression, anxiety • Impaired wound healing • Cushingoid appearance • Weight gain, central obesity, peripheral edema • Hypertension • Hyperglycemia • Thinning of the skin, acne, hirsutism • Dyspepsia	• Nausea, vomiting, anorexia • Alopecia • Stomatitis • Fatigue • Skin and nail changes • Abdominal pain, diarrhea • Rash	• Nausea, vomiting, anorexia • Fevers, chills • Alopecia • Rash • Abdominal pain, diarrhea	• Nausea, vomiting • Fever • Headache • Abdominal pain, diarrhea, constipation, dyspepsia • Hypertension • Cough, dyspnea • Peripheral edema	• Infusion reaction • Fever, rigors, night sweats, fatigue • Arthralgia, myalgia • Headache • Dizziness • Peripheral edema • Pruritis, rash, urticaria • Abdominal pain • Cough, rhinorrhea, epistaxis

of PACNS, including lymphocytic angiitis and ABRA, and patients with inflammation predominantly confined to small cortical and leptomeningeal vessels, tend to have more benign courses and better response to treatment. In the Mayo Clinic cohort of 58 biopsy-confirmed cases of PACNS, severe disability or death (mRS score of 4–6) was observed in 0/13 patients with lymphocytic angiitis, 7/35 (20%) with granulomatous angiitis, and 3/10 (30%) with necrotizing angiitis [1]. There were no reported differences regarding rates of relapse in patients with biopsy-proven angiitis compared to those who were diagnosed by cerebral angiogram without confirmatory pathology.

5 Conclusions

In this chapter, we summarize and outline common clinical presentations, differential diagnoses, diagnostic techniques, and treatment options for PACNS. Given the rarity of PACNS and diversity of processes leading to secondary CNS angiitis, thorough workup should include use of available imaging modalities, CSF analysis, serum testing for systemic inflammatory and infectious diseases, and brain biopsy when required. Given the paucity of cases of PACNS, experience with disease course and treatment is guided by relatively small cohorts which have only recently been published. Larger national and international cohort studies are needed to provide more complete evidence regarding treatment efficacy and outcomes.

References

1. Salvarani C, Brown RD, Christianson T et al (2015) An update of the Mayo Clinic cohort of patients with adult primary central nervous system vasculitis: description of 163 patients. Medicine (Baltimore) 94:e738. https://doi.org/10.1097/MD.0000000000000738

2. Salvarani C, Brown RD, Calamia KT et al (2007) Primary central nervous system vasculitis: analysis of 101 patients. Ann Neurol 62:442–451. https://doi.org/10.1002/ana.21226

3. Arthur G, Margolis G (1977) Mycoplasma-like structures in granulomatous angiitis of the central nervous system. Case reports with light and electron microscopic studies. Arch Pathol Lab Med 101:382–387

4. Thomas L, Davidson M, McClusky RT (1966) Studies of PPLO infection: the production of cerebral polyarteritis by mycoplasma gallisepticum in turkeys, the neurotoxic properties of the mycoplasma. J Exp Med 123:897–912

5. Reyes MG, Fresco R, Chokroverty S, Salud EQ (1976) Viruslike particles in granulomatous angiitis of the central nervous system. Neurology 26:797–799

6. Hajj-Ali RA, Calabrese LH (2014) Diagnosis and classification of central nervous system vasculitis. J Autoimmun 48–49:149–152. https://doi.org/10.1016/j.jaut.2014.01.007

7. Salvarani C, Brown RD, Calamia KT et al (2011) Rapidly progressive primary central nervous system vasculitis. Rheumatology (Oxford) 50:349–358. https://doi.org/10.1093/rheumatology/keq303

8. Scolding NJ, Joseph F, Kirby PA et al (2005) Abeta-related angiitis: primary angiitis of the central nervous system associated with cerebral amyloid angiopathy. Brain 128:500–515. https://doi.org/10.1093/brain/awh379

9. De Boysson H, Zuber M, Naggara O et al (2014) Primary angiitis of the central nervous system: description of the first fifty-two adults

enrolled in the french cohort of patients with primary vasculitis of the central nervous system. Arthritis Rheumatol 66:1315–1326. https://doi.org/10.1002/art.38340

10. Ducros A (2012) Reversible cerebral vasoconstriction syndrome. Lancet Neurol 11:906–917. https://doi.org/10.1016/S1474-4422(12)70135-7

11. Nagel MA, Cohrs RJ, Mahalingham R et al (2018) The varicella zoster vasculopathies: clinical, Csf, imaging and virologic features. Neurology 70:853–860. https://doi.org/10.1212/01.wnl.0000304747.38502.e8.The

12. Gaa J, Weidauer S, Sitzer M et al (2004) Cerebral vasculitis due to Treponema pallidum infection: MRI and MRA findings [2]. Eur Radiol 14:746–747. https://doi.org/10.1007/s00330-003-2015-4

13. Melica G, Brugieres P, Lascaux A-S et al (2009) Primary vasculitis of the central nervous system in patients infected with HIV-1 in the HAART era. J Med Virol 81:578–581. https://doi.org/10.1002/jmv

14. Katchanov J, Siebert E, Klingebiel R, Endres M (2010) Infectious vasculopathy of intracranial large- and medium-sized vessels in neurological intensive care unit: a clinico-radiological study. Neurocrit Care 12:369–374. https://doi.org/10.1007/s12028-010-9335-4

15. Al-Araji A, Kidd DP (2009) Neuro-Behçet's disease: epidemiology, clinical characteristics, and management. Lancet Neurol 8:192–204. https://doi.org/10.1016/S1474-4422(09)70015-8

16. Al-Araji A, Sharquie K, Al-Rawi Z (2003) Prevalence and patterns of neurological involvement in Behcet's disease: a prospective study from Iraq. J Neurol Neurosurg Psychiatry 74:608–613. https://doi.org/10.1136/jnnp.74.5.608

17. Hirohata S (2008) Histopathology of central nervous system lesions in Behcet's disease. J Neurol Sci 267:41–47. https://doi.org/10.1016/j.jns.2007.09.041

18. Wludarczyk A, Szczeklik W (2016) Neurological manifestations in ANCA-associated vasculitis – assessment and treatment. Expert Rev Neurother 7175:861–863. https://doi.org/10.1586/14737175.2016.1165095

19. Suppiah R, Hadden RDM, Batra R et al (2011) Peripheral neuropathy in ANCA-associated vasculitis: outcomes from the European Vasculitis Study Group trials. Rheumatology 50:2214–2222. https://doi.org/10.1093/rheumatology/ker266

20. Gwathmey KG, Burns TM, Collins MP, Dyck PJB (2014) Vasculitic neuropathies. Lancet

Neurol 13:67–82. https://doi.org/10.1016/S1474-4422(13)70236-9

21. Salvarani C, Brown RD, Hunder GG (2012) Adult primary central nervous system vasculitis. Lancet 380:767–777. https://doi.org/10.1016/S0140-6736(12)60069-5

22. Salvarani C, Giannini C, Miller DV, Hunder G (2006) Giant cell arteritis: involvement of intracranial arteries. Arthritis Care Res 55:985–989. https://doi.org/10.1002/art.22359

23. McCormick HM, Neubuerger KT (1958) Giant-cell arteritis involving small meningeal and intracerebral vessels. J Neuropathol Exp Neurol 17:471–478

24. Gilden D, White T, Khmeleva N et al (2015) Prevalence and distribution of VZV in temporal arteries of patients with giant cell arteritis. Neurology:1–8

25. Pisapia DJ, Lavi E (2016) VZV, temporal arteritis, and clinical practice: false positive immunohistochemical detection due to antibody cross-reactivity. Exp Mol Pathol 100:114–115. https://doi.org/10.1016/j.yexmp.2015.12.007

26. Everett CM, Graves TD, Lad S et al (2008) Aggressive CNS lupus vasculitis in the absence of systemic disease activity [5]. Rheumatology 47:107–109. https://doi.org/10.1093/rheumatology/kem264

27. Ramos-Casals M, Nardi N, Lagrutta M et al (2006) Vasculitis in systemic lupus erythematosus. Lupus 85:95–104. https://doi.org/10.1097/01.md.0000216817.35937.70

28. Ramos-Casals M, Solans R, Rosas J et al (2008) Primary Sjögren syndrome in Spain. Medicine (Baltimore) 87:210–219. https://doi.org/10.1097/MD.0b013e318181e6af

29. Caballol Pons N, Montalà N, Valverde J et al (2010) Isolated cerebral vasculitis associated with rheumatoid arthritis. Jt Bone Spine 77:361–363. https://doi.org/10.1016/j.jbspin.2010.02.030

30. Niemelä RK, Hakala M (1999) Primary Sjögren's syndrome with severe central nervous system disease. Semin Arthritis Rheum 29:4–13

31. Ma WL, Li CC, Yu SC, Tien HF (2014) Adult T-cell lymphoma/leukemia presenting as isolated central nervous system T-cell lymphoma. Case Rep Hematol 2014:917369. https://doi.org/10.1155/2014/917369

32. Ponzoni M, Ferreri AJM, Campo E et al (2007) Definition, diagnosis, and management of intravascular large B-cell lymphoma: proposals and perspectives from an international consensus meeting. J Clin Oncol 25:3168–3173.

https://doi.org/10.1200/JCO.2006.08.2313

33. Song DK, Boulis NM, McKeever PE, Quint DJ (2002) Angiotropic large cell lymphoma with imaging characteristics of CNS vasculitis. Am J Neuroradiol 23:239–242

34. Schaafsma JD, Hui F, Wisco D et al (2016) High-resolution vessel wall MRI: appearance of intravascular lymphoma mimics central nervous system vasculitis. Clin Neuroradiol:1–4. https://doi.org/10.1007/s00062-016-0529-9

35. Berlit P (2009) Isolated angiitis of the CNS and bacterial endocarditis: similarities and differences. J Neurol 256:792–795. https://doi.org/10.1007/s00415-009-5018-5

36. Calabrese LH, Mallek JA (1988) Primary angiitis of the central nervous system. Report of 8 new cases, review of the literature, and proposal for diagnostic criteria. Medicine (Baltimore) 67:20–39

37. Birnbaum J, Hellmann DB (2009) Primary angiitis of the central nervous system. Arch Neurol 66:704–709

38. Alhalabi M, Moore PM (1994) Serial angiography in isolated angiitis of the central nervous system. Neurology 44:1221–1226. https://doi.org/10.1212/WNL.44.7.1221

39. Ferris EJ, Levine HL (1973) Cerebral arteritis: classification. Radiology 109:327–341. https://doi.org/10.1148/109.2.327

40. Vollmer TL, Guarnaccia J, Harrington W et al (1993) Idiopathic granulomatous angiitis of the central nervous system. Diagnostic challenges. Arch Neurol 50:925–930. https://doi.org/10.1001/archneur.1993.00540090032007

41. Kadkhodayan Y, Alreshaid A, Moran CJ et al (2004) Primary angiitis of the central nervous system at conventional angiography. Radiology 233:878–882. https://doi.org/10.1148/radiol.2333031621

42. Mossa-Basha M, Hwang WD, De Havenon A et al (2015) Multicontrast high-resolution vessel wall magnetic resonance imaging and its value in differentiating intracranial vasculopathic processes. Stroke 46:1567–1573. https://doi.org/10.1161/STROKEAHA.115.009037

43. Obusez EC, Hui F, Hajj-Ali RA et al (2014) High-resolution MRI vessel wall imaging: spatial and temporal patterns of reversible cerebral vasoconstriction syndrome and central nervous system vasculitis. AJNR Am J Neuroradiol 35:1527–1532. https://doi.org/10.3174/ajnr.A3909

44. Ahn S-H, Lee J, Kim Y-J et al (2015) Isolated MCA disease in patients without significant atherosclerotic risk factors: a high-resolution magnetic resonance imaging study. Stroke 46:697–703. https://doi.org/10.1161/STROKEAHA.114.008181

45. Ryoo S, Cha J, Kim SJ et al (2014) High-resolution magnetic resonance wall imaging findings of moyamoya disease. Stroke 45:2457–2460. https://doi.org/10.1161/STROKEAHA.114.004761

46. Aviv RI, Benseler SM, Silverman ED et al (2006) MR imaging and angiography of primary CNS vasculitis of childhood. Am J Neuroradiol 27:192–199. 27/1/192 [pii]

47. Pomper MG, Miller TJ, Stone JH et al (1999) CNS vasculitis in autoimmune disease: MR imaging findings and correlation with angiography. Am J Neuroradiol 20:75–85

48. Powers WJ (2015) Primary angiitis of the central nervous system. Diagnostic criteria. Neurol Clin 33:515–526. https://doi.org/10.1016/j.ncl.2014.12.004

49. Salvarani C, Brown RD, Calamia KT et al (2011) Primary central nervous system vasculitis presenting with intracranial hemorrhage. Arthritis Rheum 63:3598–3606. https://doi.org/10.1002/art.30594

50. Salvarani C, Brown RD, Calamia KT et al (2008) Primary central nervous system vasculitis with prominent leptomeningeal enhancement: a subset with a benign outcome. Arthritis Rheum 58:595–603. https://doi.org/10.1002/art.23300

51. Miller DV, Salvarani C, Hunder GG et al (2009) Biopsy findings in primary angiitis of the central nervous system. Am J Surg Pathol 33:35–43. https://doi.org/10.1097/PAS.0b013e318181e097

52. Torres J, Loomis C, Cucchiara B et al (2016) Diagnostic yield and safety of brain biopsy for suspected primary central nervous system angiitis. Stroke 47:2127–2129. https://doi.org/10.1161/STROKEAHA.116.013874

53. Alrawi A, Trobe JD, Blaivas M, Musch DC (1999) Brain biopsy in primary angiitis of the central nervous system. Neurology 53:852–855. https://doi.org/10.1212/WNL.53.4.852

54. Salvarani C, Brown RD, Calamia KT et al (2008) Primary central nervous system vasculitis: comparison of patients with and without cerebral amyloid angiopathy. Rheumatology 47:1671–1677. https://doi.org/10.1093/rheumatology/ken328

Chapter 22

Cerebral Venous Thrombosis

Sameen Jafari, Catherine Albin, and Saef Izzy

Abstract

Cerebral venous thrombosis (CVT) is a relatively uncommon but serious neurologic disorder that is potentially reversible with prompt diagnosis and appropriate medical care. The clinical presentation is extremely variable and nonspecific, and imaging is critical in diagnosis. Many risk factors for CVT have been reported, most of which overlap with those of peripheral venous thromboembolism. Patients can be diagnosed with magnetic resonance imaging, CT venography, or catheter angiography. The management of CVT patients includes treatment of associated conditions, anticoagulation with parenteral heparin, prevention of recurrent seizures, and surgical decompression in patients with large venous infarcts/hemorrhages with impending herniation. After the acute phase, patients should be anticoagulated for up to 3–12 months based on etiology. In patients who develop clinical and radiological signs of impending herniation, decompressive surgery can be both lifesaving and result in a good functional outcome. The prognosis is otherwise favorable in most cases, especially compared to arterial stroke, although a significant proportion of patients do suffer from chronic symptoms.

Key words Anticoagulants, CT venography, Cerebral vein, Cerebral venous thrombosis, Decompressive surgery, Dural sinus thrombosis, Hemicraniectomy, Intracranial hypertension, MR venography, MRI, Pregnancy, Prognosis, Seizures, Thrombectomy, Thrombolysis

1 Introduction

Cerebral venous thrombosis (CVT) is an uncommon but potentially devastating cause of neurological deterioration, commonly affecting a younger patient population. Timely diagnosis and management can lead to favorable outcomes, and therefore CVT should always be considered in the differential for new onset of neurological symptoms and signs. This chapter will discuss the pathophysiology, risk factors, clinical presentation, diagnosis, medical and surgical management options, and prognosis of CVT with an emphasis on the importance of individualizing treatment on a case-by-case basis.

Fawaz Al-Mufti and Krishna Amuluru (eds.), *Cerebrovascular Disorders*, Neuromethods, vol. 170,
https://doi.org/10.1007/978-1-0716-1530-0_22, © Springer Science+Business Media, LLC, part of Springer Nature 2021

2   Pathophysiology

2.1   Etiology

The broad precipitants of CVT, similar to thrombosis found in any venous system, include those initially described in Virchow's triad:

- Stasis of blood flow.
- Endothelial injury.
- Hypercoagulability.

However, unlike in the extracerebral vascular systems where local organ pathology can account for a significant number of vascular thromboses, local pathology in the brain, head, and neck only accounts for about 1/3 of all cases of CVT [1]. Of more importance are hundreds of other systemic and local risk factors that may give rise to cerebral thrombosis. As reported in the International Study on Cerebral Vein and Dural Sinus Thrombosis (ISCVT) cohort, which is the largest multicenter CVT study, 44% of patients had more than one cause or predisposing factor for developing CVT [2].

Unique complications can develop from a thrombosis in the cerebral venous system due to importance of maintaining a stable pressure-volume relationship in the skull. Two theories have been posited to explain the development of intracranial hypertension following a CVT and thus the clinical symptoms seen in this population (Fig. 1).

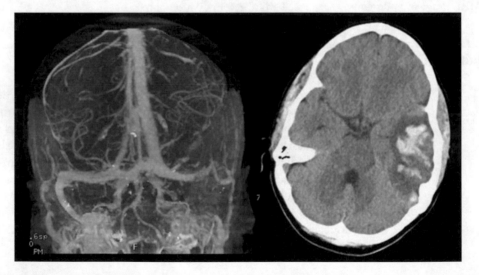

Fig. 1 Computed tomography (CT) imaging of a patient presenting with sudden-onset headache and underlying malignancy. CT of the venous system (left) shows the presence of a left transverse sinus and internal jugular vein thrombosis. CT of the head (right) shows an area of intracranial hemorrhage in the left temporal area as a result of thrombosis

The first theory proposes that a thrombosis in the cerebral veins increases venous pressure, reduces capillary perfusion pressure, and thus results in an increase in the blood volume. This increase in recruited blood volume is locally directed to the site of thrombosis causing high pressure, disruption of the blood-brain barrier, and vasogenic edema. This will decrease both cerebral perfusion pressures and cerebral blood flow leading to local ischemia and cytotoxic edema [3]. This has been validated by changes seen in diffusion-weighted and perfusion-weighted magnetic resonance imaging studies in patients with CVT [4]. The second theory proposes that the increased pressure caused by thrombosis of a major cerebral sinus disrupts CSF absorption which then results in hydrocephalus [3].

2.2 Anatomy: Location

Venous drainage of the brain parenchyma is accomplished by both a superficial and a deep system of sinuses located in the dura. The superficial system drains the superficial surface of the cerebral hemispheres into the cortical veins. The superolateral surface is drained by the superior sagittal sinus (SSS). Blood from the posteroinferior aspect drains into the transverse sinus, which also receives blood from the SSS, and drains into the internal jugular veins [5]. The deep white matter and basal ganglia are drained by the inferior sagittal sinus, internal cerebral veins, and basal vein. The confluence of these veins forms the great vein of Galen, which becomes the straight sinus and drains into the transverse sinus.

Based on the data from the ISCVT, the most commonly involved sinus is the superior sagittal sinus, which is involved in 62% of cases with CVT [5] (Fig. 2).

2.3 Risk Factors

Risk factors for CVT thrombosis can be broadly thought about in terms of transient risk factors versus permanent risk factors and also local factors versus systemic factors. Many patients that develop CVT have more than one risk factor.

2.3.1 Transient Risk Factors

Of the transient factors, oral contraceptive use and puerperium are consistently reported to impose significant risk. In some series of patients with CVT, 70% of women identified using oral contraceptive, and another 14% were peripartum [6]. Indeed, the calculated odds ratio of oral contraceptives in a meta-analysis was estimated at 5.59 (95% confidence interval [CI] 3.95–7.91; $P < 0.001$) [7]. Pregnancy has been reported as accounting for 5–20% of all CVTs in high-income countries [8]. The reason why female hormones play such an important role in the pathology of CVT is not understood, but it has been hypothesized that there is impairment of protein S activity resulting in a prothrombotic state [9]. Other important transient risk factors to consider are hypercoagulability associated with malignancy, dehydration, and inflammatory systemic disorders (Table 1). Transient disruptions to local tissue in the brain, head, and neck which directly or indirectly alter venous

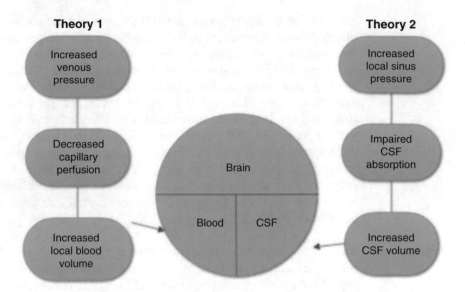

Fig. 2 Theories by which CVT can cause increased pressure in the cranial cavity by either increasing blood volume (theory 1) or CSF volume (theory 2) and leading to a common pathway of raised intracranial pressure causing clinical symptoms and eventual brain tissue damage [3]

Table 1
Risk factors for cerebral venous thrombosis

Transient (modifiable) risk factors	Long-term risk factors

• Estrogen containing contraceptive (oral, transvaginal, patch)
• Puerperium
• Dehydration
• Systemic inflammatory disorder (i.e. DIC)
	• Genetic prothrombotic states, including:
	  – Prothrombin G20210A mutation
	  – Factor V Leiden
	  – Protein C deficiency
	  – Protein S deficiency
	  – Antithrombin III deficiency
	  – Resistance to activated protein C
	  – Antiphospholipid antibodies
	  – Anticardiolipin antibodies
	  – Hyperhomocysteinemia
	• Inflammatory disorders: Sarcoid, IBD
	• Hypercoagulability associated with malignancy
	• Vasculitis
	• Nephrotic syndrome
	• Hematologic disorders

• Intracranial infections • Head and neck infections • CNS surgery or trauma • Head and neck surgery or trauma	• CNS tumors

Abbreviations: *DIC* disseminated intravascular coagulopathy, *IBD* inflammatory bowel disease, *CNS* central nervous system [2, 6–8, 10, 11] circulation such as parenchymal disease (tumor or encephalitis), trauma, surgery, and infection are also consistently reported as attributable risks.

2.3.2 Permanent Risk Factors

The most important permanent risk factors are genetic thrombophilias which were a noted risk factor in nearly a quarter of patients in a large series [2]. Interestingly, prothrombin gene mutations were more than twice as common in patients with CVT than those with other venous thromboembolism (VTE), suggestive of an increased role in the prothrombin gene in cerebral pathology [10]. Factor V Leiden is also commonly discovered, likely because it is the most prevalent cause of inherited thrombophilia [7]. Hyperhomocysteinemia has also been noted in a minority of patients who develop CVT [5, 11]. Other important causes of long-term or permanent risk factors include medical conditions such as nephrotic syndrome, inflammatory bowel disease, and hematologic disease (Table 1). Locally, central nervous system neoplastic disorders have been noted to cause both local and geographically remote thromboses [12].

3 Clinical Presentation

The clinical presentation of CVT is widely variable depending on which sinus is involved—for example, while patients with cavernous sinus thrombosis are likely to demonstrate cranial neuropathy III, IV, or VI, those with superior sagittal sinus thrombosis would be more likely to show bilateral motor deficits. There are however several epidemiologic and symptomatic features that should raise concern for CVT (Table 2).

Patients with CVT are three times more likely to be female, and the average age of presentation is 35–39 years (range 16–86) in most large series, which is much younger than patients with arterial strokes [2, 5, 13]. Symptoms of increased intracranial pressure (ICP) are often, but not always, present. Of these, headache is the single most commonly reported symptom, present in 40–95% of patients [5, 12, 14], due to raised ICP caused by distension of the occluded sinus. [6]

Table 2
Incidences of common clinical presentation of aseptic CVT [2, 5, 12]

Headache	40–95%
Seizure	27–47%
Papilledema	28–41%
Altered mental status	15–39%
Stupor or coma	14–15%
Focal motor deficits	5–43%
Visual symptoms	5–13%

An important subgroup of patients that merit special discussion are those with infective cerebral venous sinus thrombosis. Unlike other causes of CVT, the cavernous sinus is the most commonly involved sinus in infective CVT, resulting from sinus infection, dental abscesses, otitis media, or orbital infections. These patients are often acutely ill and appear toxic; with fever almost always present [15]. Given the predisposition of infection in the cavernous sinus, patients will often present with proptosis, diplopia, cranial neuropathies, and chemosis along with headache and papilledema due to increased intracranial pressure. *Staphylococcus aureus* is implicated in 60–70% of cases of infective CVT [16].

Up to 39% of patients with CVT present with changes in mental status, either in the setting of increased ICP, involvement of the deep venous drainage system causing bilateral thalamic infarcts, or a possible postictal state. Close neuromonitoring and expanding the workup to look for these complications are important next steps in management.

4   Management: Diagnostic Approach

4.1   *Noninvasive Imaging*

Traditionally, conventional digital subtraction angiography was the gold standard for diagnosis of CVT. However, with advances in imaging techniques, MRI/MRV has largely replaced the need for invasive imaging.

- *CT*: Non-contrast CT is generally useful in detecting parenchymal findings that raise suspicion for CVT (Table 3) and ruling out other pathologies. CT may also be normal or may only reveal hyperattenuation signal in the region of the thrombosed sinus, which is both insensitive and nonspecific.

- *CTV*: Although insensitive, contrast-enhanced CT may show the more specific "empty-delta" sign, in which there is a hypodense core in the sinuses—representing thrombus—surrounded by contrast-enhanced dural collateral venous flow [1].

- *MRI*: MRI is both more sensitive at visualizing the thrombosed vessel, as well as more sensitive to the resulting parenchymal changes

Table 3
Parenchymal findings associated with CVST [1, 2, 18]

No parenchymal lesion seen (30–37%)
Bilateral lesions (18%)
Lesions:
Infarct
Subarachnoid hemorrhage
Parenchymal hemorrhage
Edema (both vasogenic and cytotoxic edema)

as a consequence of the thrombosis. However, a limitation of MRI in the acute period (3–5 days) is that the thrombosed sinus is hypointense on T2 and isointense on T1, making it insensitive for detection of thrombus. The susceptibility-weighted images are helpful in demonstrating the thrombosis as a hypointense signal, which can aid in acute diagnosis [17]. Intracerebral hemorrhage (ICH) related to CVT is typically ill-defined and subcortical, with surrounding edema out of proportion to the suspected age and size of the ICH.

- *MRV*: MRV is useful in demonstrating a flow void; however, it can be difficult to discern hypoplasia versus thrombosis without also reviewing the MRI series; therefore, these studies should be reviewed in combination. Remember, absence of the anterior segment of superior sagittal or transverse sinuses can be a normal variant and can resemble a thrombus on MRV. Ordering MRV with gadolinium or CTV allows better visualization of smaller veins and sinuses and is more sensitive for ruling out CVT.

4.2 Invasive Imaging

Given the improved diagnostic yield with combined MRI/MRV, invasive imaging is rarely performed for diagnosis despite being the gold standard.

- An advantage of angiography is the ability to provide a therapeutic intervention at the time of diagnosis.
- Conventional angiography can also be useful in ruling out the misdiagnosis of dural arteriovenous fistula, which can appear similar on noninvasive imaging [19].
  - However, it is important to remember that around 40% of dural arteriovenous fistula may have co-existing CVT, making cerebral angiogram a very useful diagnostic and therapeutic tool in these complex cases.

4.3 Laboratory Testing

Unfortunately, there are no laboratory tests that sensitively diagnose CVT.

- D-Dimer has been investigated as a test to rule in or rule out CVT, and it is neither sensitive nor specific.
  - In a series of patients with confirmed CVT and the only clinical symptom being headache, over a quarter had a normal D-dimer [20].

The role of laboratory testing is more important once the diagnosis of CVT is made and is used to establish the cause and often includes screening for infectious etiology, occult malignancy, and thrombophilias. This should be considered even when one provoking factor has been found as patients often have more than one risk factor for CVT.

5   Management: Treatment Approach

5.1   Acute Phase

Acute treatment should be initiated and monitored in an inpatient setting [5]. Management goals in the acute phase include:

- Preventing clot propagation.

- Treatment of clinical symptoms.

- Monitoring for the risk of complications such as venous infarction, intracerebral hemorrhage, raised intracranial pressure, or hydrocephalus [21].

5.1.1   Treatment of Thrombosis (Fig. 3)

- *Anticoagulation.*

  - The mainstay of treatment, similar to any thrombosis in the venous system, is to start anticoagulation.

  - Several series and a meta-analysis of two randomized controlled trials show that treatment with either intravenous dose-adjusted heparin or weight-based low molecular weight heparin (LMWH) showed a treatment difference in favor of anticoagulation [5, 21, 22].

  - Intravenous dose-adjusted heparin is normally preferred compared to weight-based LMWH due to the following advantages:

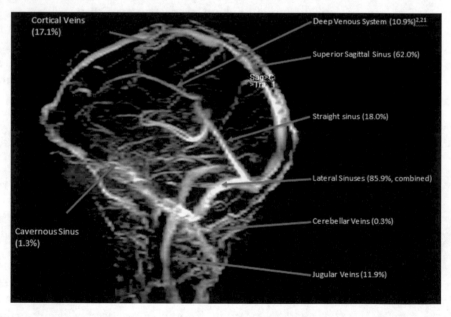

Fig. 3 MRV imaging of the cerebral venous system, labelled with percentage of the most common location of CVT as seen in patients in ISCVT [5]

- Ability to closely monitor dose.

- Target a specific therapeutic level 80–100 s (2–2.5 times greater than baseline APTT) [5].

- A shorter half-life (APTT should normalize in 1–2 h when heparin stopped) [21].

- Availability of a reversal agent, protamine, which can be used if there is a concern for new or worsening symptoms.

- Current evidence in CVT does not show any clear benefit of using LMWH over intravenous heparin, unlike that seen in literature of treatment for deep venous thrombosis or pulmonary embolus [5, 22].

  However one randomized controlled trial showed decreased mortality with LMWH which may be due in part to more sustained therapeutic anticoagulation with LMWH than with intravenous heparin in their patients (found to have suboptimal anticoagulation 61.3% of the time with unfractionated heparin) [23].

  Therefore, initiating treatment with LMWH can be considered in stable patients with low risk factors for bleeding.

- Although the presence of intracerebral hemorrhage (ICH) with CVT on admission is a risk factor for poor outcome (all deaths in one trial occurred in patients with CVT and ICH), this is not a contraindication for treatment with anticoagulation [5, 21, 22].

  Studies suggest that acute anticoagulation does not seem to worsen underlying CVT-related ICH [5, 22].

- There is no evidence for using aspirin over anticoagulation for the treatment of CVT [5].

- *Thrombolysis.*
  - Thrombolysis can include systemic or local thrombolytic treatment with and without mechanical thrombectomy [5].

    Evidence for this therapeutic option is mainly from uncontrolled series or case reports which show improved recanalization rates, but it is also associated with an increased risk of extra- and intracranial hemorrhage [5, 22].

  - There may be a role for thrombolysis in 9–13% of patients with CVT who have poor outcomes despite anticoagulation [5].

  - Another possible indication for this therapy is a rapid decline in neurological exam (i.e., coma).

- *Corticosteroids.*
  - A case-control study from ISCVT showed that corticosteroids worsen outcomes in CVT patients and they should only be used if there is an underlying steroid-responsive condition, i.e., vasculitis [5, 24].

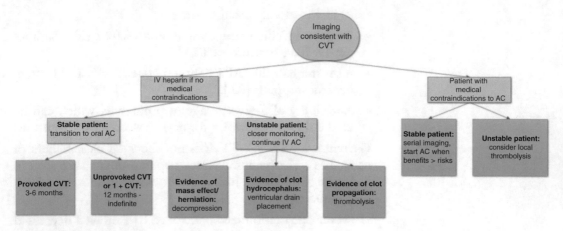

Fig. 4 Proposed algorithm for managing CVT. Anticoagulation is recommended if there are no medical contraindications (i.e., history of gastrointestinal bleeding, thrombocytopenia, etc.). Note that the presence of ICH on presentation with CVT is not a contraindication. If patient is stable or improves, then plan to transition to an oral anticoagulant, and duration of therapy will be dependent on the underlying etiology/risk factors. If the patient is unstable, the goal should be to determine a possible etiology and escalate treatment as described. Patients with medical contraindications to anticoagulation should be closely monitored and may be eligible for local thrombolysis if unstable or, if stable, conservative management (hydration, serial imaging) [5, 22]

5.1.2 Complications

- *Cerebral Edema and Herniation: Decompressive Surgery.*
  - The most frequent cause of early death in CVT (within 30 days of symptoms) is from transtentorial brain herniation, commonly due to expansion/re-bleeding of an existing ICH (Fig. 4).
  - Decompressive hemicraniectomy can be considered in patients with ICH extension, increased cerebral edema, and/or impending herniation [22, 25].
    The role of this therapy is lifesaving. It should be considered early in the course as literature shows that decompressive surgery in patients with CVT led to a favorable outcome (modified Rankin score less than or equal to 3) [5].

- *Seizures: Anti-epileptic Medications* (Fig. 5).
  - Seizures are a common finding in patients with CVT, especially in the setting of a cortical venous infarct.
    27–47% of CVT patients present with seizures, and 2% develop them during hospitalization [5].
  - A common practice is to start anti-epileptic medication in patients who either present with seizures or have seizures during their hospitalization.
  - The presence of a supratentorial parenchymal lesion along with CVT was associated with early seizures (occurring <2 weeks from presentation) [5, 22, 26].

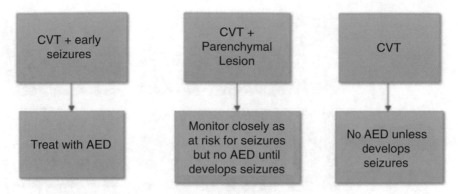

Fig. 5 Treatment of seizures in CVT. Abbreviation: anti-epileptic drugs (AED)

However, there is no sufficient evidence to recommend pro-phylactic anti-epileptic treatment in patients with supra-tentorial lesions who have no history of seizures.

They should be closely monitored with a low threshold to start anti-epileptic medication if seizures begin [5, 22, 26].

- *Hydrocephalus: Ventricular Drain Placement.*

  – Obstructive hydrocephalus is an uncommon complication and usually occurs when thrombosis of the internal cerebral veins results in a ventricular hemorrhage.
  This is usually seen in neonates but can occur at all ages [5].

- *Intracranial Hypertension: Serial Lumbar Punctures and Acetazolamide.*

  – 40% of patients can present with isolated intracranial hyper-tension and require visual field testing and fundoscopy to monitor for impending vision loss [5].

  – Serial lumbar punctures can help normalize intracranial pres-sure but are a temporary measure and can be difficult to coordinate in patients on long-term anticoagulation [5, 22].

  – Acetazolamide has been used in an outpatient setting for chronically raised intracranial pressure [5, 22].

5.2 Chronic Phase

5.2.1 Oral Anticoagulation

- *Length of Treatment* (Fig. 6).

  – When patients demonstrate stability on intravenous anticoa-gulation, it is reasonable to switch to oral anticoagulation.

  – Length of treatment will depend on the underlying etiology and risk of recurrence [5, 22].

  – Thrombophilias can be divided into mild or severe.

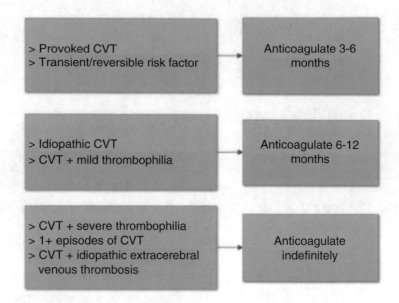

Fig. 6 Length of anticoagulation treatment increases with risk of recurrence of CVT

> Mild thrombophilia = protein C or S deficiency and hetero-zygous factor V Leiden or prothrombin G20210A mutation [22].

> Severe thrombophilia = antithrombin deficiency, homozy-gous factor V Leiden or prothrombin G20210A muta-tion, antiphospholipid syndrome, or a combination of two or more thrombophilic conditions [22].

– Recurrence rate for CVT has been quoted between 1.3% and 2.2% over a 16-month follow-up period [2, 28].

– Consider hematology involvement for patients with complex comorbidities or indeterminate thrombophilia results to help determine the best treatment duration to minimize both the adverse risks of unneeded anticoagulation with the risk of recurrent CVT [5].

• *Oral Anticoagulation Agent.*

– The most common oral anticoagulants used are vitamin K antagonists, commonly warfarin, with an INR goal of 2–3 [5, 22].

> Warfarin treatment, can have significant lifestyle implications for younger patients, and therefore novel oral anticoagu-lants may be considered. These have been shown to be non-inferior to warfarin in the treatment of deep vein thrombosis and pulmonary embolism [27], which may be extrapolated to the CVT population.

Retrospective uncontrolled studies with small sample sizes have examined novel oral anticoagulants (a factor Xa inhibitor [28] and a direct thrombin inhibitor [29], respectively) compared to vitamin K antagonist after initial treatment with parenteral heparin. The studies found similar clinical outcomes and recanalization rates between the novel agents and vitamin K antagonists without evidence of significant bleeding, but larger studies are needed [28, 29].

- *Remote Seizures and Epilepsy.*
  - Remote seizures are classified as seizures occurring >2 weeks from CVT diagnosis and occur in 5–32% of patients, while epilepsy is considered to be more than one remote seizure and occurs in about 5% of patients [22].

    Therefore, it is reasonable to continue anti-epileptic medication for at least 1 year from CVT and then consider withdrawal if stable [22].
  - Risk factors for remote seizures and epilepsy include early seizures, hemorrhagic lesions, and paresis, and the presence of any or all of these can assist in clinical decisions about when to taper AED (anti-epileptic drugs) [22, 26].
- *Underlying Etiology.*
  - Whenever possible, the underlying etiology should be treated to prevent propagation of clot or recurrent CVT.

    This can include antibiotics for CNS infection, removal of prothrombotic agents (oral contraception), and treatment of any underlying malignancy.

---

6 Prognosis

Most data is from retrospective studies including a multicenter study from the USA but also case series from other countries. Caution is needed when interpreting this data as different countries have different healthcare resources affecting the quoted rates of mortality and dependence. Three important outcomes are shown in Table 4 [2, 5, 22].

- Neurological Worsening.
  - This was seen in 23% of patients and is a broad category encompassing depressed consciousness, mental status disturbance, new seizure, worsening new focal deficit, increase in headache, or visual loss. One third of these patients were found to have a new parenchymal lesion on imaging. A risk factor for neurological worsening was depressed

Table 4
Outcomes in CVT [2, 5, 22, 25]

Outcome	Percentage	Risk factors
Neurological worsening	23%	– Depressed consciousness on admission
Early death	3–15%	– Coma on admission – Thrombosis of deep cerebral vein – Posterior fossa lesion – Right hemispheric ICH
Long-term outcomes		
Complete recovery Dependence Death	79% 5.1% 8.3%	– CNS infection – Any malignancy – Thrombosis of deep cerebral vein – ICH on admission – GCS < 9 – Age > 35 years – Male sex – Mental state disturbance

consciousness on admission [2, 5, 22]. Patients presenting in coma with deep venous thrombosis and ICH on imaging are at high risk of a poor outcome and need to be managed carefully to allow the best neurological recovery possible.

- Early Death.
  - Early death was classified as death within 30 days from symptom onset and quoted to be between 3% and 15% across different studies. In ISCVT, early death occurred in 3.4% of patients. Risk factors for early death are detailed in Table 4. Late death was mostly due to the underlying etiology of CVT, for example, malignancy or CNS infection. [2, 5, 22]

- Long-Term Outcomes.
  - These figures are all from ISCVT and were assessed at 16 months. 79% of patients made a complete recovery, 5.1% were dependent (modified Rankin score of 3 or higher, i.e., requiring assistance for activities of daily living and ambulation), and 8.3% were deceased (including early and late deaths). Risk factors for poor long-term outcomes (death and dependence) included CNS infection, any malignancy, thrombosis of deep cerebral veins, ICH on admission, Glasgow Coma Scale of <9, age < 35 years, male sex, and mental state disturbance [2, 5, 22].

7   Conclusion

Early clinical recognition and rapid diagnosis and management strategies are essential to reduce the mortality and morbidity of this complex disease. CVT patients commonly require multifaceted management decisions by a multidisciplinary team while they are closely monitored in well-equipped neurocritical care units. There is a clear need for more evidence-based data to optimize early detection and better management of CVT patients in order to improve their favorable outcomes and reduce neurological morbidity.

References

1. Gokhale S, Lahoti SA (2014) Therapeutic advances in understanding pathophysiology and treatment of cerebral venous sinus thrombosis. Am J Ther 21(2):137–139

2. Ferro JM, Canhao P, Stam J, Bousser MG, Barinagarrementeria F (2004) Prognosis of cerebral vein and dural sinus thrombosis: results of the international study on cerebral vein and dural sinus thrombosis (ISCVT). Stroke 35:664–670

3. Filippidis A, Kapsalaki E, Patramani G, Fountas KN (2009) Cerebral venous sinus thrombosis: review of the demographics, pathophysiology, current diagnosis, and treatment. Neurosurg Focus 27(5):E3

4. Yoshikawa T, Abe O, Tsuchiya K, Okubo T, Tobe K, Masumoto T et al (2002) Diffusion-weighted magnetic resonance imaging of dural sinus thrombosis. Neuroradiology 44:481–488

5. Saposnik G, Barinagarrementeria F, Brown RD Jr et al (2011) American Heart Association Stroke Council and the Council on Epidemiology and Prevention. Diagnosis and management of cerebral venous thrombosis: a statement for healthcare professionals from the American Heart Association/American Stroke Association. Stroke 42:1158–1192

6. de Bruijn SFTM (2001) Clinical features and prognostic factors of cerebral venous sinus thrombosis in a prospective series of 59 patients. J Neurol Neurosurg Psychiatry 70(1):105–108

7. Dentali F (2006) Thrombophilic abnormalities, oral contraceptives, and risk of cerebral vein thrombosis: a meta-analysis. Blood 107(7):2766–2773

8. Bousser M-G, Crassard I (2012) Cerebral venous thrombosis, pregnancy and oral contraceptives. Thromb Res 130:S19–S22

9. Tchaikovski SN, Rosing J (2010) Mechanisms of estrogen induced venous thromboembolism. Thromb Res 126:5–11

10. Wysokinska EM, Wysokinski WE, Brown RD et al (2008) Thrombophilia differences in cerebral venous sinus and lower extremity deep venous thrombosis. Neurology 70:627–633

11. Coutinho J et al (2013) Cerebral venous thrombosis and thrombophilia: a systematic review and meta-analysis. Semin Thromb Hemost 39(8):913–927

12. Raper DMS et al (2016) Geographically remote cerebral venous sinus thrombosis in patients with intracranial tumors. World Neurosurg 98:555–562

13. Azin H, Ashjazadeh N (2008) Cerebral venous sinus thrombosis—clinical features, predisposing and prognostic factors. Acta Neurol Taiwan 17(2):82–87

14. Sparaco M, Feleppa M, Bigal ME (2015) Cerebral venous thrombosis and headache – a case-series. Headache 55(6):806–814

15. Ebright JR, Pace MT, Niazi AF (2001) Septic thrombosis of the cavernous sinuses. Arch Intern Med 161(22):2671–2676

16. Khatri IA, Wasay M (2016) Septic cerebral venous sinus thrombosis. J Neurol Sci 362:221–227

17. Bousser M-G, Ferro JM (2007) Cerebral venous thrombosis: an update. Lancet Neurol 6(2):162–170

18. Kumral E, Polat F, Uzunköprü C, Çallı C, Kitiş Ö (2011) The clinical spectrum of intracerebral hematoma, hemorrhagic infarct, non-hemorrhagic infarct, and non-lesional venous stroke in patients with cerebral sinus-venous thrombosis. Eur J Neurol 19(4):537–543. https://doi.org/10.1111/j.1468-1331.2011.03562

19. Simon S, Yao T, Ulm AJ, Rosenbaum BP, Mericle RA (2009) Dural arteriovenous fistulas masquerading as dural sinus thrombosis. J Neurosurg 110(3):514–517

20. Crassard I, Soria C, Tzourio C et al (2005) A negative D-dimer assay does not rule out cerebral venous thrombosis: a series of 73 patients. Stroke 36:1716–1719

21. Einhaupl K, Bousser MG, de Bruijn SF, Ferro JM, Martinelli I, Masuhr F, Stam J (2006) EFNS guideline on the treatment of cerebral venous and sinus thrombosis. Eur J Neurol 13 (6):553–559

22. Einhaupl KM, Villringer A, Meister W, Mehraein S, Garner C, Pellkofer M, Haberl RL, Pfister HW, Schmiedek P (1991) Heparin treatment in sinus venous thrombosis [published correction appears in *Lancet*. 1991; 338:958]. Lancet 338:597–600

23. Misra UK, Kalita J, Chandra S, Kumar B et al (2012) Low molecular weight heparin versus unfractionated heparin in cerebral venous sinus thrombosis: a randomized controlled trial. Eur J Neurol 19:1030–1036

24. Canhao P, Cortesao A, Cabral M, Ferro JM, Stam J, Bousser MG, Barinagarrementeria F, ISCVT Investigators (2008) Are steroids useful to treat cerebral venous thrombosis? Stroke 39:105–110

25. Canhao P, Ferro JM, Lindgren AG et al (2005) Causes and predictors of death in cerebral venous thrombosis. Stroke 36:1720–1725

26. Ferro JM, Canhao P, Bousser MG, Stam J, Barinagarrementeria F, ISCVT Investigators (2008) Early seizures in cerebral vein and dural sinus thrombosis: risk factors and role of antiepileptics. Stroke 39:1152–1158

27. Van der Hulle T, Kooiman J, den Exter PL et al (2014) Effectiveness and safety of novel oral anticoagulants as compared with vitamin K antagonists in the treatment of acute symptomatic venous thromboembolism: a systematic review and meta-analysis. J Thromb Hemost 12(3):320–328

28. Shulman S, Kearon C, Kakkar AK et al (2009) Dabigatran versus warfarin in the treatment of acute venous thromboembolism. N Engl J Med 361:2342–2352

29. Geisbüsh C, Richter D, Herweh C, Ringleb PA, Nagel S (2014) Novel factor Xa inhibitor for the treatment of cerebral venous and sinus thrombosis – first experience in 7 patients. Stroke 45:2469–2471

Correction to: Cerebrovascular Disorders

Fawaz Al-Mufti and Krishna Amuluru

Correction to:
Fawaz Al-Mufti and Krishna Amuluru (eds.), *Cerebrovascular Disorders,*
Neuromethods, vol. 170,
https://doi.org/10.1007/978-1-0716-1530-0

The original version of this book was inadvertently published with incorrect affiliation for one of the volume editors. The affiliation has been updated in the book as follows:

Fawaz Al-Mufti, MD
Associate Chair of Neurology for Research, New York Medical College,
Associate Professor of Neurology, Neurosurgery and Radiology, New York Medical College,
Director of the Neuroendovascular Surgery Fellowship,
Neuroendovascular Surgery (Interventional Neurologist) Attending
Medical Director of Neurocritical Care, Westchester Medical Center at New York Medical College,
Valhalla, NY, USA

The updated online version of this book can be found at
https://doi.org/10.1007/978-1-0716-1530-0
https://doi.org/10.1007/978-1-0716-1530-0_1
https://doi.org/10.1007/978-1-0716-1530-0_6
https://doi.org/10.1007/978-1-0716-1530-0_18
https://doi.org/10.1007/978-1-0716-1530-0_19

Fawaz Al-Mufti and Krishna Amuluru (eds.), *Cerebrovascular Disorders*, Neuromethods, vol. 170,
https://doi.org/10.1007/978-1-0716-1530-0_23, © Springer Science+Business Media, LLC, part of Springer Nature 2021

INDEX

Fawaz Al-Mufti and Krishna Amuluru (eds.), *Cerebrovascular Disorders*, Neuromethods, vol. 170,
https://doi.org/10.1007/978-1-0716-1530-0, © Springer Science+Business Media, LLC, part of Springer Nature 2021

Printed in the United States
by Baker & Taylor Publisher Services